Design and Analysis in Chemical Research

Sheffield Analytical Chemistry

Series Editors: J.M. Chalmers and R.N. Ibbett

A series which presents the current state of the art of chosen sectors of analytical chemistry. Written at professional and reference level, it is directed at analytical chemists, environmental scientists, food scientists, pharmaceutical scientists, earth scientists, petrochemists and polymer chemists. Each volume in the series provides an accessible source of information on the essential principles, instrumentation, methodology and applications of a particular analytical technique.

Titles in the Series:

Inductively Coupled Plasma Spectrometry and its Applications
Edited by S.J. Hill

Extraction Methods in Organic Analysis
Edited by A.J. Handley

Design and Analysis in Chemical Research
Edited by R.L. Tranter

Spectroscopy in Process Analysis
Edited by J.M. Chalmers

Design and Analysis in Chemical Research

Edited by

ROY L. TRANTER

Statistics and Data Evaluation Manager
QA Compliance Group
Glaxo Wellcome Operations
County Durham, UK

CRC Press

First published 2000
Copyright © 2000 Sheffield Academic Press

Published by
Sheffield Academic Press Ltd
Mansion House, 19 Kingfield Road
Sheffield S11 9AS, England

ISBN 1-85075-994-4

Published in the U.S.A. and Canada (only) by
CRC Press LLC
2000 Corporate Blvd., N.W.
Boca Raton, FL 33431, U.S.A.
Orders from the U.S.A. and Canada (only) to CRC Press LLC

U.S.A. and Canada only:
ISBN 0-8493-9746-4

All rights reserved. No part of this publication may be reproduced, stored in a retrieval system or transmitted in any form or by any means, electronic, mechanical, photocopying or otherwise, without the prior permission of the copyright owner.

This book contains information obtained from authentic and highly regarded sources. Reprinted material is quoted with permission, and sources are indicated. Reasonable efforts have been made to publish reliable data and information, but the author and the publisher cannot assume responsibility for the validity of all materials or for the consequences of their use.

Trademark Notice: Product or corporate names may be trademarks or registered trademarks, and are used only for identification and explanation, without intent to infringe.

Printed on acid-free paper in Great Britain by
Bookcraft Ltd, Midsomer Norton, Bath

British Library Cataloguing-in-Publication Data:
A catalogue record for this book is available from the British Library

Library of Congress Cataloging-in-Publication Data:
Design and analysis in chemical research / edited by Roy Tranter.
    p.   cm. -- (Sheffield analytical chemistry : v. 3)
  Includes bibliographical references and index.
  ISBN 0-8493-9746-4 (alk. paper)
  1. Chemistry, Analytic --Statistical methods.   2.Experimental
  design. I. Tranter, Roy. II. Series.
QD75.4.S8D48   1999
543'.07'2--dc21
                                                          99-28562
                                                              CIP

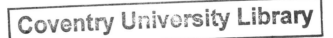

Preface

Within the chemical sciences, statistics has the reputation of being hard and usable only by mathematicians or masochists. It also has the reputation of either telling us what we already know or making predictions that are wrong. So why should we bother with it?

Both of these reputations are undeserved and often stem from a dry theoretical statistics course or from experience of the statistics used publicly for essentially political aims. But 'real' statistics is quite different. It is the aim of this book to show that it is essentially an extension of the logical processes used by chemists every day, and that its use can, and does, bring greater understanding of problems more quickly and easily than the purely intuitive or "let's try and see" approaches.

For this we must be careful to distinguish between the tools we use to make the statistical calculations—the equations, algorithms and software—and the thought processes that allow us to decide which is the best tool to use and which is the best method for interpreting the results of its use. The latter is best described as *statistical thinking*. It encompasses the tools but extends the context to include awareness and appreciation of the sciences of measurement, experimentation and logic. It is the philosophy of rational data analysis and interpretation.

Statistics is a mathematical subject and it does involve equations—we cannot get away from this—but the equations are generally no more difficult than those routinely used in spectroscopy, kinetics, structure-activity relationships, molecular modelling or any other chemical system requiring computation. Many of the statistics equations (and concepts) are much simpler! What is different about statistics is that it is all about handling variability and uncertainty. Most chemists are more comfortable with certainty and, for many, the concept of error is associated with mistakes and poor work.

Simply accepting that all measurements have some uncertainty associated with them is the first, major step in coming to terms with statistics. The second is accepting that uncertainty can be measured and handled in a quantitative way. Once we can get a handle on uncertainty, we can control it. Although it is not possible, in any practical sense, to remove uncertainty, we can certainly reduce it and its effects, and so increase our confidence in the chemical truths that we discover.

Statistics is a broad subject and it has very wide application in the sciences, engineering, financial and behavioural arenas. It can be presented in many ways, all good in their own contexts. Here we have chosen to concentrate on principles and interpretation rather than on formal derivation and proof. References are given for those wanting to get into the latter. All we need to be aware of here is that all the methods we describe are well established, well documented and widely accepted. This does not mean that they give the "truth"—only that their properties are well understood and that they will give consistent and interpretable results when used appropriately.

You are not entering an equation-free zone. Far from it. But the equations needed to understand and/or to implement a particular tool are given with explanation and interpretation to help you come to terms with them.

One of the best ways of understanding how or where to use statistics is through applications. We have included many examples of actual or possible use from a wide area. The context of research chemistry is used throughout, but many of the good, easy-to-explain examples come from analytical and process chemistry, where quantitative measurement and interpretation are the norm. However, remember our objective of establishing principles. The principles in these examples apply to all branches of research chemistry, including organic and inorganic synthesis and molecular design, as well as the more obvious topics of physical chemistry and chemical physics.

In a book of this size, it is impossible to cover the whole of the vast subjects of statistics and chemometrics. So we have chosen to cover the basic statistical methods that underpin the *statistical thinking* approach. These are chapters 1-8. Chapters 9-13 describe the tools that are frequently used in chemical situations where a quantitative model is needed to describe or test a relationship between variables. Chapter 2 focuses on data quality, as we can have no confidence in statistical results if we have no confidence in the data, no matter how impressive the calculations might be.

These days, the computation of statistics is a trivial task. However, the easy availability of good hardware and software does mean that it is very much easier to do statistical computation without actually understanding what is being done. The corollary is that you have more time to develop that understanding and to consider the interpretation of the results that you have calculated. If you do get stuck or want advice on how to get away from the 'black box' approach, there are many quite friendly and helpful statisticians out there.

Finally to the authors. They are all people well known and well respected in their areas. Some are professionally trained statisticians and some are chemists with a deep understanding and appreciation of statistics. All are practical users of statistics and have wide experience of using statistics and the principles of *statistical thinking* in many areas of chemistry. I am greatly indebted to them for the time and effort they have put in to writing their chapters, and for the patience they have shown to me as editor of this volume. You will find much to appreciate in their work. Enjoy it and apply it!

Roy Tranter

Contributors

Dr M. Robert Alecio	Director, Positive Probability Limited, 9 Church Street, Isleham, Ely, Cambridgeshire CB7 5RX, UK Email: Robert@Alecio.freeserve.co.uk
Mr Anthony G. Ferrige	Director, Positive Probability Limited, 9 Church Street, Isleham, Ely, Cambridgeshire CB7 5RX, UK Email: gfv63@dial.pipex.com
Marion Gerson	Director, Centre for Quality Engineering, University of Newcastle upon Tyne, Newcastle upon Tyne NE1 7RU, UK Email: m.e.gerson@ncl.ac.uk
Mrs Sonya Godbert	Senior Consultant Statistician, Statistical Services, Glaxo Wellcome Research and Development, Park Road, Ware SG12 0DP, UK
Professor Theodora Kourti	Department of Chemical Engineering, McMaster University, Hamilton, Ontario L8S 4L7, Canada Email: kourtit@mcmaster.ca
Professor Olav M. Kvalheim	Department of Chemistry, The University of Bergen, Allegate 41, N-5007 Bergen, Norway Email: olav.kvalheim@kj.uib.no
Dr Ivan Langhans	CQ Consultancy, Kapeldreef 60, B-3001 Heverlee, Belgium Email: CQ@CQConsultancy.be
Dr Willem Melssen	Department of Analytical Chemistry, Katholieke Universität Nijmegen, Postbus 9010, 6500 GL Nijmegen, The Netherlands Email: willem@sci.kun.nl

Dr Max A. Porter	Statistics Manager, Glaxo Wellcome UK International Actives Supply, North Lonsdale Road, Ulverston, Cumbria LA12 9DR, UK Email: map41247@GlaxoWellcome.co.uk
Dr John M. Thompson	School of Mathematics and Statistics, University of Birmingham, Edgbaston, Birmingham B15 2TT, UK Email: dr.jmthompson@cwcom.net jmt@for.mat.bham.ac.uk
Dr Roy L. Tranter	Statistics and Data Evaluation Manager, QA Compliance, Glaxo Wellcome UK International Product Supply, Harmire Road, Barnard Castle, County Durham DL12 8DT, UK Email: rlt48033@GlaxoWellcome.co.uk

Contents

1 Statistical thinking—The benefits and problems of a statistical approach   1
M. PORTER

   1.1 Introduction   1
   1.2 Where to go   1
   1.3 What is statistical thinking?   2
       1.3.1 Statistics   2
       1.3.2 Building in variability—Why all processes are subject to variability   3
       1.3.3 A model of the inductive/deductive processes   4
       1.3.4 Mathematical and statistical models   7
   1.4 Types and causes of variability   7
       1.4.1 Common and special causes   7
       1.4.2 Bias, systematic and random variation   8
       1.4.3 Example—Sampling from and assaying a bulk chemical   9
       1.4.4 Impacts of variability on decision making and the design of investigations   10
       1.4.5 Sampling—Why and how?   11
   1.5 Probability   12
       1.5.1 Measuring risk and uncertainty   12
       1.5.2 Rules of probability   12
       1.5.3 Random variables, distributions and statistics   14
       1.5.4 Practical and statistical significance   16
   1.6 Carrying out a statistical investigation   19
       1.6.1 Steps in an investigation   19
       1.6.2 Prospective and retrospective studies   23
       1.6.3 Designing investigations   24
       1.6.4 Pilot study   25
       1.6.5 Validation and good statistical practice   26
       1.6.6 Benefits and disadvantages   26
   1.7 Skills, experts and systems   29
       1.7.1 When and how to get help   29
       1.7.2 What you can do for yourself   29
       1.7.3 Local experts   30
       1.7.4 What a statistician can contribute   31
       1.7.5 Using statistical software   32
   References   32

2 Essentials of data gathering and data description   34
R. TRANTER

   2.1 Introduction   34
   2.2 Where to go   35
   2.3 The data cycle   35

	2.4	Data planning and design		36
		2.4.1 Measurement systems		36
		2.4.2 Experiment design		38
		2.4.3 Randomisation		39
		2.4.4 Data collection		41
		2.4.5 Digitisation		42
		2.4.6 Recording and reporting numbers		44
		2.4.7 Rounding		46
		2.4.8 Significant figures		47
	2.5	Data description		48
		2.5.1 Simple numerical checks		48
		2.5.2 Trend plots and control charts		49
		2.5.3 Scatter plots		52
		2.5.4 Box and whisker plots		53
		2.5.5 Cusum plots		54
		2.5.6 Histograms and data distributions		58
		2.5.7 Normality testing		60
		2.5.8 Outliers and discordant values		61
		2.5.9 Studentised range test for a single discordant value		64
		2.5.10 Grubb's test for a single discordant value		65
		2.5.11 Dixon's test for a single discordant value		65
	2.6	Data preprocessing		66
		2.6.1 Smoothing—moving average, Savitsky-Golay, EWMA		66
		2.6.2 Integration and differentiation		70
		2.6.3 Principal component analysis		75
		2.6.4 Other transformations		81
	Bibliography			82
3	**Sampling**			**85**
	J. THOMPSON			
	3.1	Introduction		85
	3.2	Where to go		85
	3.3	What is sampling?		86
	3.4	Sampling—The Pandora's box of chemical/biological research/development		87
		3.4.1 The peanut problem		88
		3.4.2 Risks in diagnosis		89
		3.4.3 Whose point of view?		90
	3.5	Aims and objectives of sampling		91
		3.5.1 Observational studies		91
		3.5.2 Invasive, noninvasive, remote and indirect sampling		92
		3.5.3 Sampling for process/quality control/improvement and for environmental regulation		93
		3.5.4 What kinds of physical sampling can be done?		95
		3.5.5 Sampling strategies		98
		3.5.6 Statistical/chemometric aspects		99
	3.6	Statistical sampling strategies		100
		3.6.1 Random sampling		100
		3.6.2 Systematic sampling		102
		3.6.3 Stratified sampling		102
		3.6.4 Sequential sampling		102

	3.7	Sample size estimation using the concept of the power of a statistical test	103
		3.7.1 Power and risks	103
		3.7.2 Limit of detection and limit of quantitation	104
		3.7.3 Other points	106
		3.7.4 Software implementations	106
	3.8	Sampling in the context of deterministic versus probabilistic assessment of compliance with a standard or threshold—use of the receiver operating characteristic (ROC) curve	107
	3.9	Problems associated with behaviour of granular and other materials—their effects on designing sampling schemes and on estimating sampling reliability	109
	3.10	Sampling for process control and quality management	109
	3.11	Economic aspects of sampling designs	109
	Bibliography and References	110

4 Interpreting results   113
M. GERSON

4.1	Introduction	113
4.2	Where to go	113
4.3	The objectives of experimentation and data collection	114
4.4	Setting up a known system on which to experiment	116
	4.4.1 Important assumptions about the ε_{ij} term	117
4.5	Experimenting on the known system	119
	4.5.1 Estimating the difference between two measurements	119
	4.5.2 A confidence interval for the effect of changing the temperature	121
	4.5.3 A revised confidence interval for temperature effect	122
	4.5.4 Confidence intervals for variability—Standard deviation	125
	4.5.5 Confidence intervals for variability—Ratio of two variances	127
	4.5.6 Who needs confidence intervals?	130
	4.5.7 Single-sided and double-sided intervals	131
4.6	Deciding on the size of a simple comparative experiment	132
	4.6.1 A simple approach	132
	4.6.2 A more general method	133
4.7	Reducing the amount of work to be done	135
4.8	Hypothesis testing and significance levels	136
4.9	Some more applications of confidence intervals	138
	4.9.1 Equivalence studies	138
	4.9.2 Paired comparisons	140
	4.9.3 Analysis of variance for estimating different sources of variability	142
4.10	Appendix	144
	4.10.1 Calculation of approximately Normally distributed random numbers	144
References	144

5 Robust, resistant and nonparametric methods   145
J. THOMPSON

5.1	Introduction	145
5.2	Where to go	146
5.3	Some simple and useful concepts	147
5.4	Looking at continuous measurements variables	148

	5.4.1	Some initial comments about the shapes of data distributions for continuous variables	148
	5.4.2	Estimating the location of a single set of data using arithmetic and geometric means	149
	5.4.3	Estimating the spread of a single set of continuous data	153
	5.4.4	Estimating confidence intervals for location estimators for a single set of data	157
	5.4.5	Estimating confidence intervals around spread estimates for a single set of data	158
5.5	Outlier tests for single sets of data		161
	5.5.1	Exploratory data analysis methods for outlier detection based on the fourth spread	161
	5.5.2	Exploring the shape of a single set of data	161
5.6	Robust and resistant methods in evaluating data transformations and the value of transformation		166
5.7	Randomness in a data set		168
5.8	Nonparametric methods for comparing locations of paired data sets		169
5.9	Nonparametric methods for comparison of two sets of unpaired data		173
5.10	Nonparametric goodness-of-fit tests to specific distributions		174
5.11	Nonparametric comparisons of the effects of one factor on more than two sets of data		175
	5.11.1	Kruskal–Wallis one-way analysis of variance (ANOVA) by ranks, including multiple comparisons methods	175
	5.11.2	Exploratory one-way ANOVA	176
	5.11.3	Cross-classified ANOVA designs—nonparametric and resistant/robust methods for two-way ANOVA and more complex designs	177
5.12	Estimating functional relationships or making paired comparisons with robust and nonparametric regression methods for two variables		182
	5.12.1	Tukey's three-group resistant line regression	183
	5.12.2	Theil–Kendall regression using the median of pairwise slopes	184
	5.12.3	Hettmansperger's rank regression methods	184
	5.12.4	Rousseeuw's least median of squares (LMS) and least trimmed squares (LTS) regression methods	185
	5.12.5	Tukey's biweight regression method	185
References			186

6 Experiment design—Identifying factors that affect responses 188
S. GODBERT

6.1	What is design of experiments?		188
6.2	Where to go		190
6.3	Terminology		190
6.4	Getting started		194
6.5	Two-level designs		197
	6.5.1	Full factorial	197
	6.5.2	Example: Determining the best storage conditions using a full factorial	197
	6.5.3	Blocking	203
	6.5.4	Fractional factorial	206
6.6	Confounding		209
	6.6.1	Resolution	209

	6.6.2 Example: Crystallisation study to determine the factors affecting particle size uniformity using a resolution III screening design	210
	6.6.3 Irregular fraction designs	213
	6.6.4 Plackett–Burman designs	213
	6.6.5 Taguchi designs	214
	6.6.6 D-Optimal designs	215
	6.6.7 Robustness designs	216
	6.6.8 Centre points	216
	6.6.9 Example: Determining robustness using a fractional factorial	216
	6.6.10 Example: Assessing process deviation ranges using a highly fractionated factorial design	221
	6.6.11 Other designs	225
	6.6.12 Mixed-level designs	225
6.7	Data analysis and interpretation	225
6.8	Choosing a design	227
6.9	Effect of not following the design exactly	230
6.10	Other potential problems	231
6.11	Screening designs in context	232
	6.11.1 Example: Summary of DOE performed during the development of a beta-lactam	233
Bibliography and References		236

7 Designs for response surface modelling—Quantifying the relation between factors and response   237
I. LANGHANS

7.1	Introduction	237
7.2	Where to go	237
7.3	The basics	238
	7.3.1 What is response surface modelling?	238
	7.3.2 What has this got to do with experimental design?	240
7.4	Soft modelling using polynomials	240
	7.4.1 Of true models and their approximations	240
	7.4.2 Relation between the objective of a study and the complexity of the polynomial	244
7.5	Designs for response surface modelling	248
	7.5.1 General considerations	248
	7.5.2 Central composite design	249
	7.5.3 Box–Behnken designs	252
	7.5.4 Choosing a design	252
	7.5.5 Case study	253
7.6	Doing the experiments	253
7.7	Analysing the data	255
	7.7.1 General considerations	255
	7.7.2 A step-by-step look at the analysis of designed data	255
	7.7.3 Case study	260
7.8	A closer look at the properties of RSM designs	265
	7.8.1 Prediction error	265
	7.8.2 Maximum prediction error	265
	7.8.3 Average prediction error	266
	7.8.4 Uniform precision	266

		7.8.5	Rotatability	267
		7.8.6	Estimation of the individual effects	268
		7.8.7	Robustness towards missing or 'wild' responses—replicated axial designs	269
		7.8.8	Blocking	269
	7.9	A glimpse of what else is out there		270
		7.9.1	Optimal designs	271
		7.9.2	Small composite designs	272
		7.9.3	Constrained regions	272
		7.9.4	Mixture problems	272
		7.9.5	Categorical variables	274
		7.9.6	Principal properties or multivariate designs	274
		7.9.7	Space-filling designs	275
		7.9.8	Power considerations—what are the smallest effects you can estimate?	275
		7.9.9	Robustness modelling	276
		7.9.10	Multiresponse optimisation	276
	7.10	Further reading		277
	7.11	Appendix: a case study in the use of response surface modelling		277
	Bibliography			278

8 Analysis of Variance. Understanding and modelling variability 279
M. PORTER

	8.1	Introduction		279
	8.2	Where to go		280
	8.3	Preliminaries		281
		8.3.1	Variables and factors	281
		8.3.2	The types and causes of variability	281
		8.3.3	Properties of estimates	282
	8.4	Variability in data		283
		8.4.1	A model for total variability in a data set	283
		8.4.2	Models, estimates and the analysis of variance	285
		8.4.3	Alternatives to least squares and the analysis of variance	288
	8.5	Modelling variability in simple linear regression		291
		8.5.1	Residuals and estimation	292
		8.5.2	Analysis of variance	292
		8.5.3	Assessing the appropriateness of the model	294
		8.5.4	Variability in the predictor	295
	8.6	One-way or fully randomised ANOVA		296
		8.6.1	Fixed and random effects	296
		8.6.2	Estimation of fixed effects	297
		8.6.3	Analysis of variance	298
		8.6.4	Estimation of random effects	299
		8.6.5	Allocation of experimental material	301
	8.7	Two-way, two-factor or randomised blocks ANOVA		301
		8.7.1	Blocking	301
		8.7.2	Crossed classifications	302
		8.7.3	Nested classifications	304
	8.8	A general approach		305
		8.8.1	A general model and analysis	306
		8.8.2	Classification of model components	307
		8.8.3	Generalised linear models	308

		8.8.4 Software for general linear models	308
	8.9	Examples of two-factor analyses of variance	308
		8.9.1 Randomised blocks design	309
		8.9.2 Replicated two-factor design	310
		8.9.3 Nested Factors	311
	References		313

9 Optimisation and control   314
T. KOURTI

9.1 Introduction	314
9.2 Terminology	314
9.3 Where to go	315
9.4 Why process control and optimisation?	316
9.4.1 Process control	316
9.4.2 Optimisation	318
9.5 Statistical process control (SPC)	318
9.5.1 Univariate Shewhart charts	319
9.5.2 A general model of Shewhart charts	321
9.5.3 Univariate Cusum charts	323
9.5.4 Univariate EWMA charts	325
9.5.5 Multivariate charts for statistical quality control	327
9.5.6 Hotelling's T^2 and chi-squared multivariate charts	328
9.5.7 Multivariate Cusum charts	330
9.5.8 Multivariate EWMA	332
9.5.9 Multivariate control charts based on latent variables	333
9.5.10 Principal component analysis (PCA) for multivariate monitoring	334
9.5.11 Partial least squares (PLS) for multivariate monitoring	336
9.5.12 Control charts based on latent variables	337
9.5.13 Fault diagnosis	339
9.5.14 Multiway data	340
9.5.15 Multiblock data	340
9.5.16 Issues in latent variable analysis and SPC	341
9.5.17 Other applications of multivariate charts	344
9.6 Optimisation of processes	345
9.6.1 General ideas	345
9.6.2 Optimise functions of many variables by changing one variable at a time	348
9.6.3 Multiresponse optimisation	349
9.6.4 Procedure for optimisation with empirical models	350
9.6.5 Sequential methods	354
9.6.6 Process optimisation using historical data	359
9.6.7 Product design with latent variables	360
References	361

10 Grouping data together—Cluster analysis and pattern recognition   365
W. MELSSEN

10.1 Introduction	365
10.2 Where to go	365
10.2.1 Visualisation of data	365
10.2.2 Similarity and finding clusters and groups in the data	366

	10.2.3 Using known groupings to predict membership of new samples	367
	10.2.4 Transforming data for better models	367
	10.2.5 A brief route through the chapter	368
10.3	Visualisation and mapping	369
	10.3.1 Principal component analysis	369
	10.3.2 Nonlinear mapping (NLM)	373
	10.3.3 Kohonen self-organising feature map neural network	376
	10.3.4 Parallel coordinates	379
10.4	Clustering of multivariate measurements	381
	10.4.1 Single, average and complete linkage	383
	10.4.2 Ward's clustering method	386
	10.4.3 Forgy's clustering method	388
10.5	Classification	391
	10.5.1 Linear discriminant analysis	391
	10.5.2 Soft independent modelling of class analogy	396
	10.5.3 Multilayer feed-forward neural networks	400
	10.5.4 Validation of the classification model	405
10.6	Appendix A. Transformation and scaling of the data	408
	10.6.1 No transformation or scaling	408
	10.6.2 Range scaling	409
	10.6.3 Mean centring	409
	10.6.4 Autoscaling	409
	10.6.5 Other transformations	410
	10.6.6 Example	410
10.7	Appendix B. Measures of (dis)similarity	413
	10.7.1 Minkowski distance	413
	10.7.2 Mahalanobis distance	415
	10.7.3 Correlation coefficient	416
	10.7.4 Which to use?	416
Bibliography		418

11 Linear regression — R. TRANTER — 421

11.1	Introduction	421
11.2	Where to go	422
11.3	Some terminology	422
11.4	Cause and effect	424
	11.4.1 General	424
	11.4.2 Relationships between variables	426
11.5	Correlation, covariance, r and R^2	428
11.6	Regression	432
11.7	Simple linear regression	433
	11.7.1 Linear models	433
	11.7.2 Linear least squares	434
	11.7.3 Assumptions	435
	11.7.4 Checking the results	436
	11.7.5 An example	437
	11.7.6 Standard errors of estimates	439
	11.7.7 The ANOVA table	441
	11.7.8 Lack of fit	443

	11.7.9 Confidence and prediction intervals	445
	11.7.10 Calibration lines and their use	446
	11.7.11 Weighted linear regression	448
	11.7.12 Errors in the x values	449
	11.7.13 What about a zero intercept?	450
11.8	Multiple linear regression	452
	11.8.1 Defining the model	453
	11.8.2 Interpreting the results	455
	11.8.3 Stepwise regression	459
	11.8.4 Principal components regression and partial least squares	461
11.9	Nonlinear regression	461
	11.9.1 Linearisation by data transformation	462
	11.9.2 Linearisation by series approximations	463
	11.9.3 Residuals surface methods	464
	11.9.4 Neural networks	466
	11.9.5 Genetic algorithms	468
11.10	Appendix: Summary of equations for linear regression	469
	11.10.1 Algebraic equations for simple linear regression	469
	11.10.2 Matrix equations for linear regression	471
Bibliography		472

12 Latent variable regression methods   473
O. KVALHEIM

12.1	Introduction	473
12.2	Where to go	473
12.3	Aims of regression analysis	474
12.4	Multiple linear regression	476
12.5	Use of generalised inverse to circumvent the colinearity problem	477
12.6	Principal component regression	477
12.7	Latent-variable regression methods	481
12.8	Relationship between regression coefficients in LVR and MLR	483
12.9	Residual standard deviation and leverage for outlier detection	484
12.10	Analysis of historic data from an industrial process	485
	12.10.1 The problem	485
	12.10.2 The data	485
	12.10.3 Weekly production, the response	486
	12.10.4 Exploratory analysis by PCA	486
	12.10.5 Regression analysis, MLR	488
	12.10.6 Regression analysis, PCR	488
	12.10.7 Regression analysis, PLS	491
	12.10.8 Revealing the lurking variable	494
12.11	Mixture design, spectroscopy and PLS regression for predicting concentrations	495
	12.11.1 The data	497
	12.11.2 Exploratory analysis of spectra, PCA	497
	12.11.3 Regression analysis, PCR	499
	12.11.4 Influence of unmodelled interferents	502
	12.11.5 Reducing the influence of unmodelled interferences by wavelength selection	503
	12.11.6 Correcting for unmodelled interferents	504
12.12	Additional hints for spectral profiles as input variables in industrial applications	505

	Acknowledgement	505
	References	506

13 Data reconstruction methods for data processing — 507
A. G. FERRIGE and M. R. ALECIO

13.1	Introduction	507
13.2	Where to go	508
13.3	Information content of data	509
13.4	The basics of data reconstruction methods	510
	13.4.1 The traditional approach to data processing	510
	13.4.2 Data reconstruction methods	512
13.5	Theory of data reconstruction techniques	515
	13.5.1 Quantified results	518
	13.5.2 Separating signals from noise	519
	13.5.3 The significance of enforcing positivity	519
13.6	Practical issues	519
	13.6.1 Unrealistic noise assessments	519
	13.6.2 *Model* mismatch	520
	13.6.3 Underfitted peaks	521
13.7	Data reconstruction applications	521
	13.7.1 Synthetic data	522
	13.7.2 Standard 1-D deconvolutions and reconstructions	526
	13.7.3 Peak detection	527
	13.7.4 Undersampled data	530
	13.7.5 Correcting backgrounds (low frequencies)	533
	13.7.6 Variable peak width	534
	13.7.7 Pattern recognition	536
	13.7.8 *Model* optimisation	538
	13.7.9 Electrospray charge deconvolutions	540
13.8	Concluding remarks	543
	References	545

Index — 546

1 Statistical thinking—The benefits and problems of a statistical approach
M. Porter

1.1 Introduction

Statistical thinking is a data rational, structured approach to scientific investigations such as problem solving, decision making, estimation and modelling reality. The components of the statistical approach described in this chapter provide a basis for the practical methods covered in this book. The benefits and problems of using this approach are also discussed. This approach is being recognised by more and more organisations as the correct approach.

Statistics can be defined as *the science of using data to improve the odds of correct decisions*. Almost all decisions are uncertain due to the variability inherent in the available data and to the impossibility of obtaining all relevant data. However, it is important that decisions are made from the available data, or at least are not in conflict with those data. Such decision making can be regarded as data rational, which is an essential feature of statistical thinking. This links to the quality-management maxim that 'you can't manage what you don't measure'. The types and causes of *variability* are described.

The mathematics of *probability, random variables* and *distributions* provides the foundations for applied statistics. These are described in a simple nonmathematical way to facilitate the understanding that is required to be able to apply statistical methods. This leads on to understanding the types of errors that can be made in making decisions and to understanding statistical and practical significance.

A process for carrying out *statistical investigations* is described. The importance is stressed of understanding the purposes of the investigation, of planning, of understanding the sampling and measurement systems, and of designing the investigation, as well as of choosing the correct statistical method.

The final section describes what you can do for yourself and from where you can get help.

1.2 Where to go

The list shows where information on aspects of statistical thinking can be found in this chapter.

Information needed	Go to Section(s)
Benefits and disadvantages of approach.	1.6.6, All
Design of investigations.	1.4.4, 1.6.1, 1.6.3
How and when to get *help*.	1.7.1
Induction and deduction	1.3.3
Importance of the *measurement system*.	1.3.2, 1.61
Measuring *probability*, risk and uncertainty.	1.5.1, 1.5.2
Randomisation. Why and how.	1.4.5
Importance of the *sampling system*.	1.3.2, 1.61
Statistical and practical *significance*.	1.5.4
Structured approach to *statistical investigations*	1.6.1
Why should you use *Statistics*?	1.3.1, 1.6.6, All
Links of *Statistics* to the scientific method.	1.3.3
Causes and types of *variability*.	1.4, 1.3.2
Effects of *variability* on decision making.	1.4, 1.5.3, 1.54
Why are all processes subject to *variability*?	1.3.2
What *you* can do for yourself.	1.7

1.3 What is statistical thinking?

1.3.1 Statistics

The linguistic roots of the word 'statistics' are the same as those for 'state', and most people would recognise the dictionary definition of Statistics as the branch of political science dealing with the collection, classification and discussion of facts. Even in this restricted sense, an understanding of statistics can be seen as an essential skill next in importance only to literacy and numeracy. The world is full of data that need to be interpreted: the causes of illness and of clusters of illnesses, the risks of different medical treatments and of no treatment, the possible and likely outcomes of political decisions, and so on. However, this dictionary definition is not adequate to fully describe Statistics. Your working environment is also full of data and using data is almost certainly a significant part of your job, as it is for most scientists, engineers and managers. You may wish to use data to make decisions, to estimate quantities and to model the real world. Even if you may not wish to do so, your customers and regulators such as the US Food and

Drugs Administration, Health and Safety Inspectorate and Radiological Protection may be requiring you to do so. Thus, you need a working understanding of Statistics and of statistics.

It is implicit in the definition of Statistics as the science of using data to improve the odds of correct decisions that decisions have to be made under conditions of uncertainty owing to variability in the available data and to the impossibility of obtaining all relevant data. All data are affected by variability in the processes that produce the data; see Sections 1.3.2 and 1.4. It is important that decisions are made from the available data, or at least are not in conflict with those data. Such decision making can be regarded as data rational, which is an essential feature of statistical thinking. No process can guarantee that the right decision will be made. However, the right process will improve your chances of making correct decisions, of obtaining best estimates of quantities of interest and of successfully modelling relationships.

Within the subject of Statistics, *statistics* are any quantities calculated from data. Thus, calculated statistics may be used to estimate parameters of interest, to aid decision making or to build models, or to measure how well these have been done. All scientists work with data and calculate quantities from the data, even if they do not think of these actions as Statistics. Thus, the choice is not of using Statistics, but of using or of not using best statistical practice in order

- To obtain the best estimates.
- To improve the chances of making correct decisions.
- To build good models.
- To assess the results against the data.
- To avoid bias from unconsidered causes of variation.
- To make most efficient and effective use of resources.

1.3.2 Building in variability—Why all processes are subject to variability

All measurements are subject to variation or error (see Section 1.4 for a more detailed discussion). That is, if two apparently identical measurements are made at different times, the two results will not be identical. Thus, a single sample measurement of the purity of a bulk chemical is very unlikely to give the true purity of the batch. The term *true* is used to represent the purity of the batch if all of the batch could be assayed with an error or random variability-free method. This true assay cannot, in practice, be measured but has to be estimated from sample data, and these estimates used to make decisions such as sentencing the batch. If two separate samples are taken from the bulk, they will probably give different results. If the same sample is assayed on two occasions, different results will probably be obtained. The likelihood of a single measurement being within some interval of the true value will be dependent upon the sampling and measurement systems. In some cases the measuring system may be precise enough for results to be effectively without error, though this is rare.

This variability is one reason why statistical methods are needed to allow us to obtain reliable estimates of parameters, to increase the chance of making the correct decision or to provide reliable models that can be applied to the real world. Statistical methods are also required if you wish to estimate the variability in the system. The ubiquitous presence of variability is also a reason why the properties of the sampling and measurement systems must be understood and why investigations should be designed so that the required specific information is obtained reliably.

Some of the potential causes of variability are:

- Heterogeneity in a material, both between batches and within batches
- Variability in a measuring device
- Material left behind in a tube, ampoule, container or vessel
- Differences in operating conditions, e.g. in actual temperature profile
- Differences between operators in technique, eyesight or training
- Changes in ambient conditions.

1.3.3 A model of the inductive/deductive processes

The process of scientific investigation is examined in this section using a description that is given in Figure 1.1. This description shows how statistical thinking fits into the standard scientific method.

Scientific investigations are responses to real-world problems, such as:

- How to increase the yield of a chemical process
- How to establish that, within defined operating conditions, a chemical process at plant scale will give product of acceptable quality
- How to measure the purity of a bulk material.

Either explicitly or, more usually, implicitly, such questions can only be asked with reference to some theory or *model of the problem* that aims to describe reality. As this is not the place for deep discussions on the philosophy of science, this link between the real world and theory will be called *abstraction*. Benefits of a statistical approach are that the model can be made explicit (though this is not always done) and that the model can be validated against data. See Section 1.6 for the phases in a statistical investigation. Then obtaining data tests the model or theory and *action* is dependent upon the outcome of this test.

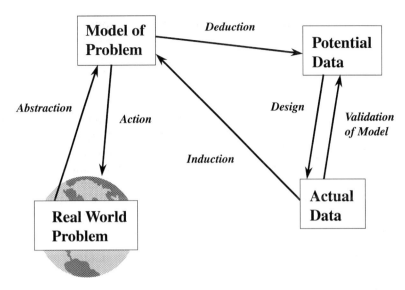

Figure 1.1 Model of the inductive/deductive process.

Thus, the problem 'How to increase the yield of a chemical process' may be answered by:

- Defining current usual or 'standard' operating conditions and measuring the yield under these conditions.
- Listing operating conditions that may give an improved yield, such as changes in amounts of materials, temperatures and times, and combinations of these.
- Operating the process and measuring the yield at each of these experimental conditions.
- Selecting the operating conditions (standard or experimental) that give the best yield (*induction from data, leading to action*).

However, as discussed in Section 1.3.2, all processes and data are subject to variation. In the example, the actual operating conditions for a particular run may vary from those planned. Possible causes are the accuracy of scales or temperature control, and operator errors. Some factors that affect the yield may not controlled or may not be controllable. Ambient conditions, mixing, temperature profiles and inhomogeneity of inputs are typical examples of such factors. Measurement of the output may be variable, for example due to sampling or the accuracy of scales. If the yield is dependent upon the batch of an input material, even measuring the yield under current standard operating conditions may be difficult. Consequently, *induction* from data back to the model by selecting the operating conditions (standard or experimental) may not be possible because the true best yield may not be identifiable or because the best yield may itself depend upon operating conditions. The design of investigations is discussed further in Section 1.6.3 and in Chapters 6 and 7.

How can the risks inherent in *induction* from limited data that is subject to variation be reduced, i.e. how can the odds of making good decisions be improved? The first step is to *explicitly* model the problem. A formal mathematical model is not necessary (see Section 1.3.4), but the model needs to include not only the effects of inputs and operating conditions upon the response, but also the components of variation (see Section 1.4). The behaviour of potential data (measurements that could be obtained under different circumstances or in the future) can then be *deduced* from the model. It follows that:

- Statistics can be chosen that allow the hypotheses about the model to be tested or parameters in the model to be estimated, i.e. to allow *induction*. Later chapters present methods for testing hypotheses and estimating parameters for particular models.
- The investigation can be *designed* so as to give acceptable probabilities of making the correct decision (or acceptable risks of making errors, see Section 1.5.4) or to give estimates of acceptable precision.

Mathematical Statistics is the science of this deductive process of developing appropriate statistics for given models and understanding their properties.

Fortunately for the practical application of Statistics, little mathematics beyond standard arithmetic and the use of tables is usually required. Well-known standard statistical methods can be applied in an easy and routine manner, often by using a statistical computer package (see Section 1.7.5), so long as the assumptions (i.e. the model on which it is based) of the standard method are valid for the problem. Any analysis should include checks on the assumptions of the methods used. In the example given above for a single experimental condition, the mean yield could be tested against the mean yield from historical data from standard operating conditions using a two-sample *t*-test so long as the assumptions of the test apply.

Once data has been obtained, it can also be used in *validation of the model* by comparing the calculated values of appropriate statistics with the expected distribution of those statistics for the potential data or by graphical methods. Outliers or discordant values in data can be identified as the abnormal values when the majority of values fit with the model.

1.3.4 *Mathematical and statistical models*

George Box is credited with saying that 'all models are wrong, but some are useful'. Useful models are those that provide insight into the real world or lead to useful predictions. Mathematical models are good at providing such insights (though not the only way) and very good at leading to predictions that can be tested. Thus, it is not surprising that core to Statistics is Mathematical Statistics, which is involved with the mathematics of variability. However, mathematics must be servant to application.

1.4 Types and causes of variability

1.4.1 *Common and special causes*

As shown in Section 1.3.2, all measurements are subject to variation. That is, two apparently identical measurements made at different times will not give identical results. There are two types, causes or classes of variation or variability in data, *special* and *common*, either of which may be systematic or random. They have different properties and it is important to understand how each can affect your measurements.

Special causes of variation are effects, factors, variables or predictors known to affect the measurement. Special causes include the effects that you wish to estimate, model or make decisions about. They also include nuisance effects that should be estimated or eliminated by good experimental design so that they do not *bias* the investigation, any estimates formed or decisions made. Special causes can be systematic or random. There are statistical methods appropriate for systematic and random effects. Examples of special causes are:

- Concentration of reagent
- Batch of intermediate
- Ambient temperature
- Laboratory undertaking the assay
- Operator assessing colour by visual inspection.

Common causes of variation are random, uncontrolled or uncontrollable effects. They are often called errors and they too can be systematic or random. The use of the word error in this context does not mean a wrong value or a mistake. The measurement has been made correctly, but the measured value is different from the *true value* because of the variability inherent in the measurement method. All measurements are subject to common cause variation.

Common cause variation is sometimes called *noise*, and the process of establishing special effects can be considered as separating signals from this background noise. One of the purposes of statistical process control (SPC) is to identify special causes of variability, to eliminate them and, thus, reduce variability in the results obtained. In the laboratory, as special causes of variability are eliminated each measurement becomes more reliable, costs are improved owing to reduction in waste, less repetition of work and lower replication, and the opportunities for improvement in the process are increased. See Deming (1986) for a fuller discussion.

1.4.2 Bias, systematic and random variation

Systematic errors are present when a sequence of measured values consistently deviate from the true, or expected, value. This is also called *bias*. Systematic errors are very difficult to detect during data analysis unless the investigation is designed to detect them. For this reason, you should try to identify possible sources of systematic error that might affect your measurements. Then design the investigation to estimate their effects, eliminate their effects by pairing or blocking, or minimise their effects by randomisation.

Examples of causes of systematic error are

- A reference calibration factor is wrong (e.g. the molar absorbtivity value for an UV measurement, or the molarity of a standard sodium hydroxide solution used in titration).
- Mobile phase for chromatography made up wrongly.
- Sampling of granular material preferentially picks up fine granules.
- Emptying a tube of powder or liquid leaves a portion behind.
- Different operators make the measurements.

All measurements are subject to random variation or error. For example, in using a stopwatch to record the duration of an event, the response time of the individual operating the watch will affect the results. There will also be variability in the response times for an individual. The impact of random variation is dependent on its magnitude relative to other causes of variability in the measurements. Thus, it is only possible to establish effects that are larger than the random error in the measurement process. However, often the size of effects of practical consequence is little greater than the random variability in the measuring system. Fortunately, replication and averaging can be used to reduce the random variability and, thus, provide valid estimates of effects.

Random errors cause the measured values to vary without any particular pattern of deviation from one result to the next. There will, however, be an overall pattern or statistical distribution of deviations, which can be specified by parameters such as an error standard deviation.

Systematic and random errors can, and do, occur together, so random error may be seen as an inconsistent pattern about a biased value.

You need to understand the causes of random error in measurements and have estimates of their magnitudes, so that your investigation can identify genuine effects and provide reliable estimates of the information you require. Some rules of thumb about random errors are:

- Random errors in weights from an analytical balance are orders of magnitude less than in volumetric dispensing.
- Time measurements made with an electronic clock can be regarded as having zero random error.
- An estimate of overall random error in a result can be obtained by adding the variances of the individual, independent contributors to the error (e.g. net weight variance = gross weight variance + tare weight variance). This can be used as an approximation; there are formal ways of carrying out the calculation.

Reduction in the random error in a result (improving the precision of the result) can be achieved by reducing the contributions of the sources of variation. You can do this by eliminating some altogether or by controlling some more effectively (for example by replacing a volumetric measurement by a weight, or by using an automated syringe instead of a manual pipette) or by averaging multiple measurements. The costs of reducing random error can be balanced against the benefits from improved information.

1.4.3 Example—Sampling from and assaying a bulk chemical

The effects of variability from independent sources are additive. Consider the process of obtaining the assay of a blended bulk material. First, a representative sample of material is taken from the bulk. How well this sample represents the bulk depends on the variability between separately taken samples, i.e. upon how well the material is blended and how well the samples are taken (for example, so that fine granules are not preferentially picked). It is important that sampling procedures are well designed.

The analyst then takes separate weighings of small quantities of material from a sample bottle and makes up solutions. These provide true replicate assays, with each step of the assay subject to error dependent upon the skill of the analyst, the quality of the equipment available and the homogeneity of the material in the sample bottle. Differences between analysts and between laboratories may also affect the results obtained in, for example, multisite trials.

Finally, the response of the measuring device to the same solution will vary randomly depending upon uncontrolled or uncontrollable factors and on the quality of the equipment. Especially within a run, the last will probably be the smallest component of variability. Thus, reproducibility experiments, which measure only the variability between replicates using the same solution run close together, are of limited value in that they measure only this smallest component.

Errors in data manipulation and rounding will then be added to the variability in the result obtained. It is important that such errors be avoided or minimised by good operating technique (see also Chapter 2).

1.4.4 Impacts of variability on decision making and the design of investigations

You wish to assess the purity of material produced by an experimental production process against the standard process. A single experimental batch has been produced along with a single control standard batch, with all other factors that might affect the purity kept constant. A sample is taken from each batch using a reliable sampling method. Each of these samples is assayed for purity just once. The results are experimental 97.8% and standard 97.5%.

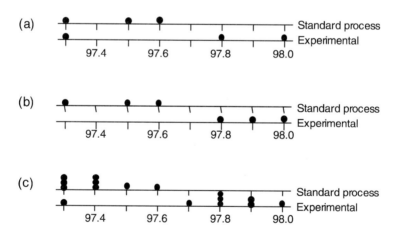

Figure 1.2 Comparison of experimental against control.

Can you say that the experimental process gives better purity than the standard? No, because you have no measure of the variability of the assay method, nor of the batch homogeneity. One way of obtaining a measure of variability is to obtain replicate samples. If three samples are obtained for each batch, the results might be as in Figure 1.2(a). The question is still quite hard to answer. If the results are as in Figure 1.2(b), the answer is clearer. Higher replication, as in Figure 1.2(c), helps further and indicates that the result of 97.3% for the experimental process looks suspect, a possible *outlier*.

In this book methods are given for

- Measuring variability (e.g. how large is the variability of the assay method?)
- Estimating the effects of variability (e.g. how does the variability affect the estimate of the difference in mean purity between experimental and standard?)
- Making decisions in the presence of variability (e.g. are the batches homogeneous?)
- Deciding how many results are needed (e.g. is one experimental batch sufficient?)
- Designing investigations that make efficient and effective use of the available material
- Identifying genuine abnormal results (outliers or discordant values) from natural variability.

1.4.5 Sampling—Why and how?

Randomisation is used because of common cause variation in measurements. Systematic common cause errors may be present in measurements, but are unknown, cannot be predicted and eliminated, are too costly to measure and eliminate, or are not expected to cause an unacceptable loss of precision. Thus, allocation of material should always be randomised within any structure used so as to estimate or eliminate special cause nuisance variation. This is equivalent to tossing a coin for service in a game of tennis. Randomisation effectively converts systematic errors into pseudo-random error.

Randomisation does not remove the cause of systematic variation; it only converts its effect into random error. This random error then contributes to the overall common cause variation in the result in the same way as any other random error. In addition, there are always likely to be unknown or unmeasurable causes of random variation affecting measurements.

Randomisation does not mean haphazard allocation. A randomisation scheme is planned and has a clear objective. The process of randomisation is discussed more fully in Chapter 2 (Section 2.4.3).

1.5 Probability

1.5.1 Measuring risk and uncertainty

All measurements have been shown to be subject to variability. Estimation, decision making and modelling all involve induction from such variable data using a model. The validity of such inferences can also only be tested by data. Thus, all estimates, decisions and models are subject to risk and uncertainty. Fortunately, probability theory and mathematical statistics provide ways of describing risk and uncertainty.

To understand and to measure risks and uncertainty, you will need to understand some basic statistical concepts. A nonmathematical understanding of these concepts is essential to understanding and correctly applying the statistical methods presented in this book. Statistical ideas are, however, formalisations of good practice in science, both in the laboratory and in undertaking plant investigations, and so you should be able relate them to your experiences.

This section gives outlines of the basic statistical concepts required in later chapters. The descriptions are, however, accurate and conform to ISO 3534 'Statistics—Vocabulary and Symbols'. Some of the concepts cannot easily be fully described in a brief outline and a better understanding may be obtained by studying the relevant chapters of undergraduate texts on probability and statistics.

The concepts introduced are *probability, random variables, data distributions, statistics* and *statistical significance.*

1.5.2 Rules of probability

There have been many attempts to define probability. The early studies considered games of chance for which the probability of an event is the ratio of favourable outcomes to the total number of outcomes assuming that all outcomes are equally likely. An application is calculating the chance of five of your numbers coming up in a lottery. This definition can be applied, for example, in assessing sampling schemes to check sterility for large batches of ampoules that have been autoclaved. However, this definition does not apply to most situations. In other areas such as actuarial science, probabilities are taken as the relative frequency of the event in a large number of cases. This definition also cannot be universally applied. Probabilities can also be taken as subjective measures of belief. Thus, probabilities do not need to be defined in any rigorous way, but can be treated a properties of the event that can be estimated and manipulated. For example, having decided on an initial belief in the probability of an event, you can see how this is changed by data as it is obtained and, thus, modify your beliefs.

What are the properties of probabilities? These are listed below together with a series of Venn diagrams (Figure 1.3) to illustrate the rules,

- The probability $P(A)$ of an event A is a number not less than zero and not greater than 1.
- If an event includes all possible outcomes, then $P(A) = 1$.
- If A and B are two events, then

$$P(A \text{ or } B) = P(A) + P(B) - P(A \text{ and } B)$$

where $P(A \text{ or } B)$ is the probability of A or B or both occurring, $P(A)$ is the probability of A occurring, $P(B)$ is the probability of B occurring and $P(A \text{ and } B)$ is the probability of A and B occurring together. (As A and B both contribute an equal amount to the overlap region in the **A or B** Venn diagram, we need to subtract one of the contributions to get the correct probability.)

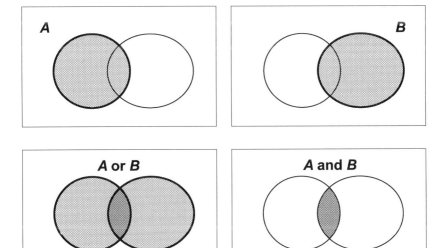

Figure 1.3 Venn diagrams illustrating $P(A)$, $P(B)$ and the construction of $P(A \text{ or } B)$ and $P(A \text{ and } B)$.

- Thus, if A and B cannot occur together, $P(A \text{ or } B) = P(A) + P(B)$. (Such events are said to be *mutually exclusive*—there is no overlap in the **A or B** Venn diagram.)
- If A and B are two events, $P(A \text{ and } B) = P(A)P(B|A)$, where $P(B|A)$ is the probability of the event B happening given that the event A has happened.

- As $P(A)P(B|A) = P(B)P(A|B)$ we can write

$$P(B|A) = P(A|B)P(B)/P(A)$$

  This is one form of Bayes' Theorem, which can be used to transform *prior* belief ($P(B)$) in an event before data has been collected, into posterior belief $P(B|A)$ having obtained data A.
- Two events A and B are *independent* if $P(B|A) = P(B)$ (and $P(A|B) = P(A)$). Thus, $P(A \text{ and } B) = P(A)P(B)$ for independent events.

As an example, consider a laboratory with several HPLC systems. Failure of the mobile phase pump in one system is independent of a similar failure in any other system. If we are interested in the probability of simultaneous failure of, say, three pumps, we simply multiply together the probabilities of failure for the three individual pumps. So, if the probability of failure for any one pump is 10^{-4}, the probability of the simultaneous failure of three pumps is $10^{-4} \times 10^{-4} \times 10^{-4} = 10^{-12}$.

As another example, consider the presence of two impurities in a material. The presence of each may or may not be independent of the other, but which case applies depends upon the process or processes that created them. If they are not independent, we need an estimate of $P(B|A)$ in order to work out the probability that both impurities appear together.

Mutually exclusive events are often obviously so. For example, if a laboratory has five HPLC systems and we have an unlabelled chromatogram from that laboratory, we know that the chromatogram can have been produced by only one of the systems. The probability that it was produced by a specific system, say system X, is the probability that that particular system would be chosen, $P(X)$. The probability that it was produced by either of two specific systems, say X or Y, is the sum of the probabilities that each of the systems would chosen, $P(X) + P(Y)$.

1.5.3 *Random variables, distributions and statistics*

A (probability) *distribution* is a mathematical function that gives the probability that a variable takes a given value or belongs to a set of values. Such variables are usually called *random variables*. Thus, for example, if a large batch of ampoules includes $100p\%$ that are defective, the probability distribution of the random variable 'number N of defective ampoules in a sample of 10 drawn at random' is given by the binomial distribution,

$$P(N = n) = {}^{10}C_n \cdot p^n \cdot (1-p)^{10-n}, \qquad n = 0, 1, 2, \ldots, 10.$$

It is not necessary for you to understand the mathematics of distributions, unless you wish to calculate probabilities. However, you must recognise that distributions and probability calculations are the basis of statistical methods. Often the assumptions of a method include statements about the distributions of data that need to be true for the use of the method to be valid. If you are not sure that a method is appropriate for this application, you should consult an expert.

Some of the most frequently used distributions are the following.

Normal (Gaussian): Random variation such as variability in assays obtained for the same sample using the same method and conditions; symmetric and bell-shaped, depends on mean and standard deviation.

Binomial: Numbers of items with some characteristic in a finite sample taken from large ('nearly' infinite) group, e.g. defective ampoules in a sample from a large batch.

Poisson: Numbers of randomly occurring relatively rare events in a finite sample, e.g. extraneous particles in a sample of bulk material.

The mathematics of probabilities and distributions is deductive, but science and statistics are inductive. Practical questions are inductive, for example:

- How do I estimate the proportion of defective ampoules in the batch from the sample data?
- How big a sample do I need to take to estimate the proportion of defective ampoules so that the estimate is likely to be within an acceptable distance of the true value?
- In routine monitoring how many ampoules should I test so that the outcome (no defectives in the sample) implies an acceptably low risk of no more than a few defectives in the batch?
- Given that there are 5 defective ampoules in a sample of 10 from a large batch, how many are there likely to be in the batch? Is it likely that only 1 in 1000 in the batch is defective?

In the last of these examples, deductive mathematics can be used work out the probability of 5 defectives in 10 ampoules if $p = 0.001$ as less than 1 in a billion, so that it appears unlikely that $p = 0.001$. The mathematics of probabilities can be used to assess whether particular values are likely and hence define what range of values is likely.

We see that statistical methods based on the theory of probabilities can be used to find likely answers to inductive questions. The probability of the answer being incorrect can also be estimated. Thus, a guide is provided to decision making, though it is a guarded guide.

As all measurements are subject to variability, the set of possible values that they could have taken is a random variable. This variable is usually called the *population*, its distribution is called the *population distribution* and its properties are called the *population parameters, true values* or *population statistics.*

Similarly, as a *statistic* is a value calculated from data, the set of possible values it could have taken is also a random variable. The distribution of possible values for a statistic can then be used to set limits on likely values (*confidence limits*), make decisions (*significance testing*) or develop models for the population parameters. For example, you may wish to test the hypothesis that the experimental process population average assay is higher than that for the standard process using the sample means.

Fortunately, you rarely need to develop the mathematics and can usually apply methods developed for standard situations. However, you need to check that the methods and their assumptions apply to your application. If in doubt, consult a statistician.

1.5.4 Practical and statistical significance

The purpose of making measurements or carrying out experiments is often to gather the information that will allow some decision to be made. For example, assume that the assays on a standard solution obtained by a laboratory on a production site are to be compared with those obtained by the development laboratory. The analysts wish to answer the question 'Are the assays the same?'

Even if a single laboratory did two runs, identical results would be unlikely because of the random variation in the measurements. This means that the question really needs to be reformed as: 'Do the results from the two laboratories come from the same population of results or are they sufficiently different as to indicate that they come from different populations?'

There is a decision to be made between these two possibilities based on data which will be called the *Null Hypothesis* H_0 (no difference) and the *Alternative Hypothesis* H_1 (there is a difference). In truth, the Null Hypothesis is either true or false; unfortunately, we do not know which. The decision to reject or not reject H_0 will be made on data obtained by measurement. As the data are subject to variability, wrong decisions may be made. In practice, decisions are based on the value of *test statistics* calculated from the data. In the example, some samples could be split into two and one of each pair assayed by each of the laboratories. The mean of the difference in assay for the pairs could be used as the test statistic. The observed value of the test statistic is compared with a *critical value*, which is derived from

- The distribution of the possible values of the statistic when the Null Hypothesis H_0 is true
- The distribution of the possible values of the statistic when the Alternative Hypothesis H_1 is true
- The risks that are acceptable to the experimenter (see below)
- The number of results or replication for each laboratory.

The result is said to be *statistically significant* if the observed value is 'bigger' than the critical value. In the example, the mean difference is statistically significant if it is larger than some value. Again fortunately, you rarely need to do any mathematics to get the critical values but can use tables or a statistical package.

In designing tests, the numbers of results or replications must be sufficient to give acceptably low risks of the two types of error:

- Saying that there is a difference between the laboratories when in truth there is no difference (*Type I error*)
- Failing to detect differences that are of practical consequence (*Type II error*).

The two types of error are summarised in Table 1.1.

Table 1.1 Types of error

	H_0 is false	H_0 is true
Decide to reject H_0	√	Type I error
Decide not to reject H_0	Type II error	√

√ = correct decision made.

The probability of a Type I error is called the *significance level* of the test and, in the example, it measures the risk of saying that there is a difference between the laboratories when there is not. This is usually denoted by α or p, and often called the *P-value* of the test. A frequently used value for α is 0.05, that is accepting a 5% risk of making a Type I error. The value $(1 - \alpha)$ is the risk of not making a Type I error and is sometimes referred to as the confidence level of the test.

The probability of a Type II error is usually denoted by β. In the example this is the risk of saying that there is no difference between the laboratories when there is in truth a difference (of practical consequence). The value $(1 - \beta)$ is called the *power* of the test and it is the probability of correctly rejecting the Null Hypothesis when it is in fact false. As the Alternative Hypothesis is not single value (H_1: difference $\neq$ zero), the power is also not a single value, but a function (the power function) of the possible values of the difference from zero. Thus, in designing a test, you have to define the power required at a particular value of the alternative. This may be expressed as: 'The power of the test is at least 90% if the difference in the averages is 0.2 (the power will be greater than 90% if the difference is greater than 0.2)'.

The required number of samples (replication) can be derived using these considerations. In practice, the cost, amount of available material or other resource may apparently limit or fix the level of replication. In such a case, the power of the test should be checked. If the test does not give adequate power for testing differences of practical significance, you are unlikely to obtain useful information and should consider aborting the test (and, thus, saving resources). If the power is greater than needed, resources could be released to other projects.

So far, this discussion about decision making has looked at the statistical aspects. It can be that a statistical test will reject the Null Hypothesis, for example showing a significant difference between laboratories, when the difference is actually very small. Statistical tests can be too sensitive if sample sizes are too large or if the variability in the data is very small. The latter can occur in validation and analytical transfer exercises when the method variability is small. To counter this, it is acceptable to define a value of *practical significance*. This is the value below which you believe any difference between data sets has no practical significance and to which it is meaningless to apply a statistical test. For example, your practical experience may tell you that an assay cannot be measured more accurately than to the closest 0.1%; thus, any difference less than this should be of no practical consequence or significance. So, if a statistically significant difference of less than 0.1% is found using this assay during a carefully controlled exercise to compare samples of material produced under different reaction conditions, the finding can be discounted as not having practical significance.

Table 1.2 shows the four outcomes for statistical/practical significance. If the outcome falls in the 'agreement' cells Yes/Yes and No/No, there is no problem. Similarly, for the statistical significance = Yes and practical significance = No cell, there is no problem because the effect is not large enough to justify action (other than to recognise that the experiment was possibly too large and that resources may have been wasted). The problem

Table 1.2 Practical and statistical significance

		Practical significance	
		Yes	No
Statistical significance	Yes	√	No problem
	No	A problem	√

√ = agreement.

occurs when the outcome falls in the statistical significance = No and practical significance = Yes cell. There are two solutions. You could obtain more data. In this case, if the effect is genuine it will be confirmed and become statistically significant, and if it is not genuine it will lose practical significance. Alternatively, you can take a gamble, based on the measured risks and likely payback.

If you apply this approach it is important that you define and justify the practical significance value at the same time as the Null and Alternative hypotheses, that is, *before* you gather the data. The setting of practical significance value, after gathering data and finding that you should reject the Null Hypothesis even though the effect is small, cannot be justified.

1.6 Carrying out a statistical investigation

A process is given for carrying out a statistical investigation using a formal, planned and systematic approach. The next section gives the steps in such an investigation and is followed by sections expanding on particular topics: prospective and retrospective studies, designing investigations, and pilot studies. The requirements for statistical validation are discussed in Section 1.6.5. The final section looks at benefits and disadvantages.

1.6.1 Steps in an investigation

A formal, systematic and planned approach to investigations usually rewards the effort required, by leading to reduced project costs, shorter times, better and more appropriate information and more secure conclusions. There are many systematic approaches to investigations, all of which roughly follow and elaborate on the Deming or 'PDCA' cycle of *PLAN-DO-CHECK-ACT* well known in continuous improvement (Oakland, 1993). Statistical investigations can be represented in this cycle, as shown in Figure 1.4.

PLAN is split into five phases: define objectives; select sampling and measurement systems; assess data properties; select statistical methods; and design investigation. Planning is the most important part of an investigation and needs to be well structured. Adequate time and resources should be allocated to planning. Although phases may be taken out of order and may be repeated as information becomes available, it is rarely wise to ignore any of them. *DO* consists of a single phase: collect data. *CHECK* consists of three phases: assess data, calculate results and interpret results. *ACT* consists of a single phase: review and report. In

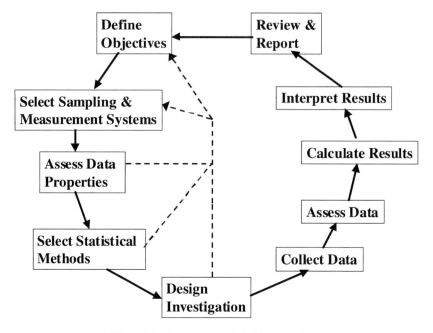

Figure 1.4 Steps in a statistical investigation.

practice many of the phases can be completed quickly and for routine tasks a standard operating procedure is all that is required.

Each of the phases will now be discussed.

PLAN—Define objectives
This covers the statements of aims, objectives and purposes. What is to be done? Why the work is being done? Who is the customer? By when must it be completed? What information is required, and how precisely? Who else is involved? What are they doing and are there overlaps? What risks are acceptable? What resources are available? What is the endpoint? How will it be recognised? What outcomes are possible?

In this phase, decisions are made on what can and should be done; roles are defined; related work and work carried out previously are reviewed; resources are allocated; and impacts on other projects are reviewed. A plan may be drawn up for reviews, meetings and communicating results.

For example, the aim of a project is to improve the purity of an intermediate in order to improve the yield at the next stage in the process. Resources for 20 laboratory-scale runs are available and the work must be completed in a month. Also production costs must not be increased.

The required outcome is an improved process acceptable to the production manager.

The process of determining objectives is as important to a small-scale, one-person experiment as it is to a large project involving many people.

PLAN—Select sampling and measurement systems
This phase is often not explicitly considered, but the sampling and measurement systems impact directly upon the likelihood of meeting the objectives. For example, poor sampling or a low-precision measurement system may cause genuine differences to be hidden in noise. This was discussed in more detail in Section 1.4. The aims, objectives and purposes are turned into data requirements. Some or all of the data required for some studies may already be available in databases.

PLAN—Assess data properties
This phase is coupled with selecting sampling and measurement systems, because these define the data that can be obtained, the accuracy and precision of that data, and at least some of the causes of variation in the data. It also leads directly into the next phase, selecting appropriate statistical methods, because these are defined by the properties of the data. The assessments should be both statistical and practical.

Many questions need to be answered (either explicitly or implicitly). What types of variables describe the data: qualitative or quantitative, discrete or continuous, finite or infinite? What values are possible? Are any of the variables linked, functionally dependent or correlated? How precisely can measurements be made? How can this precision be measured? Are external measurements of precision available, or do we have to measure precision within the investigation? What statistical distributions apply? What models apply? What components of these models have to be estimated? What hypotheses have to be tested statistically?

Thus, in the example started above, purity is measured by HPLC. The HPLC assay is a quantitative variable affected by sampling and analytical method variability. These may have been estimated during validation. Otherwise their combined effects will have to be estimated within the study. The underlying distribution of this variability is probably Normal or Gaussian. The laboratory runs can be assumed independent.

PLAN—Select statistical methods
The choice of methods will flow from the previous phase. The appropriate methods will also depend upon whether the investigation is prospective or retrospective, see Section 1.6.2. The assumptions and requirements of methods should be examined.

In this phase, assessments should be made of the Type I and Type II errors, and power curves for alternative methods. These assessments lead on to specifying the replication required. Gardiner and Gettinby (1998) describe 'statistical conclusion validity' as ensuring that sample sizes are large enough to ensure adequate power, i.e. an acceptable probability of detecting true effects of practical significance (see Section 1.6.5).

Decisions will be made on whether to undertake a pilot study (see Section 1.6.4). For larger studies, software is identified that can carry out the analysis.

Returning to the example, as assay variability is Normally distributed, parametric tests can be used such as the t- and F-tests, and the analysis of variance. However, as the number of runs is small, it may not be valid to assess many ideas empirically. In the design phase, factorial designs can be considered that will use the runs efficiently and effectively.

PLAN—Design investigation

This is discussed in more detail in Section 1.6.3. The design should also include protocols for recording results, for carrying out calculations and for computer data storage of results, and for recording abnormal processing—unusual batch temperatures, delays, suspect HPLC results, etc. Computer data entry is a common source of error and procedures should be set up to ensure that the data used in the analysis is valid. For large data sets, the use of two people entering the data independently and computer-based cross-checking will avoid most errors. For small data sets or when double entry is not possible, it is wise to check or audit the data independently. Operators, technicians and others involved may also need training.

DO—Collect data
Follow the protocol.

CHECK—Assess data

This will usually split into two parts: first, inspection of the data before any calculation, estimation or hypothesis testing; and second, inspection of residuals after calculation. The residuals are the differences between the observed values and their estimates from the underlying model. The use of residuals for assessing the appropriateness of models is discussed further in Chapter 5. Validate and check the results; audit against original records.

Preliminary analysis of data may check the following:

- Trends and unusual patterns in data, using plots against batch or run number, or statistical process control (SPC) charts
- Unusual or discordant values, using plots or distribution tests

- Invalidating correlations, using scatter plot matrices of predictor variables
- The distributions of the variables, for example by Normal probability plots
- That the assumptions of the chosen methods are valid.

It may then be necessary to review the plan for analysis, obtain more data or review the design. It may be possible to estimate components of variability and likely impacts on decisions. Similar analyses may be carried out on the residuals.

CHECK—Calculate results
Calculate the estimates, confidence limits, P-values and other statistics specified in the Select Statistical Methods phase, modified as appropriate by observations made during data collection and assessment. Check the results obtained.

CHECK—Interpret results
Assess the statistical and practical significance of results. Look for explanations of, and implications from, the results. Assess how well the aims, objectives and purposes have been met. Identify gaps, including gaps in carrying out what was planned, for example missing data. It may be necessary to rectify the latter before reporting. Draw up conclusions and recommendations from work completed. Decide on ways of presenting results in graphs, tables, words or equations (see Chapman and Mahon, 1986).

ACT—Review and report
Present and discuss results, outcomes, conclusions and recommendations formally and informally, by presentations and written reports as appropriate. If some of the objectives have not been achieved, identify what has to be done to achieve them, e.g. by obtaining additional data, by specifying a new project or by modifying objectives. Identify other actions and further work, and who is to carry them out. Identify and report lessons from carrying out the task, including practical constraints such as difficulties in sampling, limits of measurement systems, reliability of data sources and properties of data obtained, e.g. statistical distributions.

1.6.2 Prospective and retrospective studies

Many statistical investigations are retrospective in that the data has already been collected for other purposes such as routine reporting of production performance; examples might be measurement of impurities in order to assess batch quality against specification or recording of processing times, temperatures or quantities to check compliance with the manufacturing guide. Such data contain much information on the process and can be used with care for process monitoring, assessing changes, trouble-shooting and formulating hypotheses using statistical process control (SPC) techniques, regression analysis and other statistical methods.

However, the available data and knowledge on what actually happened limit retrospective studies, which may not give as reliable answers to questions as would properly designed prospective studies. For example, it is not possible to randomise retrospective studies properly, nor is it possible to ensure that all other factors were kept equal. Thus, retrospective studies may best be used for formulating ideas for testing by prospective studies. Using historical data to set action and warning limits on typical product quality, against which current performance is tested in order to identify change, is a typical example. Such limits only remain valid for as long as the historical data provides an acceptable description of process performance. However, SPC can be useful in identifying special causes of variation, the effects of which can then be estimated in a designed experiment.

1.6.3 Designing investigations

For a prospective investigation, the design turns the aims, objectives and purposes into data requirements. It also takes account of the sampling and measurement systems, and of the properties of the data collected.

In designing an investigation, account should be taken of six components of the system:

1. The properties of the response or responses to be measured, estimated or optimised. How accurately can they be measured? What factors and variables may influence the level and variability of each response? What type of variable is each response? Is a response a continuous variable such as the yield of a reaction or a count such as the number of particles in a 1 g sample, or defined by a range such as percentage of damaged tablets in a sample? The properties may include correlations or relationships between the responses.
2. The properties of the qualitative factors ('treatments') or quantitative variables ('predictors'), the effects of which on the response the experiment will explore. How well can these be controlled? How accurately can they be measured? What factors and variables

may influence the level and variability of each? What types are they? What are their likely effects on each response? Do their effects on any response interact? What questions about these effects are to be answered?
3. The properties of the qualitative factors ('blocks') or quantitative variables ('covariates'), which effect the responses and which need to be eliminated, fixed or structured so as to not bias conclusions about the responses. These may be called nuisance factors. Included in this list are the known properties of the material available, which may be used for blocking or as covariates. Other possible blocking factors are technicians, shifts, days, HPLC runs and batches of inputs. How well can these be controlled? How accurately can they be measured? What factors and variables may influence the level and variability of each? What types are they? What are their likely effects on each response? Do their effects on any response interact? Do they interact with any of the treatments or predictors?
4. In addition to the known special causes of variability in (2) and (3), there will be unknown or unknowable common causes of variability (see Section 1.4). The effects of the latter can only be minimised by randomisation.
5. The properties of the chosen statistical tests, which will define the replication required to achieve adequate power.
6. The amounts of experimental material and other resources available within the budget. Other work may be competing for these resources. If the replication is too low for validity of statistical conclusions, there is little value in carrying out the experiment.

The first step in designing an investigation should be to list the responses and to list against each the treatments, predictors, blocking factors, covariates and other possible influencing factors and variables. Against each identify the likely variability or range of values, and the likely effects on the responses and how these are to be estimated or eliminated.

The design of experiments is discussed in Chapters 6 and 7.

1.6.4 Pilot study

A pilot study may be needed to understand the properties of the selected sampling and measurement systems. For example, to measure the impact of sample size on the measurement system or to measure the impact of potential causes of variability on the data obtained. The aim of the pilot study should be to define: the data that can be obtained; the accuracy, precision and distribution of that data; and to measure at least some of the causes of variation in it. This allows the appropriateness of possible statistical methods to be explored. The assumptions and requirements of methods can be examined. Better estimates can be made of the likely sizes of effects of interest. The required replication can be better estimated and power assessments made. Alternatively, the study can be aborted, and resources saved, if variability is too large, effects are too small or resources are too limited for statistical conclusion validity. The software can be tested.

1.6.5 Validation and good statistical practice

Gardiner and Gettinby (1998) describe four types of statistical validation. These fit well with the process described above. Their four types and their fit to the phases of the statistical investigation process are as follows.

1. *Statistical conclusion validity* is defined as ensuring that there is sufficient data to allow decisions to be made with acceptable probabilities of Type I and Type II errors (see Section 1.5.4). This should be extended to include all of *PLAN*-assess data properties, *PLAN*-select statistical methods, and *CHECK*-assess data phases (in Section 1.6.1) namely also ensuring validity of the methods chosen relative to the properties of the data and ensuring that checks are made on the validity of the methods chosen against the data actually obtained.
2. *Internal validity* refers to the details of the study such as randomisation, maintenance of similar conditions within groups, avoidance of bias and recording of data including abnormal events. This may perhaps better be called design validity.
3. *External validity* refers to taking reasonable precautions to ensure that the study findings will extend to the real world.
4. *Construct validity* refers to checks that the objectives of the study are answering the correct questions.

1.6.6 Benefits and disadvantages

If it is operated in a formal, systematic and planned way, there are many benefits to the statistical approach to investigations presented here. The approach leads to reduced project costs, shorter times, better and more appropriate information, and more secure conclusions. Many of these benefits come in part from adopting a systematic approach, as a nonsystematic statistical approach will give fewer advantages.

A formal, planned and systematic approach will give clear statements of aims, objectives and purposes, a plan of action and success criteria against which progress can be assessed. Furthermore within such an approach:

- Statistical design of investigations is both efficient and effective. For example, factorial experiments have been shown to be more effective than one factor at a time experimentation in identifying the conditions for maximising the yields of chemical reactions as well as being more efficient.
- Adopting a data rational approach to problem identification and solution reduces the time required. This can be summarised in two comments by chemical plant managers: 'We now let the data talk for itself' and 'We stopped arguing about what had happened and got on with solving problems'.
- Statistical methods are designed to maximise the information gained, not only by providing the best estimates but also by measuring the precision of those estimates and by not only identifying the best process for making decisions but also measuring the risks of reaching the wrong conclusion.
- Presentation of results statistically (particularly in graphs) aids understanding.

The main disadvantages are:

- Statistical methods can be applied unthinkingly. This is even encouraged by some textbooks that are little more than 'cookbooks' of formulae and procedures. Some managers, regulatory inspectors and editors of journals can also encourage this approach. Pressures on time, expectations of others and past experience can push the user towards particular statistical designs, analyses and presentations of results. This can be overcome adopting the process given in Section 1.6.1, by training or by obtaining specialist help.
- Statistical calculations are easy, particularly with the increasing availability of statistical packages and functions within spreadsheets. However, choosing the right method can be difficult. Too much trust can be put in computer-generated results (see discussion in Section 1.7.5). Again this can be overcome by adopting the process given in Section 1.6.1, by training or by obtaining specialist help.
- There are many methods available and the user may find the differences between them difficult to distinguish. This can be overcome by understanding a few methods well, in particular their limitations, and getting help whenever a task is outside this skill set.

- Users can be tempted to use the latest method or sophisticated procedures when simple methods are adequate. This temptation should be resisted.
- Users may also be tempted to rely too much on the statistical methods. For example they may present the analysis in detail without supporting comment, conclusions and recommendations. Reporting should be appropriate for the audience. Statistical detail is often best in appendices. Alternatively, they may use the methods too rigidly (see the discussion on practical and statistical significance given in Section 1.5.4.).
- Colleagues and managers may be resistant to the use of a formal, planned and systematic approach, believing incorrectly that costs and times will be increased. This can be overcome by successful and proven applications, and by reference to the wide literature giving such successes.
- Furthermore, colleagues and managers may be resistant to the use of statistics, preferring to trust to their experience, judgement and insight, partially out of lack of understanding of how the process given in Section 1.6.1 can reinforce these. This can also be overcome by successful and proven applications, and by reference to the wide literature giving such successes. The enthusiast may have to become an educator.
- One common comment is 'If an experiment needs statistics to analyse, it can't be a good experiment'. This is simply incorrect, as is shown by many examples throughout this book. In addition, few experimenters and organisations can today afford to detect and, thus, use only those changes that are large enough to meet this condition.
- Arguments based on probability theory and statistical distributions can be complex and, thus, difficult to follow, and many people look for simple answers. However, if an answer is simple, it is probably wrong. This is particularly true for modern chemical processes and systems, which are complex, subject to many causes of variation and probably partially optimised. This can be overcome by focusing on conclusions and recommendations, supported by statistical arguments essential to the argument. Slowly, statistical ideas can be introduced and colleagues educated.

Finally:

- The weaknesses of a statistical investigation should be evident (e.g. if the replication is not adequate to provide valid conclusions), whereas the weaknesses of a nonstatistical investigation are hidden.

1.7 Skills, experts and systems

1.7.1 When and how to get help

There are many ways to seek help when you feel that you do not have the statistical knowledge needed in your work, and to validate what you are doing even if you feel confident that you know what should be done. You should look to continuously improve your understanding and knowledge, expanding outwards from the current needs of your job. In some industries, it is likely that the regulators will be demanding more and more use of statistical methods, for example in the design of and analysis of data from validation. There are things that you can do for yourself, such as reading books, attending courses, developing a self-help group and learning about software. These are discussed in the next section. Within your organisation there may be people who have relevant statistical knowledge, are enthusiastic about applying Statistics and pleased to help. Such people may not be called statisticians. These are discussed in Section 1.7.3 and 1.7.4. The final section discusses software in more detail.

1.7.2 What you can do for yourself

You have already started this process by reading this book as far as this. The other chapters cover many of the statistical concepts and methods that you will need in your work. There is a great volume of other published literature, but much of it is aimed at particular groups of students with the objective of helping them pass examinations. You may already have attended a course as part of a science degree that appeared to have little relevance to the rest of the course. This was probably partially due to the failure of academic science to integrate data analysis within courses (for example, to use statistical experimental design in laboratory work), partially due to the focus of much academic Statistics on mathematics rather than applications and partially due to your inexperience. The last of these is no longer true. Statistical methods are most easily understood when you can see or seek applications in your own work.

Useful books and articles are given in the references at the end of this chapter, and throughout this book. Davies and Goldsmith (1976), though out of print, gives a thorough and fairly advanced introduction to statistical methods for the chemical industry. The books by Caulcutt and others cover less ground, and may be an easier starting point: Caulcutt (1983), Caulcutt and Boddy (1983). A software-based approach is taken by Gardiner (1997). Sprent (1988) gives a more general basic introduction to statistics. Gardiner and Gettinby (1998) discuss the design and analysis of experiments. One way of introducing statistical methods is as part of Continuous Improvement or Total Quality Management. Useful books are Oakland (1993) and Bendell (1993), though you should also read Ishikawa (1985), who introduces his seven basic techniques for quality improvement. Finally, anyone who is interested in the statistical approach, statistical thinking, quality improvement and management should read Deming (1986).

Training in Statistics and in using statistical software is available through colleges, universities and professional organisations (such as the Royal Statistical Society in Britain, the American Statistical Association and the American Society for Quality Control). Many courses focus on specific applications. Professional organisations may also provide useful contacts with statistical consultants.

One way of developing statistical skills is to set up a self-help group or network. Identifying like-minded people can reduce the burdens of learning, of trying out methods and of persuading the sceptical, and can help you avoid pitfalls. Regular meetings can be held to discuss problems and to share experiences and best practice. Vandenbroeck and Vandevyvere (1996) describe the setting up of such a skill group at Chemical Research Group Shell. They talk about the experience and the lessons learnt. Such a group can also act as a focus for arranging formal training, for developing local experts, for creating the computing infrastructure and for communicating with professional statisticians within the company, in academia or elsewhere (see below).

Section 1.7.4 outlines what professionally trained statisticians can contribute. However, to maximise that contribution, you need to be able to communicate with them. You will need to describe the process, project, task or problem in a way that they can understand, with the detail needed to define the statistical project. This can be done using the process given in Section 1.6.1. An understanding of statistical thinking and statistical concepts will help. Also good statisticians will want to understand your problems and your language, and to facilitate communication.

1.7.3 Local experts

Your interest in statistics may have been fostered by working with an enthusiast. This person is probably the most important factor in developing your knowledge. You should call on their help, not only to solve problems but also to help your learning process by suggesting books and courses. However, as with all experts, their solutions should not be taken blindly, especially if they are only a little further up the learning curve than you.

Within a department, a self-help group with dedicated professional support may be the best solution. However, professional support may not be available and, even if available, will be limited. Thus, the development of local experts will give benefits. These will undertake significant statistical training, but remain within the department. Such people will usually understand most of the statistics required for applications in the department, and will also have a deep understanding of the work of the department. Thus, local experts play a role similar to that of paramedics in being trained in relevant skills, providing initial support and reducing the burden on specialists. They will, however, not be able to undertake all that a professionally trained statistician can do. It is thus worth also gaining access to professional help, either within the company or externally. The register maintained by the Royal Statistical Society and mentioned above is a useful source of information.

1.7.4 What a statistician can contribute

Most statistical tasks will be undertaken by nonstatisticians simply because there are not enough statisticians to undertake them all. So what is the role of the statistician? Porter has outlined the role of the professionally trained or chartered statisticians in industry. Although statisticians have skills in analysis, using statistical computer packages and presentation of data, to use them only for these is to waste their skills. R.A. Fisher is quoted as saying 'To consult a statistician after an experiment is finished is often merely to ask him to conduct for a post-mortem.' Thus, involving the statisticians in design is important, as is involving them in defining the process by which the investigation is undertaken. Statisticians will bring a unique perspective that comes from being outside the area of application and from wanting to understand and model the essential features of the problem, particularly the likely sizes and significance of effects and the acceptable risks in making decisions.

Porter (1993) lists the attributes of a good statistician:

- Involvement, especially with decision making
- Good communication skills, listens and persuades, uses the customers language
- Available and approachable, has an open door
- Takes a flexible, customer focused approach
- Able to learn other subject areas and, thus, to understand and construct models
- Able to meet customer deadlines
- Good technical skills, mathematics, computing, presentation.

Many of these skills are equally important to the user of statistics.

1.7.5 Using statistical software

Although many statistical methods can be carried out using a calculator, most are more conveniently carried out using computer software. Some became possible only with the development of high-speed computers and can only be carried out using a computer.

Spreadsheets offer many of the calculations described in the book and are widely available. As the statistical functions may only be a small part of the spreadsheet package, they may not give the quality or range of analyses provided by statistics packages. Spreadsheets are a good way of capturing data and sharing them with other systems, with most statistical packages taking data from a spreadsheet. However, care must be taken to ensure that data are valid when input and remain so.

There are many statistics software packages, some of which are suitable for the nonexpert user, though some are not. You should seek expert statistical advice, as well as computing advice, before you purchase a statistics package. One consideration in selecting a package is that regulators of many industries and professions now require proof of validation of computer software.

Whatever software you use, you need to ensure that it is carrying out the calculation that you want. For example, what divisor is used in calculating the standard deviation? You should be aware that spreadsheet statistical routines and statistical packages are not all error free. You should check the results obtained whenever possible before your first use of a routine. This can be done by testing the software with small data sets, the results for which can easily be checked against manual calculation. Software manuals often include examples that can be used for validation. These examples are also often helpful in understanding methods.

References

Bendell, T., Kelly, J., Merry, T. and Sims, F. (1993) *Quality, Measuring and Monitoring*, Century, London.
Caulcutt, R. (1983) *Statistics in Research and Development*, Chapman and Hall, London.
Caulcutt, R. and Boddy, R. (1983) *Statistics for Analytical Chemists*, Chapman and Hall, London.
Chapman, M. and Mahon, B. (1986) *Plain Figures*, HMSO, London.
Davies, O.L. and Goldsmith, P.L. (1976) *Statistical Methods in Research and Production*, 4th edn, Longman, London, for ICI.
Deming, W.E. (1986) *Out of the Crisis*, MIT Centre for Advanced Engineering Study, Cambridge, Massachusetts.
Gardiner, W.P. (1997) *Statistical Analysis Methods for Chemists*, Royal Society of Chemistry, Cambridge.
Gardiner, W.P. and Gettinby, G. (1998) *Experimental Design Techniques in Statistical Practice: A Practical Software Based Approach*, Horwood, Chichester.

Ishikawa, K. (1985) *What is Total Quality Control: The Japanese Way*, Prentice-Hall, Englewood Cliffs, NJ.
Oakland, J.S. (1993) *Total Quality Management*, 2nd edn, Butterworth-Heinemann, Oxford.
Porter, M.A. (1993) The role of the statistician in industry. *The Statistician*, **42** 217-227.
Sprent, P. (1998) *Understanding Data*, Penguin, London.
Vandenbroeck, P. and Vandevyvere, P. (1996) Statistics in a new business environment: an example. *The Statistician* **45**, 287-292.

2 Essentials of data gathering and data description
R. Tranter

2.1 Introduction

The reason for applying statistical methods is to gain an objective assessment of particular information contained in the data. There are four essentials to obtaining good quality and meaningful results:

- The data are of good quality.
- The information being sought is contained in the data.
- Appropriate data analysis methods are used.
- The results of the data analysis are interpreted within the context of the source of the data.

The last three of these items are the subjects of the other chapters in this book. The first is the subject of this chapter.

Good preparation for, and proper design of, data collection are key to the whole process of successful data analysis. It cannot be stressed too strongly that the effort on planning and design, put in *before* data collection is made, will be amply repaid. Measurement systems and data recording are important parts of this process that we shall look at.

The visual inspection of all data is good practice and should always be performed before carrying out a more detailed analysis. Trends, distributions, outliers and relationships may be obvious without further treatment. Such preliminary analysis may give good indications as to whether more detailed analysis is necessary and whether it may (or should) proceed. So we look at the use of a selection of tools to display data, confirm randomisation, check Normality, identify outliers, etc. All are aimed at assessing the quality of the data before it is committed to detailed analysis.

Finally, some simple data processing before more complex data analysis might give useful information in its own right and/or it might improve the reliability or quality of results from the later processing. Some common data pre-processing techniques form the last part of the chapter.

The techniques described in this chapter are essentially univariate methods, that is, they apply to one variable at a time. They can be applied to individual variables in multivariate data and some useful information will be obtained. *However, it is dangerous to apply univariate methods to multivariate data without applying a large dose of caution at the same time.*

Ideally, multivariate data should be handled with multivariate methods. This will usually mean principal component analysis (PCA) as the first step, and then examining the principal components by methods similar to the univariate methods. Although PCA is introduced in this chapter as a data transformation method, Chapters 8, 10 and 12 illustrate principal component methods that may be used also to describe or check the quality of the data.

2.2 Where to go

Information needed	Go to Section(s)
Is there anything I can do to get good measurements?	2.4
What problems are there with getting data from instruments?	2.4.4, 2.4.5
How can I check for differences between groups in my data?	2.5.4
Is there a simple check for trends and correlations in data?	2.5.2, 2.5.3
I think there may be some step changes in my data.	2.5.2, 2.5.5
Are my data suitable for use with statistical methods requiring Normality in data?	2.5.6, 2.5.7, 2.6.4
What can I do about outliers?	2.5.3, 2.5.6, 2.5.8
What precision should I use to record data?	2.4.6, 2.4.7, 2.4.8
How do I go about getting rid of noise in my data?	2.6.1
Can I enhance the resolution between two peaks in the spectra?	2.6.2
How do I look at a large data set with many variables?	2.6.3

2.3 The data cycle

Most research activities are iterative in nature. There is an initial idea, the formation of hypotheses, testing these with data and reformulating the hypotheses to start the cycle all over again. In this way, new information is used to refine hypotheses and take them closer to the 'truth'. Occasionally, information from new data is so startling that a major step is taken, the original hypotheses are abandoned and new ones are formulated.

This data cycle is so fundamental to problem solving (and, really, research is problem solving) that we often do not appreciate that we are following the cycle or know where we are on it. This is a pity because understanding the cycle and knowing which stage is being followed generally leads to more efficient problem solving—the many formal problem-solving techniques that are available are essentially different implementations of the data cycle.

The cycle is discussed in detail in Chapter 1. Here, we are concerned with the first part of the cycle. Having identified and understood the questions to be answered, we should plan good quality data gathering and checking processes to ensure the reliability of the data being processed. Without this first step we may end up with unusual results with a high chance that they are worthless. The alarming thing is that without the planning and checks, we may not realise the worthlessness of the results!

2.4 Data planning and design

The correct design of a measurement is fundamental to producing reliable and useful results in an efficient and cost-effective way. In this context, measurement design is a combination of choice of the measurement system, the formal experiment design and the data collection and recording subsystems.

Before the measurement can be designed you must know what information is required and you should have made at least a preliminary choice of statistical data analysis to get the desired result. Both of these will affect the measurement design. As indicated in the data cycle, some iteration may be necessary to get the best design.

This section describes some of the issues that need to be considered before embarking on data measurement.

2.4.1 Measurement systems

Here, *measurement system* means the complete process of making a measurement: it includes sample preparation, the introduction of the prepared sample to the instrument, the instrument that makes the measurement and the people involved in these stages. Each of these stages may be broken into several substages and some may be more important than others in particular situations. Thus, with powdered, crystalline or granular materials, the way in which the sample is handled may be extremely important as some methods cause segregation by particle size, while the order in which solvents are added can determine whether the material of interest actually ends up in solution!

For investigations, one-off measurements and development work, the appropriateness of the measurement system for the information required must be assessed. Examples of points to note are:

- Does the system measure the parameter I want directly or indirectly (e.g. direct volume measurement or estimating volume from weight using specific gravity)?
- Is the system affected by external factors (e.g. room temperature, humidity, sunlight, magnetic fields)?
- Is the noise (precision) of the system less than the effect I want to measure?

The first two of these can be particularly insidious in their effects if not thought through properly. Indirect measures at best introduce an extra level of uncertainty in the final result that shows as a larger standard deviation estimate than would be obtained by direct measure. The worst case is that the measurement is meaningless simply because the expected indirect link was broken or modified at some stage and was not noticed (e.g. with specific gravity, the measurements were made at a different temperature from that used to estimate the specific gravity). The critical phrase here is 'not noticed'. If you need to use an indirect measure, then you should have checks in place that detect a broken or modified link.

On the whole, external factors are easily handled, providing their effects are recognised. Often the factor can be kept constant, or the measurements can be carried out in a timescale such that the factor is effectively constant. In this case the factor has no effect on the measurement. If the factor cannot be kept constant for the duration of the measurements, then the order of measurements can be randomised so that there is no correlation between groups of measurements and the way the factor varies. The factor does exert an effect on the measurements in this situation, but it has been changed from a potentially undetected bias effect into an extra variability (increased standard deviation). Randomisation is discussed in Section 2.4.3. It is possible, of course, to record the values of the uncontrolled factors at the times of making the desired measurements and to then apply some form of correction to the results based upon the values of the uncontrolled factors. This is effectively creating a model between the uncontrolled factors and the measurement of interest, but the main difficulty is that, because the factors are uncontrolled, the values they take may not allow an effective model to be generated.

Understanding noise generated by the measurement system is critical to successful measurements. Noise will come from many different sources (weighing, volumetric measurements, electrical, electronic, optical, mechanical, external factors, etc.) that combine together to give the overall error (standard deviation) in the measurement value. In general terms, if the noise is greater than the effect being sought, then it will not be possible to distinguish the effect from the noise. If the effect is two or three times larger than the noise, then it can be determined, though the level of uncertainty in the result may be unacceptably large. With the effect at least a factor of 10 greater than the noise, the effect will be determined essentially free from the noise in the measurement system. An example is given in Section 2.4.4. The effects of large measurement noise may be reduced by combining many replicate measurements (the measurement noise reduces in proportion to the square root of the number of measurements combined). There are also data processing techniques that can extract signals from large amounts of noise in certain circumstances (see Chapter 13).

2.4.2 Experiment design

There are many formal statistical tools to aid the design of experiments and these are essential to get the correct numerical values for a design. Chapters 6 and 7 discuss the techniques in detail. Here, I give only a general summary from the overall design perspective.

Experiment design is the label applied to the steps taken to ensure that

- an adequate number of measurements is made
- randomisation patterns are effective
- nonrandomised systematic errors can be estimated
- the underlying distribution of measurements is checked against the requirements of the statistical method.

Replication is an important part of experiment design and it is required to gain information about variation due to specific causes. It is determined by

- the aims of the study
- the power to detect important differences
- the variability in the data
- the particular design used.

In addition, nonstatistical considerations such as cost (financial, time and resources) have to be taken into account and these may be deciding factors. However, it must be remembered that inadequate replication renders a study unable to meet its objectives and hence wastes those costs.

Table 2.1 gives some simple examples of design considerations for particular types of information requirement.

Table 2.1 Examples of design considerations and required information

Information required	Design considerations
Purity of material isolated from a reaction	Take representative samples from throughout the material. Pool samples before analysis if interested only in the average value. Assay samples individually if interested in homogeneity of the material. Randomise the samples prior to assay but retain individual identities.
Comparison of two methods without eliminating extraneous effects	Select a range of samples to cover anticipated future applications. Divide each sample into two parts and analyse each part by a different method. Compare the individual results on the parts of the same sample.
Comparison of two methods eliminating extraneous effects	Select a range of samples to cover anticipated future applications. Submit a part of each sample to both methods of analysis on the same occasion in a random sequence (to ensure that testing by one method is not always undertaken before testing by the other method). Compare the averages and standard deviations of the results of all samples by each method.
Which of six factors has the biggest effect on a measurement	Identify the range over which each factor needs to be assessed. Determine whether all combinations of extreme factor values can be set. Work out whether all the measurements can be made as a single randomised set or whether they need to be grouped together. Can you afford to do all required measurements in a reasonable timescale?

2.4.3 Randomisation

Randomisation is used when systematic errors may be present in measurements but cannot be eliminated, and when it does not cause an unacceptable loss of precision in the result. As there are always likely to be unknown causes of variation affecting measurements, allocation of material within a design should always be randomised. Randomisation effectively converts systematic errors into a pseudo-random error. It does not remove the cause of a systematic error, it only converts its effect into a random error that contributes to the overall random error of the result in the same way as any other random error. Note, though, that random does not mean haphazard; a randomisation scheme is planned and has a clear objective.

Several randomisation schemes may have to be applied simultaneously to a measurement if it is affected by more than one systematic error.

In the common randomisation schemes that follow, only one cause of bias has been identified. The indicated scheme will randomise the effects of this bias but it will not necessarily randomise other biases that may be present and which might be important in practice.

- Mobile phase composition, column performance and flow rate in HPLC may drift or have cyclical changes over several hours. These would appear as a systematic difference between samples of the same material measured at the beginning and end of a run; putting replicates of individual samples in random assay order will randomise these effects.
- Two spectrometers may be used for an assay; allocating samples to the spectrometers in random order will randomise any bias between the spectrometers.
- Segregation of powder in a bottle may occur owing to particle size differences; taking samples from different positions in the bottle will randomise the effects of segregation.

Be careful, though: some randomisation schemes are not scientifically meaningful. For example, randomly allocating samples for assay by two different methods of analysis (UV and HPLC, for example) will randomise the bias between the methods when the results from the two methods are combined, but the bias often arises because the methods actually measure different things. (In UV, all components that absorb at the chosen wavelength are measured, while in HPLC only a single component is measured). In this situation it is not correct to combine the results of the two methods.

The process of randomisation is usually straightforward. The items to be randomised are labelled in some sequential way (1,2,3,... or A,B,C,...). A sequence of random numbers is generated and each is assigned to the ordered items in turn as it appears and ignoring any duplicate random numbers that may occur. The random number assigned to the item is the position in which the item will be used in the experiment. Two examples are given in Table 2.2, one illustrating the randomisation of 10 samples in an HPLC run and the other the allocation of 10 samples to two spectrometers (the same sample labels have been used for convenience in the table).

The first column of the table gives the sample labels A–J, and the second column a sequence of ten random numbers (0–9). The HPLC run sequence column is the order the samples should be run on the HPLC system. Thus, sample H was assigned random number 0 so it is run first. Sample E has random number 1 so it is run second, and so on.

Table 2.2 Examples of the process of randomisation

Sample label	Random number	HPLC run sequence	Label random number	Run on spectrometer A	Run on spectrometer B
A	7	H	Odd	A	B
B	4	E	Even	D	C
C	2	C	Even	E	F
D	5	G	Odd	G	H
E	1	B	Odd	J	I
F	6	D	Even		
G	3	F	Odd		
H	0	A	Even		
I	8	I	Even		
J	9	J	Odd		

When assigning the samples to two spectrometers, a simple approach is to assign all of the samples with odd random numbers to instrument A and all the even-numbered samples to instrument B. The results of this approach are shown in Table 2.2. If the samples had to be divided between, say, five instruments, then the random numbers need to be divided into five groups. Thus, numbers 0,5 (samples H and D) would be assigned to instrument 1, numbers 1,6 (samples E and F) to instrument 2, and so on.

Sequences of random numbers can be generated from many sources, including printed tables, calculators and computer programs such as spreadsheets. When using any of these sources you must remember that if you start producing the random numbers from the same point you will always get the same sequence of random numbers. If you apply this sequence repeatedly to the same type of experiment (for example, randomising samples in an HPLC run), you run the risk of not randomising some effect that does influence the results.

You should select the starting point of the random number sequence at random. With a printed table of numbers, you should choose a different page (some compilations of random numbers are large), column and position within the column each time the table is used. With calculators and software, there is usually a means of randomising the starting point (often called seeding or randomisation).

2.4.4 Data collection

While it is obvious that all the data required for a statistical analysis have to be collected and recorded, it is perhaps not so obvious that poor data collection and recording can affect the result of the analysis. This comes mainly from collecting data at an inadequate resolution, that is, the number of digits in the values is too small.

For example, the absorbance of a solution at 274 nm may be reported as 0.68326 by a UV spectrometer. Recording this as 0.7 may be sufficient if only the presence or absence of the absorbing species in the solution is required. However, if the absorbance is used as a monitor of concentration of the absorbing species, and changes in concentration of 0.1% are important, as they might be when monitoring the kinetics of a reaction, then all five digits must be recorded.

Related to this is the noise in the measurement system. The noise may arise from many sources but, typically, instrument noise usually comes from optical, electronic or electrochemical fluctuations. Often these are a function of the instrument design and there is little the user can do to modify them, apart from using a different instrument. The effect on data quality is usually easy to assess.

Let us continue with the UV measurement example. If the standard deviation of repeated measurements on the sample, without removing and replacing it in the instrument, is 0.0013 absorbance unit, then we can assess the usefulness of the instrument for certain types of measurement, again assuming the sample absorbance is reported as 0.68326. The standard deviation, 0.0013, is a measure of the variability caused by the instrument; thus, we expect the true absorbance of the sample to be somewhere in the range approximately 0.680–0.686. This is more than adequate to determine the presence or absence of an absorbing component but it is not good enough to measure changes in concentration of the order of 0.1%. For the latter the standard deviation due to the instrument has to be of the order of 0.01% of the measurement value, i.e. about 0.00005 absorbance units. There are many spectrometers that can achieve this and these are the ones to use in this type of experiment.

Finally, as measurement systems are increasingly capable of recording data automatically, it is important to check that they are not truncating or rounding values before recording.

2.4.5 Digitisation

The vast majority of instruments used to make measurements in chemical research generate digital values rather than present an analogue value on a meter, chart recorder or scale. This is very convenient for storing the values electronically and for transmitting them to computers for further processing. Usually there is no problem with this process, but an appreciation of the characteristics of digitisation can avoid some serious pitfalls, particularly if an instrument is used in a way that was not originally intended.

Most measurement sensors (the part that actually responds to the variable being measured—thermocouples, resistance thermometers, pH electrodes, photodiodes, radiofrequency antennae, etc.) generate an analogue signal and this is converted into a digitised form using an analogue-to-digital (A/D) converter. Among all the design characteristics an A/D designer has to consider, there are two major ones that the end user needs to appreciate. These are the digitisation level and digitisation frequency.

An A/D converter generates a digital representation of the input analogue signal in binary format. The digitisation level is the range of the input analogue signal accepted by the A/D converter (usually in volts) divided by the number of output binary values that can be generated by the converter. The number of binary values is 2^n, where n is the number of bits in the binary value. It represents the change in the analogue signal that will cause a change in the least significant bit of the binary value. Thus, if the input analogue range is 5 volts and the output is an 8-bit binary value, the digitisation level is $5/2^8$ volt/bit=0.0195 volt/bit. The analogue signal would have to change by 0.0195 volts to cause a 1-bit change in the output value.

The effect of this is shown in Figure 2.1 for two A/D converters with the same input range of 0–10 volts monitoring the same signal varying between 0 and 2.5 volts at the same digitisation frequency. The solid line is for a 6-bit converter and the dashed line figure for a 12-bit converter.

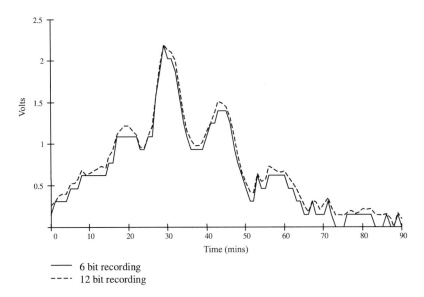

Figure 2.1 The effects of 6-bit and 12-bit digitisation on a signal.

While both converters follow the overall shape of the signal, the 6-bit converter loses more detail than the 12-bit converter—the steps in the output values are more pronounced.

All A/D converters take a finite time in which to convert the analogue signal into a binary value. This determines the highest frequency with which the converter can operate. In practice, this frequency (the so-called free-run frequency) is rarely used; instead, a lower selected frequency is used. The choice of the digitisation frequency determines the highest-frequency features that can be determined in the signal (the Nyquist frequency or limit). This effect is shown in Figure 2.2 where the intensity of transmitted light through a thin moving film is plotted as a function of the distance along the film. The low-frequency digitisation rate (100 Hz) loses nearly all of the fine structure that the higher frequency (1000 Hz) detects.

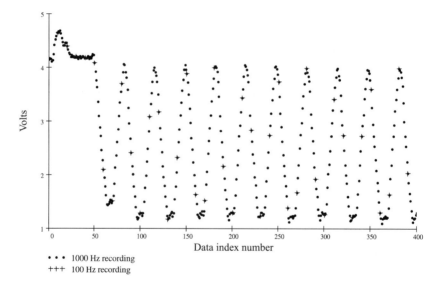

Figure 2.2 The effects of 1000 Hz and 100 Hz digitisation frequency on a signal.

The consequences of digitisation extend beyond missing fine detail in the signal. For example, the combination of high digitisation rate and coarse digitisation level can produce serious artefacts in numerical differentiation of the signal (see Figure 2.18).

2.4.6 Recording and reporting numbers

Numerical values, be they the original measurements or the results of statistical analysis, are usually stored in some form of database (a paper-based notebook, electronic file or database application file). The values are used in all sorts of calculations to derive further results. If they are put into the database with too few digits, it may mean that the calculations give limited information and, sometimes, none. There are many parallels with digitisation levels discussed in Section 2.4.5.

A very common example is impurity levels, where only a single digit may be reported. Figure 2.3 shows an example of the effects of data recovered from a database. The effect of rounding is obvious. The data have only five values (0.0, 0.1, 0.2, 0.3, 0.4) and any subtleties in the way the amount of impurity varies from sample to sample have been lost.

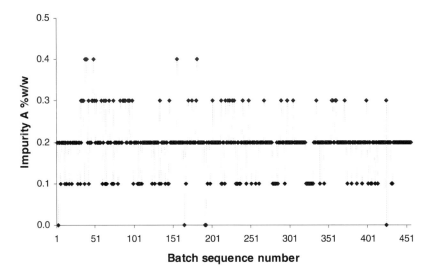

Figure 2.3 Example of data rounding on the information content of data. These data were recovered from a database where rounding had been applied at the data entry stage.

Considerations of rounding and significant figures are closely related to those for recording numbers (see also Sections 2.4.7 and 2.4.8, respectively). However, there is a clear distinction between recording and reporting numbers.

Numbers are *recorded* so that they may be used in future data analyses. If the future data analyses are not to be distorted by rounding errors, it is necessary that the numbers are recorded with at least one digit more than the expected precision of the measurement. This requirement should be applied to results of calculations where those results may be used in other calculations as well to original values.

The number of digits in a *reported* value is often not related to the precision of the measurement or to the number of digits recorded. It is implied by some contractual, legal, regulatory or aesthetic requirement. Usually, the implied number of digits required is much less than the precision of the measurement.

For example, a contractual specification may say that the content of a specific molecule in a product will not be less than 95% w/w of the label claim. An assay result of 97.2536% w/w would be recorded as 97.25% w/w (or even 97.254% w/w) but reported as 97% w/w (note that an assay result of 94.873% w/w would be reported as 95% w/w and, thus, meet specification!). Use of the reported figure in future data analyses would be inappropriate.

2.4.7 Rounding

Rounding is the process of restricting the number of digits in a given value that are displayed (recorded) or used (reported). There are several conventions in common use throughout the world and there are pros and cons for each. Whichever convention is used, it is necessary to apply it consistently to all values, particularly when comparisons of values may be made. A convention that is commonly taught in the UK, that is consistent with many regulatory documents used worldwide, and that is commonly used in computer software is

- Identify the rightmost digit in the number that is to be displayed or used. Label this digit as R.
- If the next digit to the right of R is 5, 6, 7, 8, or 9 then increase the value of R by one, if the digit to the right of R is 0, 1, 2, 3 or 4 do not change the value of R.
- Remove all digits to the right of R.
- The rounding operation at R must be carried out in a single step and not as a sequence of rounding operations starting at the extreme right of a number and working towards R.

The examples in Table 2.3 illustrate the process.

Rounding should be applied only to the final result of a calculation; rounding input data values or intermediate calculation results will cause errors and can generate grossly distorted results (see Figure 2.3 for example). If the final result is likely to be used in other calculations, for example when comparing results over a period of time, it is recommended that, on rounding, at least two digits (preferably more) are left to the right of the digit identified as the least significant from an experimental or assay point of view.

DATA GATHERING AND DESCRIPTION 47

Table 2.3 Examples of rounding using the rounding rules

Digits to be retained	R	Digits to be discarded	Result
5.34	2	49	5.342
5.34	2	94	5.343
0.002	5	1	0.0025
0.002	5	5	0.0026
0.002	4	6	0.0025
1379	2	687.88	13 793 000

Rounding is an important consideration with computer software. It is a function of the internal representation of numbers used by the software, the way the computer's hardware handles register carry, overflow and underflow during binary operations, and the way a particular calculation has been implemented by the software programmer. Discussion of the detail of these points is beyond the scope of this book, but you must be aware that poor implementations can give believable but seriously wrong results. Protection against these effects can be obtained by using one of the many test data sets that have been published to highlight problems with particular calculations. A recent compilation has been made by Butler *et al.* (1996).

2.4.8 Significant figures

The number of significant figures quoted in a result should be related to the overall precision of the measurement system used to get the result. The standard deviation of a group of measurements may be used as the estimate of precision.

As a rule of thumb, a standard deviation can be assumed to have two or three significant figures. If a standard deviation is reported with more than three digits, simply assume three. A value derived from the group of measurements, such as the average, is then quoted so that its least significant digit is in the same decimal place as the least significant digit of the standard deviation.

Some examples of rounding while maintaining an appropriate indication of precision are given in Table 2.4.

There is a caution to be applied here for numbers with many digits. For example, the average number of theoretical plates of a chromatographic column may be calculated to be 754 622.7996 with a standard deviation of 25.65. The three most significant digits in the average (754) do not change from measurement to measurement and it is not appropriate to round the average to 750 000 as it implies (falsely) a poor precision in the measurements. Instead, rounding to 754 623 indicates the actual

Table 2.4 Examples of rounding to a precision defined by a standard deviation

Average	Standard deviation (SD)	Position of LSD of SD	Result
5.342	0.091	.00x	5.342
5.342	0.91	.0x	5.34
97.456	1.2	.x	97.5
97.456	1.2783	.0x	97.46
3432.1	110	x0.	3430
3432.1	62	x.	3432

level of precision in the result based on a standard deviation with two significant digits.

2.5 Data description

This section describes some simple techniques for inspecting the data *before* applying any statistical tests. Such inspection is quick and is designed to show up any problems with the data that might distort the outcome of a statistical test. Some of these inspections should ideally be carried out while the data are being gathered as doing this allows you to repeat or check a measurement immediately if one is found to be unusual. This approach is strongly recommended.

Most of the techniques described in this section are essentially graphical and use visual inspection rather than statistical test methods. Although it is convenient to use computer software such as spreadsheets to produce nicely formatted charts, it is not essential for these methods. Hand-drawn plots on plain or ruled paper (not even graph paper!) are just as effective and often have the advantage over a computer of accessibility at the site of measurement.

This section assumes the inspection of a single data set, but the techniques apply equally well to the comparison of two or more independent data sets.

2.5.1 Simple numerical checks

Simply reading the numbers in a data set for values that do not seem to fit with what you expect from the data will trap many errors that could invalidate the set. The value of doing this as the data are recorded cannot be over emphasised.

Some typical examples are:

- The pH of 20 individual vials of a solution to be used in a designed experiment is being measured. The required range for the experiment is pH 6.0–7.0. The first recorded value is pH 8.0. Should you continue?
- The weights of kegs of drug substance are being monitored after filling. The sequence so far is (all values in kg), 21.80, 21.23, 20.99, 21.76, 21.22, 12.33, 20.92. Is there anything unusual?
- The UV absorbance of a solution is being monitored for stability over a period of time at half-hour intervals. The first seven values are, 0.6222, 0.6234, 0.6223, 0.6219, 0.6745, 0.6732, 0.6738. Do these values fit your expectations?

In the first example, the pH is clearly outside the range expected and, if it is the true pH of the solution, continuing with the experiment will give unexpected, if not meaningless, results. It is possible that only this vial has the wrong solution or that the pH electrode has failed. A small amount of investigation will quickly discover the cause of the problem and prevent a significant amount of work being wasted.

The value 12.33 in the second example is almost certainly a transcription error and it should be 21.33. If this is spotted at the time of weighing the kegs, it is a simple matter to reweigh the keg as a check. If it is not spotted until after the kegs of material have been used, then it has to be *assumed* to be a transcription error and a decision has to be taken about the acceptability of risk introduced if the assumption is accepted.

There is quite clearly a step change in the absorbance values in the third example. This certainly does not meet the expectation of a stable solution, nor does it meet the expectation of a solution that is progressively changing. A fault with the spectrophotometer, or with the way the sample is presented to it, is a likely cause of the problem and it needs to be resolved before further measurements are made.

2.5.2 Trend plots and control charts

A trend plot is a graph of the data in which the x-axis is a sequence value such as time of measurement, sample number, batch number or position in the population, and the y-axis is the value of the measurement at that time, sample, batch or position. It is most important that each data value is correctly identified with its sequence value and that the sequence value is itself meaningful in terms of the information sought.

Trend plots are good at highlighting single unusual values, drift in values and step changes in values. However, if the number of data values is small, less than 6 or 7, you should be cautious about interpreting apparent changes to indicate a trend as they could be due only to the random variations in your data!

The choice of using separated points or lines in a trend plot is purely aesthetic and you can choose the one that shows the effects of interest most clearly.

Some typical trend plots are shown in Figure 2.4.

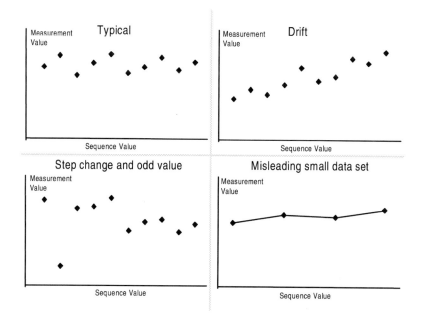

Figure 2.4 Trend plots showing different types of data problem.

A form of trend plot that is widely used to monitor production processes (and which could be used effectively in laboratories to monitor the performance of critical instruments or experimental conditions) is the *control chart*.

The simplest control chart is a trend plot of a variable with three specific values of the variable drawn as horizontal lines on the plot (Figure 2.5). The first is the expected or target value of the variable and the other two are the upper and lower action limit values placed at the target value $\pm 3s$, where s is the standard deviation of the variable estimated from at least 50 values (preferably 100 or more). The probability of a value not being between the action limits simply due to

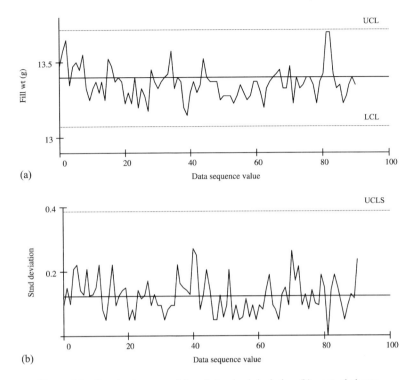

Figure 2.5 Shewhart average (a) and standard deviation (b) control charts.

random fluctuations in a Normal distribution is 0.27%. This is accepted as being sufficiently low for a value outside the action limits to be classed as atypical and for the process to be described as 'out of control'. Some form of action is called for to bring the process back under control.

Control charts of this type are called Shewhart charts, after W.A. Shewhart.

There are many forms of control chart and many refinements to their use that go beyond the scope of this chapter but that are covered in Chapter 9. However, it is worth mentioning here the range and standard deviation control charts because of their ability to show changes in the variability of data. If data values are acquired in small groups at specific times (i.e. temperature is measured at five positions in an isothermal heater block, or the diameters of 12 plastic injection-moulded components are measured), the range or standard deviation of the values within the groups can be plotted as a control chart. Again, the expected value and an action limit (only one is needed as a range or standard deviation less than zero cannot be obtained) are drawn. Changes in variability are easily seen in the plot. An example of a standard deviation control chart is included in Figure 2.5.

2.5.3 Scatter plots

Scatter plots are useful for showing relationships between two variables. It is necessary that values of the variables can be paired by some common property such as sample number, batch number, time or position. The values in one of the data sets are plotted on the x-axis and those of the other on the y-axis. Examples of data sets suitable for scatter plots are

- two methods used to assay the same samples
- drug content and weight of individual tablets
- pH of a mobile phase and elution time of a component
- temperature of a reaction and yield.

Scatter plots can show linear and nonlinear relations between the variables, even when there is quite a lot of noise in the data, unusual values and clustering of values. Scatter plots are most effective with large numbers of data values. Those with less than half a dozen values are not particularly informative unless the data have been designed for a regression analysis.

Some typical examples are shown in Figure 2.6. If there is no relationship between the variables then a fairly even scatter of points is

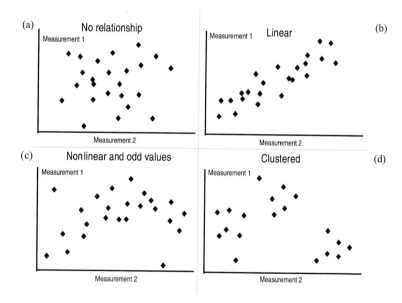

Figure 2.6 Examples of scatter plots showing particular data forms.

seen in the plot (plot (a)). A weak relationship may be difficult to distinguish from no relationship. However, a strong linear or nonlinear relationship will be quite evident (plots (b) and (c)). Plot (d) illustrates the effect of grouping or clustering in the data.

Plot (c) is interesting in that it shows two outlying points—that is two points that are some distance away from the overall trend in the data. These are quite obvious in this plot, but they would be very difficult to spot by scanning a table of numbers or by looking at each variable independently.

Many software packages allow three-dimensional scatter plots to be displayed. These are particularly useful if the package allows interactive rotation of the plot. With higher-dimensional data, many packages will draw a matrix of scatter plots for all possible combinations of pairs of variables. However, these can be overwhelming and difficult to interpret with more than about half a dozen variables. In these situations, parallel coordinate plots (Chapter 6) and principal component plots are needed (Section 2.6.3).

2.5.4 Box and whisker plots

Box and whisker plots are useful for the comparison of groups of measurements of a variable. The grouping can be, for example, by time, batch, location, temperature, distinct sets of conditions (as found with multifactor experiment designs). The plots will show

- unusual values within groups
- groups with much greater or much smaller variability then other groups
- groups with an average value much different from other groups.

When constructing a plot, the *x*-axis is marked with the group descriptors and the *y*-axis with the values of the variable. The values of the variable within a group are plotted along a vertical line above the group descriptor. A rectangular box is drawn to enclose the middle 50% of values in the group and the median value of the values in the group is marked with a horizontal line. A short horizontal line inside the box is used to mark the position of the average. Finally, vertical lines (the whiskers) are drawn from the upper and lower sides of the box to points that are 1.25 times the distance from the median to the upper box side and 1.25 times the distance from the median to the lower box side, respectively, or to the most extreme value if it is closer than the 1.25 length. Fortunately, many software packages offer box and whisker plots as an option! If the data are Normally distributed, then 60% of the data are expected to be within the range defined by the whiskers.

Box and whisker plots are a very powerful visual tool for comparing variability between groups of data. As seen in Figure 2.7, groups with unusually small or large variability are easily seen, as are differences in the location of the groups and skewness in the data. In many ways, the box and whisker plot is the visual analogue of a one-way analysis of variance (Chapter 8).

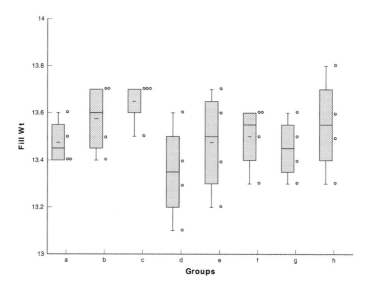

Figure 2.7 Box and whisker plot of the first few points used in Figure 2.5.

There are several variants of box and whisker plots. One is shown in Figure 2.7 where the data values are plotted in a form of histogram alongside the corresponding box. This is a box and whisker plot with a so-called dot plot. Another variant is to insert a notch in the box of the box and whisker plot at the median of the data values and to adjust the size of the notch according to the confidence interval of the median. The choice of which variant to use depends on the information to be displayed, but for most data-checking purposes the simple box and whisker plot is sufficient.

2.5.5 Cusum plots

Cusum plots are a form of trend plot designed to be particularly sensitive to changes in the average level of a variable. They are formed by taking the difference between the value of the variable and its target value, preserving the sign of the difference, and then forming the cumulative sum (Cusum) of the differences from the start of the monitoring sequence.

Thus, the Cusum, c_i, at position i in the sequence is

$$c_i = \sum_{j=0}^{i}(x_j - t)$$

where t is the target value, i is the current sequence (often time) point, x_j is the data value at sequence point j.

It is usual to scale Cusum values by dividing them by $2\sigma/\sqrt{n}$ before plotting. This has the effect of visually de-emphasising nonsignificant changes and also of allowing Cusums of different variables measured at the same time to be directly compared. Here, σ is the population standard deviation of individual measurements and n is the number of measurements used to generate the value of x_j. Often $n=1$, that is no replicate measurements have been made and x_j is not an average. Ideally σ should be determined from a large amount of historical data. In practice such an estimate is not available and σ has to be estimated from the current data.

Although σ can be estimated from the standard deviation of all of the data in the data set, it is likely to be overestimated if any step changes or trends occur in the data. These are, after all, the events we suspect as being in the data and we do not want their effects to be included in the standard deviation estimate. An estimate of σ that minimises the effect of step changes and trends is the successive difference, or process, standard deviation. This can be calculated in one of two ways from the differences between adjacent pairs of data values:

$$s = \sqrt{\frac{n\sum_{i=1}^{k-1}(x_{i+1} - x_i)^2}{2(k-1)}}$$

or

$$s = \sqrt{n}\,\frac{\sum_{i=1}^{k-1}|x_{i+1} - x_i|}{1.128(k-1)}$$

In these equations, k is the number of averages plotted and n is the number of replicates forming the average. (It should be noted that these equations for estimating standard deviation are quite general in application and are not restricted to Cusums—they are often used to estimate σ for Shewhart charts (Section 2.5.2) for example).

There are two principal uses of Cusum plots. One is in the real-time (or near real-time) monitoring of changes in a variable. In this use, the target value, t, is the set point or expected value of the variable. The purpose is to detect a significant movement of the variable from its target as soon as possible. This is accomplished with a V-mask. This use of a Cusum is described in Chapter 9.

The second use is to assess 'historic' data (data already gathered and not being used in real-time monitoring and control) for any significant changes that occur in time. This use is described here.

In this use of the Cusum, the target value is the average of the data values. This forces the first and last Cusum values to be identically equal to zero and it tends to emphasise visually the points at which significant step changes occur. Because of their shape, these points are often called breakpoints.

The value of the Cusum at each sequence point is plotted against sequence number. Table 2.5 shows a short section of data with the corresponding Cusum values. Figure 2.8 shows the corresponding Cusum plot.

If the data values are randomly scattered about the target value, the Cusum will be a random scatter around zero. However, if a step change in

Table 2.5 Calculation of a Cusum for the first part of the data shown in Figure 2.8

Sequence no.	Fill Weight (g)	Difference	Cusum
1	13.4	0.0379	0.0379
2	13.5	0.1379	0.1759
3	13.4	0.0379	0.2138
4	13.6	0.2379	0.4517
5	13.7	0.3379	0.7897
6	13.7	0.3379	1.1276
7	13.4	0.0379	1.1655
8	13.5	0.1379	1.3034
9	13.7	0.3379	1.6414
10	13.7	0.3379	1.9793
11	13.7	0.3379	2.3172
12	13.5	0.1379	2.4552
13	13.6	0.2379	2.6931
14	13.1	−0.2621	2.4310
15	13.4	0.0379	2.4690
16	13.3	−0.0621	2.4069
17	13.4	0.0379	2.4448
18	13.7	0.3379	2.7828
19	13.6	0.2379	3.0207
20	13.2	−0.1621	2.8586

The average of all of the data used for calculating the differences is 13.3621 g. The successive difference standard deviation (second equation) used for normalising the plot in Figure 2.8 is 0.1276.

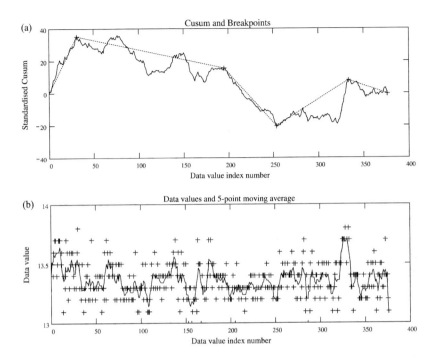

Figure 2.8 Plot of a Cusum. The first 20 points of the plot are the data from Table 2.5. Possible breakpoints are shown in the Cusum plot (a). The original data values (crosses) and a 5-point moving average (line) are shown in (b).

the average value of the values occurs, the Cusum will show a definite drift away from zero in the direction of the step change (positive for a step change above target, negative for a change below target). Because the Cusum has a much lower noise level than the original data, the significant breakpoints are readily detected.

A simple test for a significant change is to find the maximum absolute difference between the Cusum value and the value of the line joining the confirmed breakpoints closest to the suspected point (initially the first and last points of the plot are used) at the same sequence value. The original data values in the range between the two confirmed breakpoints are split into two groups at the suspected breakpoint and a one-way analysis of variance is carried out. If the between-group mean square is significantly larger than the within-group mean square (see Chapter 8), then the suspected breakpoint is confirmed as a significant breakpoint. The significant breakpoints in the example data are shown in Figure 2.8 and are joined by dotted lines to illustrate how constant the data values are between the breakpoints (if the data are constant the Cusum will follow the dashed line closely).

The ANOVA table for the first breakpoint in Figure 2.8 (that at sequence value 31) is given in Table 2.6, and shows that the averages of the data values each side of the breakpoint are very highly significantly different.

Table 2.6 ANOVA table showing that the first breakpoint (sequence value 31) in the Cusum of Figure 2.8 is highly significant

SUMMARY

Groups	Count	Average	Variance
Data 1–32	32	13.503	0.028054
Data 32–377	346	13.349	0.023087

ANOVA

Source of variation	SS	df	MS	F	F crit
Between groups	0.691989	1	0.691989	29.45	3.87
Within groups	8.834572	376	0.023496		
Total	9.526561	377			

Because of the sensitivity of the Cusum plot, it is wise to check that detected breakpoints actually have practical significance as well as statistical significance and to try to identify what has caused the change. In the case of the data used here as an example, the detected breakpoints correspond to changes in the filling control made by the operators.

Cusum plots are very good at detecting changes in average value but they are not particularly sensitive to changes in variability. However, it is possible to generate Cusum plots of ranges, standard deviations or any other statistic and to examine these for breakpoints and to use these to detect significant changes in the statistic.

2.5.6 *Histograms and data distributions*

Histograms are plots of data that show the frequency distribution of the values. There are three valuable uses of a histogram:

- Highlighting unusual values or groups of values (extreme values or clusters)
- Checking deviations of your data from the distribution you expect
- Comparing distributions of several data sets for differences in shape or position.

When checking your data against an expected distribution, the two types you are most likely to expect with measurements of a variable, such as weight, concentration, pH, are

- *Boxcar* (where all values have equal probability of occurring—it is also known as a uniform distribution)
- *Normal* (where the most frequently occurring values are close to the average).

Figure 2.9 shows some typical histograms. If your data are counts of events or items then you will expect to see a binomial or Poisson distribution (not shown).

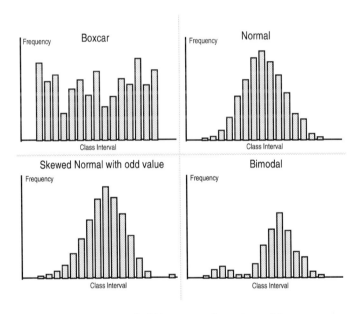

Figure 2.9 Typical histograms of experimental data.

The most likely deviations, particularly from the Normal distribution, are *skewness* or lack of symmetry (more high or low values than expected) and *bimodal* (two overlapping distributions).

Most statistical tests will tolerate some degree of skewness but, if it looks pronounced in the histogram, you may find some distortion of results and some tests may be invalid (for example the F-test).

Bimodal distributions usually indicate either that an experiment is poorly designed or that an unexpected change that has taken place during the experiment. Continuing with the planned data analysis is unlikely to give meaningful results and the experiment should be repeated after eliminating the cause of the bimodality.

When constructing a histogram from a large data set (i.e. at least 50 values), aim to have between 10 and 20 class intervals, or bins, on the variable's axis. It is usual to make them of equal width but it is not necessary to do so, and often the first and the last intervals are defined to count all values respectively above and below certain values. As the perceived shape of a histogram is dependent on the number of class intervals, their placing and their size, it is sometimes worth replotting a histogram with different conditions. Note should be made of the data precision when setting the class intervals—their boundaries should be simple multiples of the precision. Software that automatically scales the class interval axis sometimes does not do this and produces illogical intervals.

If the number of data values you have is small, say less than 20, then you should be cautious about interpreting the histogram shape in terms of Normality and bimodality. The box and whisker plot (Section 2.5.4) is an alternative to drawing multiple histograms when comparing several variables.

2.5.7 *Normality testing*

A common requirement in statistical testing is that the data have a Normal distribution. In many situations the simple visual inspection of the histogram of the data will be sufficient, particularly if a Normal probability curve is overlaid on the histogram. Some statistical tests, however, are quite strict about the Normality requirement and a more quantitative measure is needed. Many statistical packages provide such tests and these may be used. In the absence of the tests, a more searching visual check can be made using a cumulative Normal probability plot.

The basis of the Normal probability plot is dividing the area under the Normal distribution curve into n equal sized areas, where n is the number of data values being tested. The Normal deviate corresponding to the boundaries of each of these areas is then found. If the data come from a Normal distribution, then it is expected that one value will fall between each adjacent pair of deviates. This is shown visually by sorting the data into increasing order and plotting the ordered Normal deviates on the y-axis against the sorted data values on the x-axis. The y-axis is often scaled to show cumulative probability rather than Normal deviate.

Normally distributed data are expected to fall on a straight line whose slope and intercept depend on the average and standard deviation of the data. Nonlinear trends in the plot indicate deviations from Normality. Figure 2.10 shows the Normal probability plot for the data illustrated in

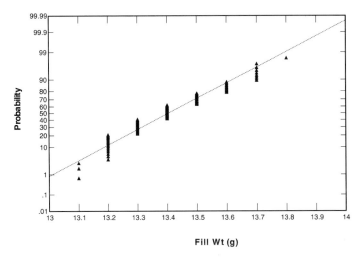

Figure 2.10 Normal probability plot.

the Shewhart plots of Figure 2.5. The data fall reasonably close to the straight line and this indicates reasonable Normality in the data. In passing, it is worth noting that the effects of rounding the data to one decimal digit show up strongly in this plot.

Again, such tests are not particularly useful if you have a small data set.

If Normality is important and you have a small data set, then you may be able to establish that the measurements you have just made come from a Normal population by looking at accumulated historical data by the same measurement or from similar measurements. In addition, most analytical methods derive benefit from the Central Limit Theorem, which states that the averages of groups of values tend towards a Normal distribution irrespective of the underlying distribution (providing that distribution is unimodal). As most analytical methods involve averaging at some stage (e.g. duplicate injections, replicate sample preparations, blending, etc.), an assumption of Normality is not unreasonable in the absence of any other information.

2.5.8 Outliers and discordant values

An *outlier* in a set of data is an observation that appears to be inconsistent with the remainder of that set of data. If on examination it is found that the outlier is truly inconsistent because of an error or special circumstance, then it should be classed as a *discordant* observation. This distinction has been introduced in order to separate outliers into those that present no useful information other than that an error has occurred (discordant values) and the remainder, which may present useful information about the underlying distribution of values.

Every effort should be made to discover what, if anything, caused an outlier (before making any classification as a discordant value). Records of the experiment, instrumentation logs, delivery records of materials, discussions with suppliers of materials or those who helped with the experiment, and so on are all useful sources of information. It is worth noting that an outlier may be a clue to some previously unsuspected and possibly interesting factor affecting your data. In these circumstances, investigation of the causes of the outlier can be rewarding.

Outliers occur for many reasons, for example:

- They may be near the extremes of the underlying distribution.
- The set of data may arise from a different distribution from that assumed.
- The outliers could be the result of sampling errors, recording errors or technical errors.

In the first two cases the outliers are consistent with their underlying distributions and should not be classified as discordant. In the third case, if the error is confirmed, then the outlier should be classified as discordant. A common situation in which the first two cases can occur is during homogeneity testing when extreme values can occur because of inhomogeneity in the material that has been sampled. Such values should not be classed as discordant (unless there is firm evidence of an error in sampling, measurement or data recording).

Identifying outliers

Plots and histograms of the data can be very revealing and should be the first stage in the search for outliers. They have many advantages over visual checks of tables of numbers (see Figure 2.9).

Usually the largest or smallest values in a set of data will come under suspicion as possible outliers, but an outlier can occur even if it is within the range of the other data. For example, when there is a trend in the data with time or some other variable, a value may lie an exceptional distance from the relationship seen with the remainder of the data but still be within the overall range of data values. Trend and scatter plots are particularly useful for identifying these outliers (see Figure 2.4 and Chapter 9).

Outliers are those values we can identify as being different from what we expect. We should remember that values may be wrong even when they are totally consistent with the remainder, but there is no statistical method to detect such values.

Dealing with discordant data
Dealing with discordant values raises many ethical, as well as legal, issues within some industries and professions. There is no simple answer to these issues beyond saying that discordant values cannot be just ignored and hidden. At the very least there should be a written statement with the experiment notes that explains what action has been taken and why. In some circumstances, the statement may have to extend to a full investigation report into the probable cause of the value.

At a statistical level, it may be appropriate to reject a discordant value and to re-analyse the remaining data. It is, however, worth checking to see exactly what effect the inclusion of the discordant value has on the final result—if it is small and there is no clear cause for the value, it can be argued that the value should be included in the analysis. This might be important with some experiment designs where the loss of data can complicate the statistical analysis and affect its interpretation.

If possible, replace the discordant value with a repeated measurement of the item that led to it. This might be feasible when the measurement is nondestructive (e.g. weighing) and the item has been retained. With destructive measurements a 'similar' item might be available (e.g. a sample from a bulk container). However, be careful that the items and the measurement system have not changed in any important way.

An alternative approach is to regard the discordant value as a missing value and to calculate a replacement value (e.g. using the average or an imputed value as a replacement). There are severe dangers of seriously biasing the results of the analysis with this approach and it should be used only when other routes have failed.

Finally, the uncertainty raised by the discordant value and the difficulty or impossibility of getting a replacement value may mean that the whole data set should be abandoned and a new set of experiments carried out.

Discordancy tests
Many discordancy tests are available, each appropriate in specific situations (see Barnett and Lewis, 1994). Three are described in the Sections 2.5.9–2.5.11 to check whether either the smallest or the largest value is discordant. They are the Studentised range, Grubb's and Dixon's tests. All are of the form

$$\text{Test statistic} = \frac{\text{difference}}{\text{variation}}$$

The *Studentised range* test uses for the variation an estimate obtained elsewhere, such as historical data. For this reason it is the preferred test.

Grubb's and *Dixon's* tests estimate the variation from the sample data. These two tests are in common usage and have similar properties for sample sizes up to 15, beyond which *Grubb's* test is preferable.

It is assumed for these tests that

- the data form a single random sample of independent observations from a Normal distribution;
- there may be only a single outlier in the data.

None of these tests should be used if more than one value is suspect. In this case, investigate the data and consider rejecting the whole sample. In particular, you must not apply these methods sequentially to remove more than one outlier from a data set. However, it is possible to carry out tests on non-Normal data, multiple data sets and data sets with more than one outlier. If you think you have one of these situations you should consult a statistician or someone experienced in these techniques (see also Chapter 5).

The following data set is used to illustrate the tests. It consists of the weights (in mg) recorded for a sample of 10 tablets from a single batch and they have been sorted into increasing magnitude:

$$102.1, 103.3, 103.5, 103.6, 103.9, 104.0, 104.0, 104.1, 104.3, 104.4$$

Label them $x_1, x_2, \ldots, x_{n-1}, x_n$ so that x_1 is the smallest value and x_n is the largest:

$$x_1 = 102.1, \quad x_2 = 103.3, \ldots, x_{n-1} = 104.3, \quad x_n = 104.4$$

The critical values referred to in the illustrations are available in many statistical texts and software packages.

2.5.9 Studentised range test for a single discordant value

Define R as

$$\frac{x_n - x_1}{s_\nu}$$

where s_ν is the independently estimated standard deviation of the values with ν degrees of freedom.

Compare R with the appropriate critical value for a sample size of n and with ν degrees of freedom for the standard deviation. If R exceeds it, then declare x_n or x_1 discordant. Note that the smallest sample size catered for by this test is $n=3$.

Suppose that for the example data set an estimate of s_ν is available from previous samples, namely $s_\nu = 0.48$ with $\nu = 60$ degrees of freedom. Then,

$$R = \frac{x_n - x_1}{s_\nu} = \frac{104.4 - 102.1}{0.48} = \frac{2.3}{0.48} = 4.79$$

The 5% critical value of R for $n=10$ and $\nu=60$ is 4.65. Since R exceeds this, one of the two values used in the test is classed as discordant. We should choose 102.1 in this example as it is farther from the average of the data set.

2.5.10 Grubb's test for a single discordant value

Define G as the maximum of

$$\frac{x_n - m}{s} \quad \text{and} \quad \frac{m - x_1}{s}$$

where m is the average (mean) of the n values and s is the standard deviation of the n values. Compare G with the appropriate critical value for a sample size of n. If G exceeds it, then declare x_n or x_1 discordant.

For the above data set, $m = 103.72$ and $s = 0.666$. Since x_1 is farther from m than is x_n we use

$$G = \frac{m - x_1}{s} = \frac{103.72 - 102.1}{0.666} = \frac{1.62}{0.666} = 2.43$$

The 5% critical value of G for $n=10$ is 2.29. Since G exceeds this, we would label 102.1 as a discordant value.

2.5.11 Dixon's test for a single discordant value

Define D as the maximum of

$$\frac{x_n - x_{n-1}}{x_n - x_1} \quad \text{and} \quad \frac{x_2 - x_1}{x_n - x_1}$$

Compare D with the appropriate critical value for a sample size of n. If D exceeds it then declare x_n or x_1 discordant. Note that the smallest sample size catered for is $n=4$.

For the above data set, $x_n - x_{n-1} = 0.1$ and $x_2 - x_1 = 1.2$, so we use

$$D = \frac{x_2 - x_1}{x_n - x_1} = \frac{103.3 - 102.1}{104.4 - 102.1} = \frac{1.2}{2.3} = 0.522$$

The 5% critical value for $n=10$ is 0.464. Since D exceeds this, we would label 102.1 as a discordant value.

2.6 Data preprocessing

The data description techniques described in Section 2.5 are very useful for detecting the unexpected in the data. Often, we expect to see a change or pattern in the data but it may not easily visible or extractable. The data preprocessing techniques in this section are a selection from the most commonly used techniques. They all aim to enhance a particular part of the information contained in the data. But remember: if the information is not in the data in the first place, these techniques will not create it.

Finally, before we look at the techniques in detail, it should be mentioned that most instrumentation used in laboratories implements some form of smoothing or other filtering algorithm on the data it collects *before* it is presented to the user. Often this is done completely transparently and the user may have no control over what is done. This is not usually a problem in normal use of the instrument, as the instrument designer generally chooses algorithms to match the type of measurement, the instrument design and its intended use. However, it is worth remembering that the 'raw' data you obtain from an instrument has already undergone a significant amount of preprocessing and you may find some unusual effects if you use the instrument near the limits of its capability or with difficult samples or outside its intended use.

2.6.1 Smoothing—moving average, Savitsky–Golay, EWMA

Smoothing algorithms aim to enhance underlying structures in the data by reducing the features that mask or confuse them. Generally, this is achieved by reducing the high-frequency 'noise' components of the data while causing only small changes to the lower-frequency structures of interest. The techniques described here are part of the very large class of filters. They are simple to understand and implement and they will work with almost any data set where there is a sequential ordering of the data, be this by time, wavelength or batch number.

Alternative forms of smoothing algorithms, and filters designed to achieve particular enhancement effects with certain types of data, rely on, for example, Kalman filters, Fourier transforms and wavelet transforms. These are not covered in this book, though some references are given in the bibliography to this chapter. All of these techniques remove information from the data to enhance, more or less selectively, the required information. Chapter 13 in this book describes techniques that can extract specific information without destroying other information in the data.

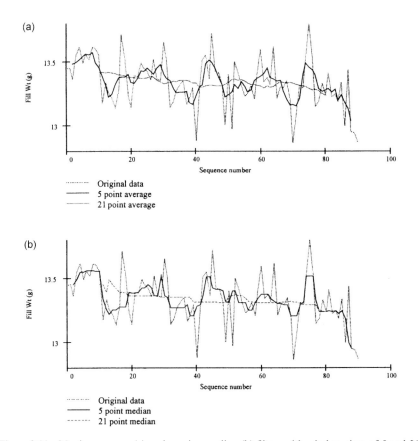

Figure 2.11 Moving average (a) and moving median (b) filters with window sizes of 5 and 21.

Moving-average and moving-median filters

The simple moving-average filter is sometimes described as a moving-window filter. This is because a value in the data set is replaced by another value calculated from data values that appear in a small window placed over the data set. Usually the window is centred on the value to be replaced. The most common calculated value used to replace the original value is the average of the values in the window, but replacement with the median value is often used, particularly when spikes or step changes occur in the original data.

The size of the window, that is the number of data values used to calculate the replacement value, determines the degree of smoothing applied to the data. For convenience, an odd number is chosen. A small number such as 3 or 5 will give a small degree of smoothing and will affect only the highest frequencies in the data. A large number such as 21 or 25 will give a large degree of smoothing, almost completely removing high frequencies but also affecting the low frequencies, including the structures you may wish to see. Middle range values (7, 9, 11) have effects between these two extremes. Which value should be used is very dependent on the application and the structures in the data.

The replacement value is calculated using the equations for average, a_i, or median, m_i, respectively,

$$a_i = \frac{1}{w} \sum_{j=-r}^{r} x_{i+j} \qquad m_i = \text{med}(x_{i-(w-1)/2}, \ldots, x_i, \ldots, x_{i+(w-1)/2})$$

where, x_i is the value to be replaced, w is the window size and $r=(w-1)/2$.

Figure 2.11 shows the effect of applying the average and median filters to the same data with window sizes of 5 and 21.

Savitsky–Golay filter

The Savitsky–Golay (S-G) filter is another window filter, developed by Savitsky and Golay in 1964. Unlike the simple moving-average filter in which all values in the window are given equal weight in the calculation of the average, the S-G filter uses a specific scheme that gives more weight to values in the centre of the window than to those at the edges. As a consequence, the S-G filter tends to give a smaller amount of distortion to the underlying low frequency structures than the simple filter for the same amount of high frequency suppression. The equation for the replacement value is that of a weighted average,

$$sg_i = \sum_{j=-r}^{r} c_j \cdot x_{i+j} \bigg/ \sum_{j=-r}^{r} c_j$$

where, x_i is the value to be replaced, w is the window size, c_j is the appropriate S-G weighting coefficient (note that the coefficients are different for different window sizes) and $r=(w-1)/2$.

Figure 2.12 shows the effect of applying S-G filters with window sizes of 5 and 21 to the same data that was used for Figure 2.11. Comparison of the different methods of smoothing shown in the figures is interesting. Although the simple moving-average and median filters give a reasonable representation of the data when the window sizes are small, the S-G filter gives much better representation at comparable window sizes.

The S-G filter is often used by instrument designers and, where a control is offered to set the level of smoothing, it selects the window size for the filter. The tables of weighting coefficients are given in the papers by Savitsky and Golay, but it should be noted that the tables in the

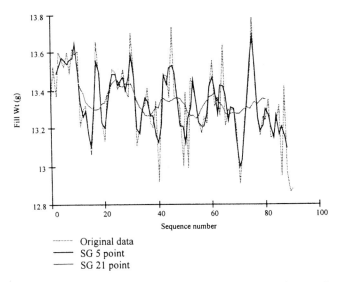

Figure 2.12 The application of S-G filters with window sizes 5 and 21 to the same data shown in Figure 2.11.

original paper contain a small number of errors that were corrected in later publications (see Steiner *et al.*, 1972).

Exponentially weighted moving average smoothing
The simple window smoothing filters use the data in the immediate vicinity of the value being replaced to calculate the replacement value. The exponentially weighted moving average (EWMA) filter uses all of the preceding values, from the start of the data series, to calculate the replacement value. As the name suggests, the replacement value is an average of preceding values that are weighted with a decreasing exponential function so that the older values have less weight than the recent values. A very common use of the filter is to smooth and monitor a process variable in real time so that predictions can be made of the next one or two data values to drive some form of real-time control of the process. This application is discussed more fully in Chapter 9. Here we shall describe the EWMA filter purely as a smoothing filter.

The equation of the filter is,

$$ewma_i = (1 - \lambda) \cdot x_i + \lambda \cdot ewma_{i-1}$$

This shows that the replacement value, $ewma_i$, for the current data value, x_i, depends not only on the current data value but also on the previous

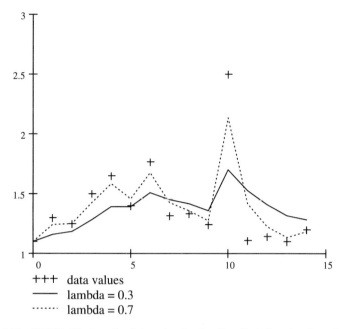

Figure 2.13 EWMA filtering of a data series showing the effect changing the value of λ.

replacement value, $ewma_{i-1}$. Expansion of this equation by repeatedly replacing the *ewma* term on the right-hand side of the expression gives

$$ewma_i = (1-\lambda) \cdot x_i + \lambda \cdot [(1-\lambda) \cdot x_{i-1} + \lambda \cdot ((1-\lambda) \cdot x_{i-2} + \lambda \cdots)]$$
$$= (1-\lambda)[x_i + \lambda x_{i-1} + \lambda^2 x_{i-2} + \cdots + \lambda^{i-1} x_1 + \lambda^i x_0]$$

As the value of λ_i approaches zero as i increases, so the contributions of the preceding data values to the replacement value decrease.

The effect of the filter is controlled by the value of λ, as shown in Figure 2.13.

2.6.2 Integration and differentiation

Integration and differentiation of data sequences are complementary ways of enhancing features in the data. Integration is the calculation of the area defined by two boundaries on the x-axis, some baseline y-value and the curve created by the data values between the two x boundaries. If one of the x boundaries is fixed at the start of the data sequence and the other is moved progressively along the data sequence, the area contained by the figure progressively increases (if the data values are positive; it decreases if the data values are negative). The plot of the area with the x value as it moves along the data series is the integral plot. The increment size for moving the boundary can be one or more data points, with small increments preserving more of the detail in the data than large increments.

Cumulative sum techniques (e.g. Cusum and Normal probability plots) are a form of integration.

Features in the data, such as peaks and humps, that are wider than the increment size appear as steps in the plot of the integral. The height of the step is proportional to the area of the feature (if all features have similar width, then the height is proportional to the height of the feature) and the slope of the riser to the step is proportional to the width of the feature (see Figure 2.14).

An important application of integration as a data processing technique is in NMR spectroscopy, where it is used to calculate the relative sizes of peaks and hence the numbers of atoms in a molecule that contribute to the peaks in the spectrum. Similar, but less common, applications are in voltammetry/amperometry, differential scanning calorimetry and electron spin resonance spectroscopy.

Differentiation is a much more widely used technique, particularly in vibrational and electronic spectroscopy. There are two main reasons for

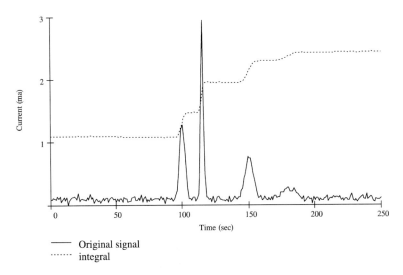

Figure 2.14 Illustration of the integration of a signal. The integral curve has been offset from its initial zero value for clarity.

its use: it can remove the effects of baseline offsets and trends, and it can enhance the resolution between overlapping peaks.

Baseline and trend removal are easily seen consequences of differentiation using a UV spectrum as an example. The absorbance of the sample at a given wavelength, λ, can be described by some function $f(\lambda)$, so that

$$A = f(\lambda)$$

If the measured spectrum is subject to both a baseline offset and a linear increase in absorbance with increasing wavelength, the measured absorbance is

$$A = \beta_0 + \beta_1 \lambda + f(\lambda)$$

where β_0 is the baseline offset and β_1 is the slope of the trend.

The first differential of this equation is

$$\frac{dA}{d\lambda} = \beta_1 + f'(\lambda)$$

It is seen that the baseline offset, β_0, has been eliminated and the trend has been converted into an offset of the first differential. A further differentiation gives

$$\frac{d^2A}{d\lambda^2} = f''(\lambda)$$

And now the effect of the linear trend has been eliminated. Figure 2.15 illustrates baseline offset and linear trend removal from a simulated UV spectrum. The effect is most clearly seen at the high-wavelength end of the spectrum.

Trends across a spectrum are rarely linear, so a second differential will not eliminate them completely. Higher-order differentials (up to about tenth order!) have been explored, but there are severe problems with this approach (described later in this section) and it is now rarely used. In practice, if a trend is sufficiently small compared with the size of the spectral features, the a second derivative effectively removes its effect, even if it is a nonlinear trend.

The other reason for using differentiation is to enhance the apparent resolution between overlapping peaks. This is also seen in Figure 2.15.

Several features should be noted. The first is that the second-derivative spectrum is inverted with respect to the zero-order spectrum: what is a peak in the latter becomes a valley in the former. This is a consequence of

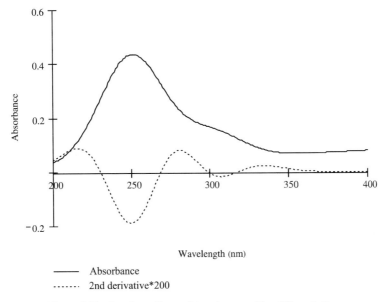

Figure 2.15 Baseline offset and trend removal by differentiation.

the derivative process—the fourth-order derivative would be inverted again, to have the same orientation as the zero-order spectrum.

The second point is that the small hump on the side of the peak in the zero-order spectrum has become a clearly defined peak in the second-order spectrum—the desired effect. This is due partly to the main part of each peak being narrowed by the differentiation and partly to the creation of side lobes on both sides of each peak that fortuitously act together between the peaks to enhance the separation of the peaks. The creation of side lobes during differentiation is one of the reasons for not going to high orders of differentiation—each level creates new lobes on the lobes that already exist and it can become difficult to distinguish between artefactual lobes and real peaks.

The third feature is the position of the peaks as determined peak maximum (even though it is displayed as a minimum in the second derivative). The positions used when calculating the curves were 250 nm and 300 nm for the large and small peaks, respectively. The position of the large peak is seen to be very close to 250 nm in both the zero-order and second-derivative spectra. The position of the small peak in the zero-order spectrum is essentially indeterminate. However, a clear peak is seen in the second derivative at about 308 nm. This is close to, but not exactly the same as, the true value. It is an effect always seen with this type of resolution enhancement and is a function of the amount of overlap, the peak widths and their relative sizes. It has to be borne in mind that peak positions determined in this way are rarely the true positions and that the reported values depend on the calculation method used.

So far we have ignored one aspect of a spectrum when discussing differentiation, and that is noise. Differentiation has the effect of enhancing noise relative to peak size. Each level of differentiation has this effect and it is one of the main reasons for not going much beyond a second derivative with real spectra. Clearly, high-frequency noise can be removed from a spectrum using one of the smoothing algorithms already discussed, but this usually leaves behind small bumps and wobbles and it is these that become enhanced by differentiation in exactly the same way that small true peaks do. The small steps caused by digitisation levels also cause problems in differentiation. They correspond to a high-frequency noise signal and generate spurious peaks in the derivative. Figure 2.16 shows the effect of noise and digitisation level on a second derivative. Although the figure uses somewhat exaggerated noise and digitisation to illustrate the points, the effects are very real and affect *all* data that have come from an A/D converter. These plots should be compared with Figure 2.15.

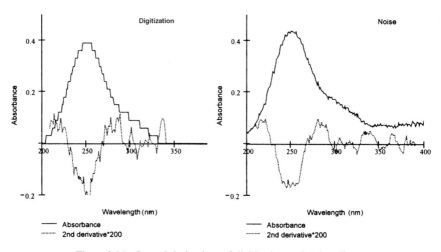

Figure 2.16 Second derivatives of digitisation and noise effects.

Finally, how is differentiation of data carried out? Numerical methods are based upon taking weighted differences between data values. The simplest form is the simple difference between adjacent values,

$$d_i = x_{i-1} - x_i$$

This method has been implemented by several instrument manufacturers, but as it is extremely sensitive to noise in the data it is not recommended for quantitative work or beyond the first derivative.

Savitsky and Golay give coefficients for their window filter to generate first and second derivatives. As the filter incorporates a degree of smoothing as well, it is reasonably robust towards noise. Although coefficients are given only up to second derivative, the filter can be applied repeatedly to data to generate higher-order derivatives. The form of the filter equation is exactly the same as that given in Section 2.6.1, though the divisor is not simply the sum of the weights. Again, the filter is implemented in many instruments.

Some statistics packages offer differentiation as a data transformation option and all mathematics software packages offer numerical integration and differentiation.

2.6.3 *Principal component analysis*

The basics of principal components analysis (PCA) are introduced here as a data preprocessing option for multivariate data. Once the principal components (PCs) have been obtained, they can be treated as ordinary variables in the graphical data description methods, particularly in two- and three-dimensional scatter plots.

A multivariate data table consists of columns, one for each variable, and rows, one for each object for which the variables have been measured (see Table 2.7).

The variables can be all of the same type, such as the particle size parameters in Table 2.7, or they can be of different types such as the other powder properties. Although many simple laboratory experiments may generate only about a dozen variables, there is in fact no restriction other than what the computer and software can reasonably handle—it is quite common for 3000–4000 wavelengths in a spectrum to be used. Similarly, there is no limit on the number of objects in the table, although it is often the case that there are more variables than objects.

When different types of variables are used, there is a potential problem in that they will almost certainly have very different numerical scales (e.g. d0.9 and Cohesivity). During PCA, those with large numerical values will dominate and effectively relegate those with small values to noise, thus losing potentially important information. This can be countered by bringing all of the variables onto the same numerical scale. The most common way to do this is by autoscaling each variable. The average value of a variable is subtracted from its values (so that each variable is centred at zero) and the differences are then divided by the standard deviation of

Table 2.7 Example of multivariate data

Batch	d0.1	d0.5	d0.9	Um	Cm	Ch	p0.5	p0.5_5	p5_10
A21	25	180	392	6.4	13.9	2.0	10.68	80.51	5.72
A22	53	178	359	14.5	12.2	0.3	10.55	84.77	3.16
A23	52	173	382	13.6	14.3	0.4	10.50	86.00	2.33
A24	44	174	343	12.8	14.4	1.4	13.00	80.66	5.75
A25	49	162	318	16.9	14.0	1.3	10.60	85.04	3.47
A26	61	170	354	17.2	13.4	1.3	11.00	86.09	2.31
A27	55	156	298	18.9	12.2	0.9	10.50	85.77	3.40
A28	59	153	270	21.8	14.4	1.1	12.10	85.13	2.58
A29	88	187	337	26.0	16.0	0.5	12.00	85.40	2.40
A30	106	219	357	30.8	10.0	0.6	12.50	83.17	4.00
A31	98	204	338	29.0	10.8	0.4	13.00	82.33	4.30
A32	99	210	358	27.5	10.1	0.3	12.50	84.20	2.96
A33	77	188	342	22.8	11.7	0.2	13.90	83.70	2.20
A34	82	199	343	23.7	11.7	0.2	13.50	81.70	4.40
A35	100	213	356	28.1	11.4	0.3	12.90	83.40	3.40
A36	94	205	339	27.8	10.3	0.2	11.50	84.60	3.60
A37	93	188	305	30.7	10.9	0.1	12.10	84.35	3.27
A38	97	180	289	33.6	6.2	0.1	11.95	84.45	3.20
A39	82	175	328	24.9	14.1	1.2	12.70	84.20	2.80
B13	73	152	237	31.0	14.8	12.5	14.30	79.50	5.60
B14	74	159	246	30.0	11.5	0.9	14.20	79.20	6.00
B15	79	152	237	33.0	15.4	2.3	14.40	79.60	5.40
C8a	116	210	302	41.3	10.0	5.0	9.50	80.60	9.30
C8b	116	210	302	41.3	10.0	5.0	9.90	81.20	8.40
C8c	116	210	302	41.3	10.0	5.0	10.20	81.00	8.30
C9	107	166	241	44.7	12.0	3.0	12.45	80.65	6.55
C10	93	157	242	38.3	16.0	3.0	11.10	80.83	7.60
C11	85	157	236	39.0	13.0	2.0	13.76	78.70	6.82
C13	95	165	264	37.0	16.0	8.0	15.05	79.46	4.89
C14	100	173	286	33.0	13.0	5.0	15.48	78.47	5.26
C15	108	162	237	44.0	13.0	6.0	16.32	78.08	4.83

d0.1, d0.5 and d0.9 are particle size parameters characterising a powder entering a process; Un, Cm and Ch are respectively the Uniformity, Compressibility and Cohesivity of the input powder; p0.5, p0.5_5, p5_10 are particle size parameters of the processed powder. The prefix letters of the batch number indicate different suppliers of powder.

the values for the variable (so the range of each variable is scaled so that ±1 corresponds to one standard deviation).

Thus, the autoscaled values are

$$a_{ij} = \frac{x_{ij} - \bar{x}_j}{s_j}$$

where, x_{ij} is the ith value for variable j, $\bar{x}_j$ is the average for column j and s_j is the standard deviation for column j.

Table 2.8 gives an example of autoscaling using the d0.9 and Cohesivity values from Table 2.7. Only the first six rows of Table 2.7 are displayed in Table 2.8, to illustrate the process of autoscaling, but the calculations have been performed using the averages and standard deviations of all of the rows (the values used are given in Table 2.8).

Autoscaling is not recommended when the variables are all of the same type, as variables with small values may truly be noise and they should not be inflated to have the same weight as variables that represent the signal. A fuller discussion of scaling is given in Chapters 9 and 10.

Table 2.8 Example of autoscaling using part of the d0.9 and Cohesivity data from Table 2.7

	d0.9	Autoscaled d0.9	Cohesivity	Autoscaled cohesivity
	392	1.7657	2.0	−0.1008
	359	1.0810	0.3	−0.6845
	382	1.5536	0.4	−0.6633
	343	0.7276	1.4	−0.3086
	318	0.2063	1.3	−0.3491
	354	0.9611	1.3	−0.3440
	298	−0.1978	0.9	−0.5037
Average	307.7085		2.2700	
SD	47.8191		2.8195	

The process of generating PCs can be visualised using just two variables, d0.1 and Uniformity given in Table 2.7. First, draw a scatter plot of the data as in Figure 2.17. Now draw a line through the data and project the data points at right angles onto the line, then rotate the line so that the distance between the two most extreme projections is a maximum. In this way the line falls in the direction of maximum variability in the data; it is said to explain, or account for, the largest amount of variation in the data. The equation of the line, if it is translated (not rotated) to pass through the origin of the plot is the first PC,

$$PC1 = b_{11}x_1 + b_{12}x_2$$

PC1 accounts for variability in the data along its length but it does not account for variability in other directions. So another line is drawn on the plot. It is rotated so that it is at right angles to the first line (it is orthogonal to PC1) and so that when the data points are projected at right angles onto it, it accounts for the maximum amount of variation in the data not accounted for by PC1. This is PC2 and it has the equation

$$PC2 = b_{21}x_1 + b_{22}x_2$$

With only two original variables there is only one direction that PC2 can take and be orthogonal to PC1, as shown in Figure 2.17. However, if there are more variables, PC2 can be at any position in a plane orthogonal to PC1, but the constraint of accounting for the maximum amount of variation then defines a unique direction.

When there are n variables, n PCs can be formed in the same way.

The length of the data projection on the lines is the corresponding amount of variance explained, or eigenvalue, of the PC.

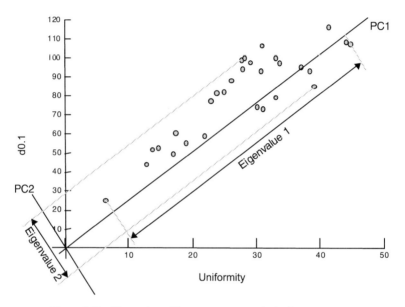

Figure 2.17 Illustration of how to generate principal components.

Clearly this graphical construction of PCs is impractical when there are many variables and computation must be used. All modern statistical and chemometrics packages include PCA as a standard option. The statistical packages generally use one of the classical eigenvector/eigenvalue methods of generating PCs, such as the Jacobi or Householder methods. These generate all possible PCs from the covariance matrix (or correlation matrix if the data have been autoscaled), **Z**, of the original data table,

$$\mathbf{Z} = \mathbf{D}^T\mathbf{D}$$

As this is done in essentially one go, the calculation can be quite lengthy if there are many variables. Chemometrics packages generally use the alternative NIPALS algorithm, which extracts and reports the PCs from the original data matrix **D** one at a time and in decreasing order of amount of variance explained.

The coefficients, b_{ij}, in the equations for the PCs are called the loadings as they describe the relative contributions, or loadings, of the original variables in the PC. Values of b_{ij} close to zero indicate that the corresponding variable has little or no contribution to that PC. Variables with large b_{ij} within the same PC are likely to be correlated if they have the same sign and anticorrelated if they have opposite signs.

The vector of values obtained by multiplying the original data matrix with one set of PC loadings gives the scores of the objects for that PC. In effect, each vector of scores is a new variable formed by the linear combination of the original variables. Thus, it is quite appropriate to treat the scores vectors as variables and, for example, draw scatter plots of them.

The PCA of the data in Table 2.7 is summarised in Tables 2.9 and 2.10 and Figure 2.18. Table 2.9 lists all of the eigenvalues, the percentage variance accounted for by each and the cumulative percentage variance explained.

Table 2.9 The eigenvalues from the PCA of the data in Table 2.7

Component no.	Eigenvalue	Percentage variance	% Cumulative variance
1	3.8872	43.19	43.19
2	2.4736	27.48	70.68
3	1.0551	11.72	82.40
4	0.7005	7.78	90.18
5	0.4979	5.53	95.71
6	0.3429	3.81	99.52
7	0.0302	0.34	99.86
8	0.0090	0.10	99.96
9	0.0037	0.04	100.00

As is often found, the first few PCs account for a high proportion (>95%) of variation in the data—in this case five. Scatter plots of these PCs allow large multivariate data sets to be examined quickly and visually for correlations, clustering and outliers. Figure 2.18 is the scatter plot of the first two PCs of the data. The plot shows some very clear clustering of the data, with one cluster approximately in each quadrant of the plot. The separation along PC1 is perhaps the clearest. Going back to the scores table for PC1 we find that negative PC1 values correspond to supplier A and positive values to suppliers B and C. The loadings of the original variables on PC1 indicate that supplier A produces material that tends to have high d0.5, d0.9 and p0.5_5 values with low d0.1, Uniformity, Cohesivity, p0.5 and p5_10 values. Compressibility, with a loading value of 0.0332, does not make much of a contribution to PC1.

The interpretation of the split along PC2 is not so simple. The scores do not fall into categories by supplier. However, the three points in the upper right quadrant correspond to three subbatches of one main batch from supplier C, while the points in the upper left quadrant correspond to a sequence of 9 consecutive batches from supplier A. The loadings for PC2 show that high positive values will be obtained with large d0.1 and d0.5 values coupled with low Compressibility, indicating that the split along PC2 is due to the properties of the input powder rather than the processed powder.

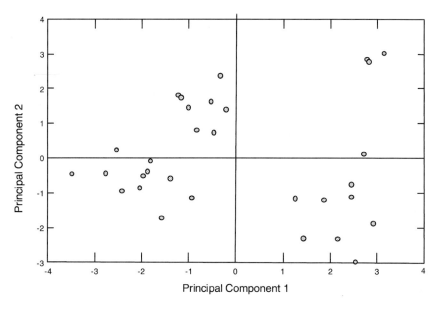

Figure 2.18 Scatterplot of the scores of the first two principal components of the data in Table 2.7 as listed in Table 2.10.

As an aside, it is worth mentioning that if these data are not autoscaled, then the first two principal components account for 98.56% of the variability in the data. PC1 is dominated by d0.9 while PC2 is dominated by the combination of d0.1 and d0.5. These are the variables with the largest numerical values. In addition, the scores plot of PC1 with PC2 does not show any clustering of the data and so the ability to detect potential problems is lost.

Table 2.10 The first five principal component loadings and scores for the data in Table 2.7

	PC 1	PC 2	PC 3	PC 4	PC 5
Loadings					
d0.1	0.3021	0.4459	0.2619	−0.0962	0.3197
d0.5	−0.1089	0.5637	−0.0107	0.4390	0.1724
d0.9	−0.4243	0.1804	−0.1019	0.5162	0.1315
Unifomity	0.4323	0.2601	0.1419	−0.2839	0.1500
Compressibility	0.0332	−0.5324	−0.1360	0.1323	0.4636
Cohesitivity	0.3764	−0.1320	−0.2583	0.2238	0.5822
p0.5	0.2487	−0.2318	0.6838	0.4181	−0.1312
p0.5_5	−0.4372	0.0861	0.0904	−0.4497	0.4065
p5_10	0.3664	0.1458	−0.5829	0.0800	−0.3031
Scores					
A21	−2.1890	−1.4352	−2.2516	1.7912	−0.9756
A22	−2.6565	−0.4445	−0.6508	−0.0782	−0.1882
A23	−3.2213	−1.0224	−0.5813	−0.0376	0.5725
A24	−0.9591	−1.6696	−0.9104	1.3796	−0.9873
A25	−1.9770	−1.4400	−0.8117	−0.7449	0.1117
A26	−2.5204	−0.8566	−0.1921	−0.4273	0.6319
A27	−1.8507	−1.0399	−0.5182	−1.4577	−0.2060
A28	−1.1561	−1.9127	0.3160	−1.3051	0.1966
A29	−1.5224	−0.4463	0.5307	−0.1776	1.3726
A30	−0.7458	2.3496	0.7325	0.7559	0.1758
A31	−0.4312	1.4064	0.7122	0.6225	−0.2563
A32	−1.3514	1.8230	0.9666	0.4606	0.2501
A33	−1.4227	0.0247	1.3816	0.5288	−0.1111
A34	−0.6946	0.5567	0.5553	1.0609	−0.5821
A35	−1.0094	1.5779	0.9200	0.7850	0.3023
A36	−1.3119	1.6917	0.3936	−0.1277	0.1439
A37	−0.7463	0.9219	0.8012	−0.7202	0.0095
A38	−0.5084	1.9237	1.1737	−1.4878	−0.9246
A39	−1.0679	−0.6047	0.6394	−0.1091	0.5959
B13	3.0962	−2.4398	−0.6167	0.5767	1.3213
B14	1.4647	−0.8768	0.4503	−0.2258	−1.7719
B15	1.8664	−1.9628	0.4787	−0.2811	−0.5118
C8a	1.9563	2.8486	−2.1290	−0.0731	0.1459
C8b	1.7428	2.7492	−1.6844	−0.1200	0.3503
C8c	1.8018	2.6949	−1.5439	−0.0162	0.3108
C9	2.4203	0.4028	0.0100	−1.1493	−0.2446
C10	2.0295	−0.9384	−1.3089	−1.1662	0.1889
C11	2.4386	−0.8271	0.0984	−0.4845	−1.1703
C13	2.7301	−1.6484	0.4922	0.7471	1.2510
C14	2.2418	−0.5821	0.9197	1.1240	−0.0984
C15	3.5538	−0.8239	1.6265	0.3570	0.0973

2.6.4 Other transformations

If it is known that the underlying distribution of the data is not Normal, but follows some simple function such an inverse, square root or logarithm, a transformation of the data using the function will generate data with an approximate Normal distribution. This will aid further processing with methods that assume Normality.

However, if it is not known, but only suspected, that the data are non-Normal, care should be exercised if only small amounts of data are available. In these situations it may be more appropriate to apply nonparametric data analysis tools rather than apply a transformation whose effects may be the opposite of those expected.

Bibliography

The topics in this chapter are covered in a very wide range of publications, both books and research papers. The small selection of references given here should help in finding further, more detailed and readable, discussions of the topics and give pointers to specific applications. The list is roughly characterised by the subject content of this chapter, with the references in the general section covering several aspects.

General bibliography

Barlow, R.J. (1995) *Statistics, A Guide to the Use of Statistical Methods in the Physical Sciences*, Wiley, Chichester.
Brereton, R.G. (1990) *Chemometrics: Applications of Mathematics and Statistics to Laboratory Systems*, Ellis Horwood, Chichester.
Caulcutt, R. (1991) *Statistics in Research and Development*, 2nd edn, Chapman and Hall, London.
Caulcutt, R. and Boddy, R. (1989) *Statistics for Analytical Chemists*, Chapman and Hall, London.
Davies, O.L. and Goldsmith, P.L. (1976) *Statistical Methods in Research and Production*, Longman, London.
Hoel, P.G. (1984) *Introduction to Mathematical Statistics*, Wiley, New York.
Kateman, G. and Pijpers, F.W. (1981) *Quality Control in Analytical Chemistry*, Wiley, New York.
Miller, J.C. and Miller, J.N. (1989) *Statistics for Analytical Chemistry*, 2nd edn, Ellis Horwood, Chichester.
Ryan, T.P. (1989) *Statistical Methods for Quality Improvement*, Wiley, New York.

Data structure

Blanco, M., Coello, J., Iturriaga, H., Maspoch, S. and Riba, J. (1990) Precision of a diode array spectrophotometer. *Analytica Chimica Acta*, **234** 395-401.
The Presentation of Numerical Values, British Standard 1957 (1953), Confirmed July 1987, British Standards Institution, London.
Butler, B.P., Cox, M.G., Ellison, S.L.R. and Hardcastle, W.A. (1996) *Statistics Software Qualification, Reference Data Sets*, Royal Society of Chemistry, Cambridge.

Green, J. (1992) Diode arrays in spectroscopy, *International Labmate*, **92** 11-16.
Hovanec, J.W., Seiders, R.P. and Ward, J.R. (1986) On the precision of a diode-array spectrophotometer. *Computer Enhanced Spectroscopy*, **3** 69-71.
Lepla, K.C. and Horlick, G. (1990) Data processing techniques for improved spectrochemical measurements with photodiode array spectrometers. *Applied Spectroscopy*, **44** 1259-1269.
Norris, K.H. (1992) A closer look at spectral noise. *Spectroscopy Europe*, **4** 26-31.
Zitko, V. (1989) A simple look at the structure of data matrices. *Trends in Analytical Chemistry*, **8** 161-162.

Data display

Box, G. and Luceno, A. (1997) *Statistical Control by Monitoring and Feedback Adjustment*, Wiley, New York.
Chatfield, C. (1989) *The Analysis of Time Series: An Introduction*, 4th edn, Chapman and Hall, London.
Tuffe, E.R. (1986) *The Visual Display of Quantitative Information*, Graphics press, Connecticutt.
Wetherill, G.B. and Brown, D.W. (1995) *Statistical Process Control, Theory and Practice*, Chapman and Hall, London.

Outliers

Barnett, V. and Lewis, T. (1994) *Outliers in Statistical Data*, 3rd edn, Wiley, New York.
Naes, T. and Isaksson, T. (1992) The importance of outlier detection in spectroscopy. *Spectroscopy Europe*, **4** 32-33.
Rius, F.X., Smeyers-Verbeke, J. and Massart, D.L. (1989) Method validation: software to plot calibration lines and their response residuals, and to detect outliers according to Cook's distance. *Trends in Analytical Chemistry*, **8** 8-11.
Thompson, M. (1989) Robust statistics—how not to reject outliers. Part 1: Basic concepts (RSC Analytical Methods Committee). *Analyst*, **114** 1693-1697.

Smoothing and filtering

Bialkowski, S.E. (1988) Real time digital filters: infinite impulse response filters. *Analytical Chemistry* (A pages), **60** 403-413.
Bialkowski, S.E. (1989) Generalised digital smoothing filters made easy by matrix calculations. *Analytical Chemistry*, **61** 1308-1310.
Bromba, M. and Ziegler, H. (1984) Variable filter for digital smoothing and resolution enhancement of noisy spectra. *Analytical Chemistry*, **56** 2052-2058.
Erickson, C.L., Lysaght, M.J. and Callis, J.B. (1992) Relationship between digital filtering and multivariate regression in quantitative analysis. *Analytical Chemistry* (A Pages), **64** 1155-1163.
Kitamura, K. and Hozumi, K. (1987) Effect of Savitzky–Golay smoothing on second-derivative spectra. *Analytica Chimica Acta*, **201** 301-304.
Moore, A.W. and Jorgenson, J.W. (1993) Median filtering for removal of low-frequency background drift. *Analytical Chemistry*, **65** 188-191.
Ratzlaff, K.L. and Johnson, J.T. (1989) Computation of two-dimensional polynomial least-squares convolution smoothing integers. *Analytical Chemistry*, **61** 1303-1305.
Savitzky. A. and Golay, M.J.E. (1964) Smoothing and differentiation of data by simplified least squares procedures. *Analytical Chemistry*, **36** 1627-1639.

Steinier, J., Termonia, Y. and Deltour, J. (1972) Comments on smoothing and differentiation of data by simplified least square procedure. *Analytical Chemistry*, **44** 1906-1909.

Terrence, A.L., Headley, L.M. and Hardy, J.K. (1991) Noise reduction of gas chromatography/mass spectrometry data using principal component analysis. *Analytical Chemistry*, **63** 357-360.

Principal components

Chatfield, C. and Collin, A.J. (1995) *Introduction to Multivariate Analysis*, Chapman and Hall, London.

Geladi, P. and Kowalski, B.R. (1986) Partial least squares regression: a tutorial. *Analytica Chimica Acta*, **185** 1-17.

Kubista, M., Sjoback, R. and Albinsson, B. (1993) Determination of equilibrium constants by chemometric analysis of spectroscopic data. *Analytical Chemistry*, **65** 994-998.

Manly, B.F.J. (1994) *Multivariate Statistical Methods. A Primer*, 2nd edn, Chapman and Hall, London, ISBN 0 412 60300 4.

Sharma, S. (1996) *Applied Multivariate Techniques*, Wiley, New York.

3 Sampling
J. Thompson

3.1 Introduction

Chemists undertaking research and development activities, whether in the academic or industrial sectors, need to take samples for various purposes, e.g. for measurement of composition, chemical, biochemical or physical properties or to investigate the effects of materials on the behaviour of physical, chemical, biological or environmental systems. Frequently, issues relating to sampling are given little detailed consideration. In many cases, this is because scant consideration was given to sampling issues in the training of chemists and the consequences of this neglect were poorly understood. This chapter will provide chemists with a basic understanding of the range of sampling issues that should be considered when planning and undertaking research and development projects. The reader will find concepts being brought together from a wide variety of disciplines, not only from areas of chemical research and development but also from disciplines unlikely to be considered obviously relevant although they have contributed to major advances in statistical sampling theory and practice. Other examples and concepts will be drawn from various practical everyday experiences of sampling that are of considerable economic or regulatory importance. It is hoped that by creating such a blend, new insights will be encouraged and greater appreciation of the need to consider very carefully the full range of sampling issues when planning or conducting research and development.

3.2 Where to go

Information needed	Go to Section(s)
Is there a difference between "specimens" and "samples"?	3.3
Statisticians seem to have a different meaning of samples from mine!	3.3
Can I be certain that a sample actually contains what I am looking for?	3.4.1, 3.4.2, 3.4.3, 3.7.1
What are random samples?	3.6.1

Information needed	Go to Section(s)
Can I get useful information from samples whose generation is not really under control?	3.5.1
What sort of sampling should I use for non-destructive testing?	3.5.2, 3.5.4
Are there specific ways to sample environmental materials or production processes?	3.5.3, 3.6.2, 3.6.3
How do I plan taking samples?	3.6
How many samples do I need?	3.6.1, 3.7.1
What issues are there when using methods close the limits of detection or quantitation?	3.7.2, 3.8
Is there any software that can help?	3.7.4

3.3 What is sampling?

What kinds of sampling can be done? To chemists, sampling often means the *physical activity* of collecting a specimen of a substance, a biological fluid, a tissue from a living or dead organism, a rock or mineral or a volume of air, water or waste. To a statistician, sampling means collecting items that may or may not be representative of a population of such items. The items may simply be numbers in a theoretical simulation or measurements on specimens or counts of the specimens themselves. To the statistician such a collection of items is a sample, whereas to the average chemist each item or specimen in the collection is a sample. To avoid confusion, 'sample' will be used in the statistical sense, unless otherwise indicated.

What do we mean by a good sample? Often, this may simply mean a large enough specimen, from which the experimenter can take smaller specimens, on which to carry out the planned measurements. But from a statistical viewpoint we need to consider whether a sample was representative of the population from which it was taken or whether it was biased in some way. As chemists, we often split samples into sub-samples, either to make replicate measurements or to make a range of different measurements or simply because the original specimen is too large to measure in its entirety. If the sample/specimen is homogeneous, this physical subdivision may not be a problem. More often than not, such an ideal situation does not apply and attention must be given to designing the subdivision so that bias is minimised, as far as is practicable, and hopefully some understanding is gained about the nature of the heterogeneity. The design of the sampling scheme should also enable us to understand the effects of the methods of preparation and preservation of a specimen prior to measurement.

What are our aims and objectives in collecting samples? In the statistical sense, sampling needs to be done in a way that preferably avoids, or at worst minimises, bias. Another aim is to gain an understanding of uncertainties or variabilities contributed by the various physical and chemical manipulations of the specimen or specimens prior to measurement, as well as by the intrinsic within-specimen and between-specimen variability.

How do these aims and objectives fit with the overall aims and objectives of our project? Whenever measurements are made, we should be concerned to gain an understanding of the uncertainties of the measurement process. In the context of sampling, it is important that the sample is of a sufficient size. From a statistical viewpoint, this means that we must have sufficient specimens to provide adequate confidence in conclusions drawn from the measurements performed on those specimens. Physically, it implies that the size of each specimen is such that there is enough material within each specimen to do all the necessary, planned measurements. In the design of a sampling exercise, careful attention should be given to the practical balance between the costs of sampling and measurement and the gaining of reliable, adequate information. In other words, sampling should be fit for the overall purpose of a project.

3.4 Sampling—The Pandora's box of chemical/biological research/development

Many examples can be found in practice in which the issue of sampling is regarded as of trivial significance, e.g. 'take a representative sample' is often all that is said and the sampling is left to an untrained operative.

I recall an experiment undertaken by Waste Regulators without prior consultation on the design. It was intended to yield information on the within- and between-batch variability of some lime sludges (filter cakes). The sampling was done by a van driver who was instructed to collect 10 specimens 'at random' from one skip containing such sludge, at each of a series of seven waste processing sites. If some areas in each skip looked different, he was to try to sample from a range of these areas within that skip. Thus, he was already departing from a random pattern, even without imposing his own interpretation of randomness into the sampling process. In the past I have often referred to this sampling protocol as a 'lucky dip'. When the uncertainties began to be assessed, the

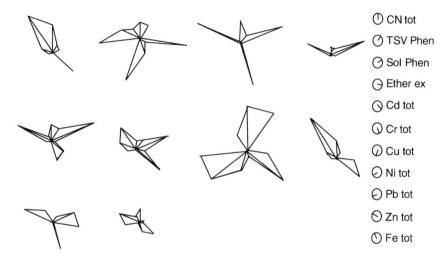

Figure 3.1 Star plots of 'lucky dip' or grab sampling from ten positions in a skip of lime sludge (filter cake) waste.

heterogeneity was found to be considerable. This may be seen in the star plot shown in Figure 3.1. Each star is made up from vectors representing different analytes (see key on diagram). The length of a vector for an analyte is indicative of the concentration in the specimen represented by a given star plot. The concentrations for each analyte are normalised relative to the maximum for that analyte in the group of 10 specimens from the skip. If all specimens had the same concentration for a given analyte, the star plots would all be regular polygons with the same overall shape and size. Here we see wide variations (orders of magnitude) in analyte concentrations between specimens. Subsequent laboratory studies on blending of a specimen of such sludge revealed not only that sampling was an important issue but that problems of sample preparation were also very important (see below).

3.4.1 The peanut problem

In a lecture at a Royal Society of Chemistry conference on 'Uncertainty and Reliability in Chemical Analysis' (December 1997), William Horwitz of the US FDA, cited the food analyst's worst sampling nightmare in which one, and only one, peanut in a lorry load is heavily contaminated with aflatoxin. If that single contaminated peanut turns up in the food analyst's sampling of the lorry load, what shall be inferred about the aflatoxin content of the load? Is the sampling protocol designed so that we can say that contamination arose from just one peanut in that lorry load? This is unlikely, as the objective probably is to assess the quality of the load, not focus on individual peanuts. Conversely, if the contaminated peanut does not appear among specimens received in the analyst's laboratory, we naturally assume that the lorry load is free from contamination.

This raises the very important issue of the 'appropriateness' not only of measurement processes but also of sampling processes. In this particular respect, we should not be concerned only with the traditionally narrow considerations of the measurement process, such as chemical sensitivity or specificity and repeatability/reproducibility. 'Appropriateness' suggests that our attention should be directed to objective criteria concerned with whether the combined sampling and measurement processes *and* the sampling strategy enable us to answer some more difficult questions.

3.4.2 Risks in diagnosis

The peanut problem is one example of such a set of difficult questions. Another commonly occurring medical example arises from carrying out a diagnostic test involving a specific chemical analysis of blood. Subsequently, a clinician might try to judge whether a patient has a particular disease, as a result of carrying out that diagnostic test and, thus, requires information about diagnostic reliability of the assay.

This is a very different question. The diagnostic sensitivity and specificity relate to the context within which the sampling and measurement are applied. They have nothing to do directly with the specificity and sensitivity of the measurement. The result of the measurement might be judged against a criterion, or even a set of criteria, that may not be completely clear-cut in that, for example, the range of measurement values for absence of disease often overlaps the range encountered in patients with the disease. In such a situation, decisions on the presence or absence of the disease may be false positive when the patient does not have the disease but the test value is above the decision threshold value. Alternatively, the decision may be a false negative (the patient does have the disease but the test value is below the threshold value). Such situations are often made worse in the context of the use of the results from a clinical chemistry laboratory by poor attention to the sampling strategy and processes that may be critical in enabling the refinement of the diagnostic capability without doing anything to refine the performance of the assay (Fraser, 1986; Walmsley and White, 1985). Thus, when researching new chemical tests for diagnostic purposes, clinical chemists have become acutely aware of the importance of such sampling strategy issues in recent years.

W.G. de Ruig and H. van der Voet (1994) have drawn attention to the problem of criteria for interpreting data, in relation to harmonisation and optimisation of analytical methods and inspection procedures. As an illustration of the problems of defining and choosing appropriate criteria, they specified criteria for judging whether there is a tower of a specific type present in a Dutch village by inspecting a set of photographs. They make the point that using different thresholds for criteria or using different combinations of criteria will inevitably produce different outcomes with regard to the probability of a false positive or false negative decision. Thus, there should be agreement on the exact specification of the criteria between the producer of the measurement data and the ultimate user of the information derived from that data. The user and producer should agree what are acceptable risks for false positives and false negatives and whether these can be satisfactorily achieved. This has major implications for the design of sampling strategies and protocols and judging their relative capabilities. This theme will be developed further in various sections of this chapter.

3.4.3 *Whose point of view?*

A practical example was encountered recently in the performance validation of a chemical ionisation mass spectrometric measurement technique for analysis of trace organic vapours in air. It used vapour standards generated from permeation tubes for four organic compounds. One series of measurements gave results from the calculations of vapour concentrations much lower than were produced by the standards generator. Other experiments in the same series had previously produced results reasonably in line with expectations of those produced by the standards generator. The chemical physicists making the measurements claimed that they were making 'absolute' measurements and so, *the standards generator must have been at fault*. Even when shown how an assessment of the measurement and sampling uncertainty budget could be derived, their conviction in the 'correctness' of their stance remained strong.

They were left to contemplate this problem.

Two weeks later, they announced that having tested the channeltron ion detector in the equipment, its sensitivity was found to have dropped sharply. They had no criteria against which to judge whether their equipment was functioning correctly and no routine protocol was built into their research programme to audit the performance of their measurement system. This drop in sensitivity, first indicated using a series of vapour standards produced using the standards generator, was further revealed by measurements of well-understood chemical reaction systems. In these experiments, reaction rate constants (previously verified by other independent methods) were measured rather than measuring the concentration of components in vapour standards produced with the standards generator.

The researchers were able to use the information gained from the reaction rate experiments to derive rough 'correction factors' for their measurements of each of the four organic vapours being studied. When their data was subsequently 'adjusted' using these correction factors, the regression of these previously doubtful results was much closer to that for previous series. This was a clear demonstration of the value of appropriate sampling of reference vapour materials produced with a reliable standards generator, the functioning of which could be independently checked. However, to these chemical physicists, it was a clear demonstration of the capabilities of their 'absolute' measurement system and they remain convinced to this day that these studies were merely validating the performance of the standards generator!

3.5 Aims and objectives of sampling

The primary aims of a well-designed sampling scheme relate to underpinning measurement reliability and, thus, to minimising bias and obtaining reasonable confidence in results while yielding valuable information about the intrinsic variability of whatever has been sampled. It is useful to relate these aims to the varied contexts in which sampling is done. One fundamental point is the sharp distinction between designed experiments and observational studies. Both are very important approaches in scientific research, but each presents different problems, especially in connection with the design of appropriate sampling schemes.

3.5.1 Observational studies

In the research or development laboratory, pilot plant or production plant, we are often able to control the important variables, and in such situations we can conduct designed experiments. In other situations, we may not know what all the important variables are but we may still be able to design experiments in a balanced way. Then we may use appropriate forms of data analysis to tease out the contributions of the variables included in the design from other uncontrolled or often unknown influences. In this chapter, the focus is on the use of concepts from other chapters directed to the solution of particular problems, such as those outlined below.

In observational studies, we are unable either to control the experimental units or to assign preset levels to variables whose effects we wish to study, or we may be confronted with combinations of such problems. We have to take things as they are. Such situations are common in biological and environmental sciences and astronomy, but less obviously so in chemical research. They may be present also in what may appear to be otherwise well-controlled experimental situations because of effects of one or more unconsidered variables. Observational studies, by their very nature, are more difficult to interpret and evaluate and they come with built-in biases, so our efforts need to be directed to minimising such biases by careful design of sampling schemes. W.G. Cochran wrote the major monograph on the planning and analysis of observational studies but regrettably did not complete it before his death. His manuscript was edited by Lincoln E. Moses and Frederick Mosteller and published in 1983 and it still remains a very important source of ideas and techniques.

3.5.2 *Invasive, noninvasive, remote and indirect sampling*

Invasive sampling generally involves physically taking a specimen for further study or inserting a measurement device into the system being sampled. In clinical or animal studies, invasive sampling may involve removing a specimen (say of arterial, venous or capillary blood or a tissue biopsy) for *in vitro* measurement or using an *in vivo* sensor (e.g. an ISFET array or a pressure transducer) or sampling system (e.g. a microdialysis probe embedded in a particular tissue, subcutaneously). Noninvasive sampling is becoming more common in clinical studies, either in a remote, noncontact mode (e.g. using near-infrared spectrometry) or in an external contact mode by placing a sensor *ex vivo* on the surface (e.g. by using a transcutaneous oxygen sensor).

Remote sampling is a form of noninvasive sampling, e.g. using satellite-based infrared spectrometers to map water distribution in soils and plants. Likewise, indirect sampling may involve remote or noninvasive measurement, as when laser beams are directed through high- or low-temperature plasma to study the spatial and temporal distribution of reactants and their physicochemical state.

Other forms of biochemical sampling of living organisms (sampling of urine, saliva, sweat or exhaled breath or other secretions or excretions) taken over a period of time may allow one to model processes within the organism, particularly relating to normal or pathological metabolic processes or to pharmacokinetic, pharmacodynamic and toxicological studies. In these cases, the temporal sequence of samples must take into account the appropriate anatomical, biochemical, physiological and pharmacological issues relevant to the sampling process. It may also be necessary to consider the time course of the physical, chemical or other processes being studied, as when we study the metabolism of a drug or the periodicity of a biochemical process, such as glucose metabolism.

When researching the use of microsensors *in vivo* in an artery, the hydrodynamics at the sampling zone of the sensor may vary, depending on the sensor's position in the artery relative to the arterial wall and whether the probe is facing into the flow or in the direction of flow. Stagnation and sensor fouling may seriously affect the ability of the sensor to take a meaningful and representative sample of measurements. Similar considerations apply in sampling flowing fluids or powders in many other situations. Such measurements are often either continuous or quasi-continuous, and then the issue of sampling frequency takes on a different dimension analogous to that of automatic on-line industrial process monitoring. Hydrodynamics and sensor fouling are also relevant to sampling in the latter context. Even in situations where sampling would seem straightforward, as in the monitoring of water purification for renal dialysis or for heat-exchanger systems, biofouling of continuous sampling and monitoring systems can cause havoc.

3.5.3 Sampling for process/quality control/improvement and for environmental regulation

Various kinds of sampling are undertaken in quality management, even in research and development projects. These include:

- Sampling for the establishment of the initial conditions for a control chart
- Sampling of the process to be controlled with the aid of the control chart
- Various kinds of acceptance sampling (both by attributes and by variables).

The issues involved in sampling for statistical process control are discussed later in this chapter, as well as in Chapter 9.

Sampling for environmental regulatory or occupational hygiene compliance monitoring involves sampling from non-Gaussian (right-skewed) distributions of pollutants. In addition, we seek to gain information on both spatial and temporal distribution of pollutants. While this might seem of peripheral interest for many areas of research and development, it should be recognised that it is of fundamental importance to those involved with ecotoxicological research on pesticides, herbicides, etc., and similar principles apply to tracing the origins and effects of impurities in chemical synthesis. In the case of discharges of chemicals into the working or external environments, the situation is analogous to sampling for process control with upper warning/action limits only. The US NIOSH (Leidel *et al.*, 1977) has issued statistical guidance on sampling strategies that takes account of the sampling distribution (in most cases approximated by log-normal, truncated log-normal or exponential distributions; see also Harvey, 1981).

In drafting new environmental regulations, there is a trend towards establishment of site-specific trigger levels and the use of statistical process control (SPC) to monitor compliance. This may be exemplified by the EU Landfill Directive (EU 6453/98 Annex A, DG1), which was approved by the EU Council of Ministers on April 26th 1999. In drafting this directive, the EU has recognised the need for a major research effort on sampling and characterisation of wastes and of sampling for environmental monitoring in order to draw up satisfactory and reliable guidelines. Additionally, there may be agreed designed discharges of pollutants during the lifetime of an industrial activity, either into the atmosphere or into ground or surface waters. Research is needed to assess the impact of such discharges, any adverse deviations from their designed and allowed extent and especially on the sampling strategies for such changing baselines, before guidance can be issued.

It is likely that sampling uncertainty will be the dominant problem, so it will be important to understand its effects on limiting the ability of monitoring programmes to detect noncompliance with the original designed environmental impacts (including the difficult issues of false negative assessments of noncompliance). This can be more difficult for sampling distributions that are skewed and elongated, as is frequently the case. It is necessary to distinguish between the sensitivity and specificity of a chemical analysis and the very different types of sensitivity and specificity of a diagnostic procedure discussed in Section 3.4 (in these cases, detection of compliance versus noncompliance, i.e. false negative and false positive assessments). Such assessments are retrospective, so it is vital to develop techniques for the design of appropriate sampling strategies to optimise operating characteristics (or statistical power) of the test systems and average run length to achieve reliable and reasonably early detection of noncompliance problems before they acquire catastrophic proportions (see below).

Other problems relate to the avoidance of cross-contamination and to the preservation of the physicochemical state of the specimen, particularly in respect of the dissolved gases and volatile/semivolatile organic compounds. In designing environmental sampling protocols, it is recommended that blank samples be included within the quality protocol for control (uncontaminated) sites and field, trip and sampling equipment

(Kulkarni and Bertoni (chapter 6) and Black (chapter 7) in Keith, 1996). Groundwater, which is collected in an anaerobic state, requires preservation of that state, not allowing loss of CO_2 or uptake of O_2, to avoid affecting the redox and pH status. Such comments also apply to the sampling of polluted surface waters, and sediments in lakes, streams, estuaries and the sea.

Other sampling problems arise in dealing with complex wastes, which may be sent from a producer to a secondary processor for reprocessing or ultimately for disposal. Some of these relate to waste disposal in landfill and involve design of sampling and analysis protocols for characterisation of the properties of the waste concerned. These involve not only the assessment of composition and its heterogeneity within and between batches but also the potential leaching characteristics under the conditions likely to be encountered in the landfill. In waste reprocessing to render materials re-usable, e.g. recovering aluminium from furnace drosses, difficulties arise because of the enormous particle size range and the variable reactivity of the dross, partially resulting from this wide range but also from compositional heterogeneity. Thus, sampling schemes need to be devised to enable the dross producer and the reprocessor to agree a fair price for the secondary material to its producer commensurate with the costs of reprocessing and a fair return on the recovered material. Such schemes have to strike a delicate balance between the costs of sampling and analysis and reliability of the information obtained. Thus, they may have some similarity to those required in the mineral-processing industries.

3.5.4 *What kinds of physical sampling can be done?*

A vast literature exists on physical sampling processes and equipment. A useful bibliography has been compiled (Thomas and Schofield, 1995) as part of the UK Department of Trade and Industry Valid Analytical Measurement (VAM) programme. A second edition of a valuable and detailed critical review of sampling systems for use in process analysis has also been published (Carr-Brion and Clarke, 1996) as part of the same programme. A useful booklet on good sampling practice (Crosby and Patel, 1995) and a critical assessment of sampling of bulk materials (Smith and James, 1981) are available from the Royal Society of Chemistry. Keith (1996) has edited an important volume on environmental sampling (now in its 2nd edition) that could be of great benefit to a wide range of readers interested in other sampling problem applications, not necessarily merely in the area of environmental science.

Sampling of bulk materials may be done on a small scale in laboratories, using both manual and automatic methods. Conventional coning and quartering may suffice with relatively homogeneous, dry, particulate materials to obtain reasonably representative specimens, but beware that, unless you have some good quality prior knowledge of the homogeneity and of the behaviour of the material, this may be a dangerous assumption. Sometimes, we take physical samples and, suspecting that there may be some heterogeneity, attempt to reduce its effects by blending several such samples into a composite physical sample. This can give us a useful 'average' view of the material but in so doing it destroys the opportunity to estimate the extent of heterogeneity.

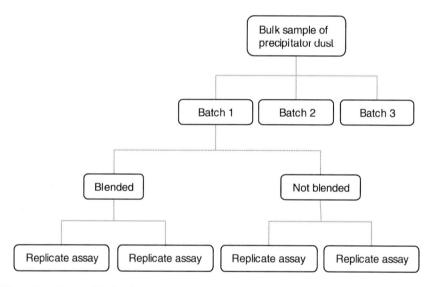

Figure 3.2 A nested design for investigating the effects of sample blending on sample variability (heterogeneity) for a precipitator dust.

This can be illustrated using some data from a nested experiment on the effects of blending versus not blending precipitator dust from an incinerator (the nested design is shown in Figure 3.2). Each of eight bulk samples of dust collected from the precipitator was split into two subsamples, of which one was blended and the other was not. Five replicate analyses for zinc content were performed on each subsample from each of the eight batches. Figure 3.3 shows the boxplots of the data for the eight sets of unblended and eight sets of blended dust assays, and one can clearly see that blending had a dramatic effect on both spread and 'average' zinc content and, as a consequence, masks the heterogeneity of this material.

With heterogeneous materials, such as contaminated soils, attempts at blending may not necessarily fully mask the heterogeneity of a bulk

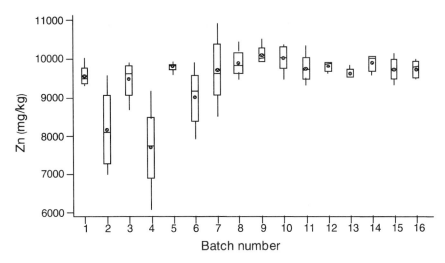

Figure 3.3 Boxplots of zinc content assay data for precipitator dusts investigated using the nested design in Figure 3.2. Batches 1 to 8 were not blended, batches 9 to 16 were blended. (Note that the dot within the box is the mean and the horizontal line within the box is the median.)

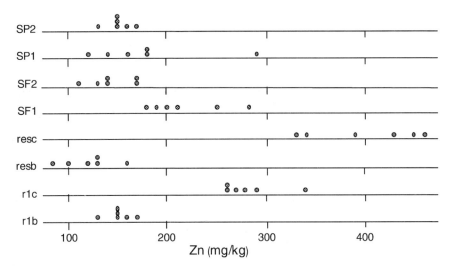

Figure 3.4 Dotplots of the zinc content in a contaminated soil, investigated using a nested design involving both an intralaboratory trial (batches SP1, SP2, SF1 and SF2) and an interlaboratory trial.

sample. Figure 3.4 shows dotplots of data from a two-stage study on waste characterisation; the first involved an intralaboratory trial in which two analysts performed six replicate assays on each of two subsamples from the bulk sample; the second stage involved two laboratories. One can see from the dotplots for zinc content in Figure 3.4 that there is a considerable heterogeneity remaining.

3.5.5 Sampling strategies

Much of the literature on sampling strategy relates to marketing/political/economic surveys and censuses (e.g. Cochran, 1977; Stuart, 1976), although many ideas are readily transferred to the domain of chemical research (Crosby and Patel, 1995; Snedecor and Cochran, 1967). Confusion often arises over what is meant by 'sample size'. To most physical scientists it is the physical size (weight or volume) of specimens but to statisticians it is the number of specimens and/or measurements. Both concepts are important and need careful consideration, especially when dealing with complex, possibly heterogeneous materials or systems.

Here, sample size discussions are about numbers of specimens or measurements. In thinking about sampling strategy we must also be concerned with issues relating to patterns of sampling, for example:

- In space and/or time
- With avoidance, or minimisation, of bias
- How easy or otherwise it may be to obtain representativeness.

Adaptive sampling strategies, in which the procedure for selecting units to be included in the sample may depend on values of the variable of interest observed during the investigation, have been the subject of intensive research (Thompson and Seber, 1996). They are useful in many real-world situations and offer us the possibility of being optimal in many situations, in the sense of giving the most precise estimates for a given sampling effort. Sometimes, we may have to analyse data from observational studies (Section 3.4) in which no design was deliberately applied. Then we may benefit from taking account of adaptive procedures in the analysis. For example, in a study of groundwater pollution, we may have to rely on measurements on specimens from existing wells, but the original selection of these may have been done taking into account water quality measurements in neighbouring areas (an informal adaptive approach).

In statistical process control of continuous production processes, for which batch sampling is inappropriate or not possible to implement, special sets of sampling inspection plans known as continuous or adaptive sampling plans were first introduced by Dodge in 1943. Continuous sampling plan techniques have been further developed during the 1960s and 1970s, especially involving incorporation of ideas from control theory (see Wetherill and Brown, 1991, chapter 14).

3.5.6 Statistical/chemometric aspects

When designing schemes for sampling that involves ultimately measuring continuous variables on the specimens taken, Normal distribution theory is often assumed, although crude correction for other distributions has been offered in the past (Deming, 1960) when doing sample size estimation. However, other approaches are possible, as discussed briefly below; they are not considered further than that brief discussion because of lack of space. The reader is encouraged to pursue the study of these, as they are important in providing sound alternatives, either when data is sampled from non-Normal distributions or when robustness and resistance are required when dealing with data with outliers.

For example, if the data population distribution is unknown, it is possible to use the appropriate nonparametric tests, which will be used in the data analysis, to model the sample size estimation (see Chapter 5 in this book and also Hettmansperger and McKean, 1998; Staudte and Sheather, 1990). In particular, we can do this using an approach based on confidence intervals for one- and two-sample tests, as well as for one-way and two-way analysis of variance (ANOVA) and factorial designs.

Another approach recognises that, by the very act of guessing when we design sampling strategies, we are actually using Bayesian ideas. Bayesian statistics explicitly uses prior estimates of uncertainty to refine the data analysis (very useful introductions to this area of statistics are to be found in Albert, 1996, whose book contains Minitab macros to perform all the computations in Berry, 1996). It is often very useful to incorporate prior knowledge explicitly by using the Bayesian approach directly in the sampling design process, especially in refining the calculation of statistical power (Carlin and Louis, 1996; Verdinelli, 1992).

If we already have some data about the population being sampled, then we might consider the use of bootstrapping to estimate the sample size (Efron and Tibshirani, 1993; Simon, 1997). If we intend to use permutation tests, then we can simulate the effectiveness and power of appropriate sample size choices (Good, 1994, 1998; Sprent, 1998; both discuss useful software for these tests).

Liggett and Inn (chapter 10 in Keith, 1996) suggest the use of various pilot experiments to evaluate alternative combinations of measurement and sampling protocols in order to assess their relative performance. This approach has much merit and has been applied successfully in industrial research and development (Yano, 1991). It might be considered a form of adaptive sampling design.

A specifically chemical problem in sampling uncertainty is that of sample recovery, especially in trace analysis. Here, the problem is both of systematic and random contributions to uncertainty and of their relative importance. This important subject was discussed in detail at the Seventh International Symposium for Harmonisation of Quality Assurance in Analytical Chemistry and the reader is referred to the proceedings (edited by Parkany, 1996) in which the use and misuse of recovery factors is examined in detail.

3.6 Statistical sampling strategies

3.6.1 Random sampling

The term 'random' in this context does not imply a haphazard 'lucky dip' into the bulk material but rather that all parts of the bulk have an equal chance of being selected. Simple random sampling is *without replacement*. In nondestructive measurement, sampling with replacement would be possible and attention then needs to be paid to preserving a genuinely random sampling process.

Random sampling of bulk liquids, powders, etc., is in reality quite difficult to achieve; much effort in designing physical sampling equipment is directed to assisting this aim. Random spatial sampling in two or three dimensions, for example of soil in a field, water or sediment in a lake or river, is even more difficult and a better approach might be the use of stratified sampling (see Section 3.6.3).

Assuming reasonable homogeneity in the bulk, the main concern would be to reduce measurement uncertainty. If the measurements have a standard deviation s and an estimated mean μ_e, the 95% confidence interval of the estimated mean is

$$\mu_e \pm t_{1-\alpha, df} \frac{s}{\sqrt{n}}$$

where α is 0.95 and relates to the two-sided probability interval for Student's t, df is the number of degrees of freedom ($df = n - 1$) with n being the number of observations in the sample (see also Chapter 4). We can use this to estimate what sample size we might need to ensure that the measured mean is within a specified confidence interval of length

$$L = 2t \frac{s}{\sqrt{n}}$$

Thus,

$$n = \frac{4t^2 s^2}{L^2}$$

is the minimum number of required observations. If we had a prior estimate of s from a series of m measurements on a similar material, then a more appropriate confidence interval for the new set of measurements would be calculated from the pooled standard deviation. The latter is obtained from that for the n new measurements and that from the m previous measurements used in estimating the required sample size.

A somewhat different problem is presented if our objective is to estimate the confidence interval for the variability, rather than that for the mean, of our measurements (see Chapter 4 of this book and also Mandel, 1964). If we consider a Normal distribution of data, then the distribution of our sample estimates, s_{exp}, of standard deviation (with degrees of freedom, $n-1$) is roughly Normal with a mean approximately equal to the standard deviation of the population, s_{pop}, and the variance of s_{exp} is roughly given by

$$V(s_{exp}) = \frac{s_{pop}^2}{2(n-1)}$$

So, the ratio s_{exp}/s_{pop} has a mean of unity and a variance given by

$$V(s_{exp}/s_{pop}) = \frac{V(s_{exp})}{s_{pop}^2} = \frac{1}{2(n-1)}$$

From Normal distribution theory, the 95% confidence interval length is given by

$$L = \frac{2 \times 1.96}{\sqrt{2(n-1)}}$$

So, our sample size for estimating the variability with 95% confidence is given by

$$n - 1 = \frac{1}{2}\left[\frac{2 \times 1.96}{L}\right]^2$$

If, for example, we want the ratio s_{exp}/s_{pop} to be within the range 0.8 to 1.2 (i.e. s_{exp} to be within $\pm 20\%$ of s_{pop}) then we have

$$n - 1 = \frac{1}{2}\left[\frac{2 \times 1.96}{0.4}\right]^2 = 48, \quad \text{so, } n = 49$$

Thus, to get our estimate to within even 20% of the population variability with 95% confidence requires a sample size considerably larger than that for estimation of the mean and, of course, we are assuming Normality. If we were to assume a log-normal distribution, then similar reasoning would enable us to devise estimates of the sample size required for that situation.

3.6.2 Systematic sampling

Suppose we wish to draw a 5% sample from a collection of items such as a production batch. We could select a random number between 1 and 20, such as 4, and then select every 20th item (4, 24, 44, etc.). This is a systematic sample. Such sampling has two advantages over simple random sampling:

- It is easier to draw (only one random number is used).
- It spreads the sampling evenly over the whole population.

It has become popular in routine sampling because of these advantages. However, there are two potentially serious disadvantages:

- If the sampling interval (e.g. a regular monthly sampling of a river for water quality) coincides with that of a periodic variation (or a multiple of that period), then the sample will be badly biased without our realising it.
- There is no reliable method of estimating the standard error of the mean (Snedecor and Cochran, 1967).

3.6.3 Stratified sampling

Three steps are involved in this type of sampling, which is generally applied in sample surveys or in sampling for attributes, in both of which we are dealing with finite populations (see Snedecor and Cochran, 1967; Cochran, 1977; Davies and Goldsmith, 1977; Desu and Raghavarao, 1990):

- We divide the population into parts, called strata.
- A sample is drawn independently and randomly from each part.
- We can estimate the population mean from the means within each stratum.

Such an approach could be taken with pelleted material or tablets, for example.

3.6.4 Sequential sampling

This approach, sometimes used in inspection plans for quality control, is based on the idea that items are drawn from a batch one by one, each item being tested before the next one is drawn. If the quality of the batch is very good or very bad, few items will need to be drawn to come to this conclusion, whereas in intermediate cases more items will need inspection to verify quality level with reasonable risk (Wetherill and Brown, 1991). Sequential sampling sometimes improves the cost and efficiency of inspection, but often the improvement is only marginal.

3.7 Sample size estimation using the concept of the power of a statistical test

3.7.1 Power and risks

When carrying out statistical hypothesis testing, errors occur because in such testing we make sharp yes–no decisions on whether to accept or reject the Null Hypothesis. Type I error occurs when the Null Hypothesis is rejected when it should not have been (see Chapter 1 for further discussion of the points in this section and Section 3.4.2 above on false positive/negative decisions and Section 3.8 on the receiver operating characteristic plot). So if we are comparing means, our Null Hypothesis may be that the population means are equal. On a small number of occasions when we make such comparisons the sample means may differ significantly, even though the population means are the same. The probability α of rejecting the Null Hypothesis is determined by our choice of significance level (e.g. 0.05 or 5%). In SPC, this corresponds to the *probability of rejection* of a production lot of acceptable quality, also known as the *producer's risk point* (Wetherill and Brown, 1990; Montgomery, 1991).

The possibility also exists that we might reject the Alternative Hypothesis when it is true. This is a Type II error, whose probability of occurrence is β, also known as the *consumer's risk point* (Wetherill and Brown, 1990; Montgomery, 1991). Sometimes, we wish to evaluate the *power* of a test, which is the probability of correctly rejecting the Null Hypothesis: $1 - \beta$. Ideally, in designing experimental or observational studies, we should think about the use of statistical tests by specifying an appropriate value of α and designing the application of the test so as to obtain a small value of β. Several factors affect β, including sample size, so one way of obtaining a low β value is by arranging to have as large a sample size as practicable, but this has to be balanced against the costs of large sample sizes.

The concept may be illustrated using the Z-test for sample means, with variance, σ^2, known. The Null Hypothesis is H_0: $\mu_{\exp} = \mu_0$ and the Alternative Hypothesis is H_1: $\mu_{\exp} \neq \mu_0$. The test statistic is calculated from

$$Z_0 = (\mu_{\exp} - \mu_0)\frac{\sqrt{n}}{\sigma}$$

H_0 is rejected if $|Z_0| > Z_{\alpha/2}$ where the latter is the upper $\alpha/2$ percentage point of the standard Normal distribution. If μ_{exp} is really $\mu + \delta$, where $\delta > 0$, then we want to check the probability of a Type II error, β, in relation to detecting the difference, δ. In terms of the standard Normal cumulative distribution, Φ, we have

$$\beta = \Phi\left[Z_{\alpha/2} - \frac{\delta\sqrt{n}}{\sigma}\right] - \Phi\left[-Z_{\alpha/2} - \frac{\delta\sqrt{n}}{\sigma}\right]$$

where σ is the standard deviation. If we have a tablet with an active drug content of 16 mg and from previous studies, we know that the standard deviation of the content is 0.1 mg and we decide to specify a Type I error probability $\alpha = 0.05$, then we reject H_0 if $|Z| > Z_{0.025} = 1.96$. Suppose we wish to find the Type II error probability for a true mean drug content of 16.1 mg, then $\delta = 0.1$ mg and if we have tested 9 tablets

$$\beta = \Phi\left[1.96 - \frac{0.1 \times \sqrt{9}}{0.1}\right] - \Phi\left[-1.96 - \frac{0.1 \times \sqrt{9}}{0.1}\right]$$
$$= \Phi(-1.04) - \Phi(-4.96)$$
$$= 0.1492$$

So the probability of incorrectly failing to reject the Null Hypothesis is 14.92% and the power of the test is $1 - \beta = 0.8508$ or 85.08%.

We can plot a set of curves of the relationship between β and the ratio, d, of the desired difference, δ, to be detected and the standard deviation, for a series of sample sizes n (see Figure 3.5) for the Z-test. These are termed operating characteristic (OC) curves or power function curves for the test. From Figure 3.5, we may see that, as sample size increases, the curve shifts to the left and becomes steeper.

During the sampling design stage, it is often helpful to investigate what the limiting detectable difference, δ_{lim}, is under various conditions. This may be evaluated using the power function curves because both δ and the standard deviation are involved in determining the abscissa of the plot as the ratio, d. As the sample size affects both σ and β, such evaluation is necessarily iterative but is very worthwhile doing.

3.7.2 Limit of detection and limit of quantitation

Another aspect of detection limits of concern to chemists is found with trace analysis or other forms of assay. The CITAC Guide (1995) on Quality in Analytical Chemistry refers to both the *limit of detection* (LOD) and the *limit of quantitation* (LOQ). Both may be linked to the

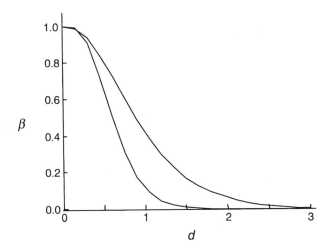

Figure 3.5 Operating characteristic or power function curves for two sample sizes of the Z-test. The steeper, left-hand curve is for a much larger sample size than the right-hand curve.

standard deviations of blank measurements, which, of course, vary with the type of specimen (both matrix and interference effects are possible). The blank standard deviation varies, as with other kinds of assay, with sample size (i.e. numbers of measurements).

A useful approach involves estimating the confidence band in the region of the intercept of the calibration curve obtained by regression analysis. However, we must not neglect the need to get a reliable estimate of the slope, otherwise we introduce unnecessary bias into the estimate of the intercept. Thus, we need to plan the regression-based experiments carefully so that we obtain a sufficiently large sample of measurements throughout the calibration to get a reliable understanding of its behaviour in the extrapolation zone. In support of such an effort, we may wish to pool data from previous series of blank assays, as well as previous calibrations, if all are conducted using the same standard operating procedures.

The reader should note the comments on regression in Chapter 5, on Robust, Resistant and Nonparametric Methods, that no regression method is perfect and we need to be very cautious in extrapolation to the intercept, even with the use of blank assay data. There we enter a zone in which data points can exert high leverage and, thus, considerable influence on the estimation of the intercept. Thus, a robust regression technique such as that described by Hettmansperger and McKean (1998) and implemented in Minitab as RREG would prove useful in estimating the intercept and its standard deviation and standard error and hence the LOD and LOQ. Two different methods for estimating the intercept are available in RREG: the default method assumes that the uncentred residuals from the regression are symmetrically distributed and provides an option to estimate the 90% confidence interval; the second method estimates the intercept as the median of the uncentred residuals and the spread is obtained using a kernel density estimation (Aubuchon, 1990).

We must be especially careful to evaluate LOD and LOQ in the same kind of matrix that will be used ultimately for the assay. If possible, the evaluation should be with conditions such that matrix and interference effects vary over a wide range so that their limiting effects may be tested. It may therefore involve a multivariate design. Such a design may be experimental, but this may not always be possible and resort to observational study design may be needed (see, for example, Thompson et al., 1994).

3.7.3 Other points

Gardiner (1997) and Gardiner and Gettinby (1998) discuss the use of power analysis in estimating sample sizes and developing sampling designs for simple experiments, as well as for completely randomised, randomised block, nested and factorial designs, with useful worked examples.

When using control charts, we may wish to know how many tests have to done (groups on a Shewhart chart or individual tests on an individuals chart) before we can detect a shift in the mean. The actual length of a run of such tests will vary randomly but we can calculate an *average run length*, $ARL = 1/(1 - \beta)$. So, as with the OC curves, we can plot a series of ARL curves to determine appropriate sample sizes for both variables and attributes control charts (Wetherill and Brown, 1990; Montgomery, 1991). We can also derive OC and ARL curves for Cusum charts (Wetherill and Brown, 1990; Montgomery, 1991) and for process capability indices (Kotz and Lovelace, 1998) and OC curves for sequential testing (Wetherill and Glazebrook, 1986).

If we allow the detectable difference, δ, defined above to tend towards zero as the sample size, n, tends to infinity, we can find the limiting value of the power function, the *asymptotic power function*, for any given statistical test. Staudte and Sheather (1990) discuss the estimation of these asymptotic power functions for various one- and two-sample tests, including t-tests, trimmed t-tests, the sign and signed rank tests and the Wilcoxon–Mann–Whitney test and their practical use in sample size calculations.

3.7.4 Software implementations

Statistical software, such as Minitab, offers the capability of performing calculations for sample size estimations for one-sample z- and t-tests and tests of proportions, two-sample *t*-tests and tests of proportions and for various ANOVA designs (one-way, two-level full and fractional factorial and Plackett–Burman). Examples of specific sample size estimation software include nQuery Advisor (Statistical Solutions, Saugus, MA 01906, USA) and SamplePower from SPSS. A useful elementary discussion of the theory underlying sample size estimation may be found in Keppel *et al.* (1992) and is helpful in guiding calculations for those without access to statistical software offering such capabilities.

For those interested in environmental sampling, Keith developed two suites of programmes: DQO-PRO (available free) and Practical QC. These may be used to determine sample sizes for various types of systematic sampling of sites using a range of grids when seeking to identify possible 'hot-spots' and for assessing 15 different types of QC samples (see Keith *et al.*, chapter 1 in Keith, 1996). The software is based on a structured approach to data quality objectives developed by the US EPA Quality Assurance Management Staff.

3.8 Sampling in the context of deterministic versus probabilistic assessment of compliance with a standard or threshold—use of the receiver operating characteristic (ROC) curve

Reference was made earlier to the problems of false positive or negative interpretations of test results. When the data distributions that we are trying to discriminate between are overlapping, as shown in Figure 3.6, we have the difficult problem of deciding on the threshold (e.g. for compliance with production standard or to assess whether land is contaminated). If we have a deterministic standard, we have problems dealing with false positives and negatives, whereas with a probabilistic approach the composite uncertainty in sampling and measurement can be used to put confidence intervals around our estimate (Ramsey *et al.*, 1995; Ramsey and Argyraki, 1997; Thompson and Ramsey, 1995; Thompson and Fearn, 1996). The reliability of a particular sampling and measurement strategy and the associated decision criteria can be assessed using the receiver operating characteristic (ROC) plot technique (de Ruig and van der Voet, 1994).

If we have sufficient information on the two overlapping distributions (or we can model them on the basis of estimates from designed experiments), then we can test the effects of varying the position of the decision threshold for the test (see Figure 3.6). The true and false positive rates can be estimated by measurement of the respective areas marked on the distribution plots in Figure 3.6. These can then be used to create a ROC

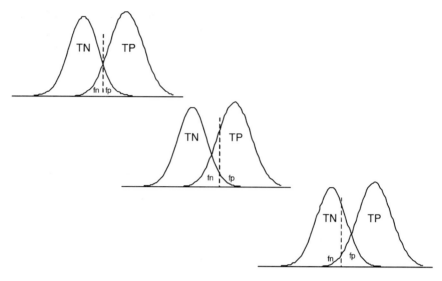

Figure 3.6 Overlapping distributions with different positions for the decision threshold. TP = true positive, fp = false positive, TN = true negative, fn = false negative.

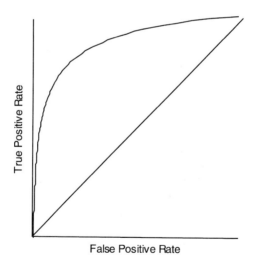

Figure 3.7 Receiver operator characteristic (ROC) plot derived from Figure 3.6.

plot as in Figure 3.7. When the ROC plot lies close to the line of identity (slope of 1), the test has poor discriminating capability (low diagnostic sensitivity and specificity). If the ROC plot curve is close to the upper left corner of the diagram, the test has high discriminating capability. Improving the design of the sampling strategy will have an impact and issues of sample size are clearly very important and highly influential in this respect.

3.9 Problems associated with behaviour of granular and other materials—their effects on designing sampling schemes and on estimating sampling reliability

Kestenbaum (1997) discussed research on the sometimes exceedingly bizarre and common behaviour of granular materials to unmix, clump, etc., when they should mix, flow, and so on (see also Flatman and Yfantis, chapter 40 in Keith, 1996). This causes significant problems industrially, not only in production but also in sampling for quality control. Experience with slurries, pastes, and so on suggests that similar problems may be found with these materials (Carr-Brion and Clark, 1996).

3.10 Sampling for process control and quality management

Here we have a two-stage process in which data must first be acquired to determine the process mean and the composite uncertainty (resulting from intrinsic process variability, together with sampling and measurement uncertainties). This enables the various limits to be estimated for control chart design and process capability indices (Kotz and Lovelace, 1998; Oakland, 1996; Wetherill and Brown, 1991). For grouped data sampled from a Normal distribution, measurements for at least 20 groups are required to get estimates of process mean and standard deviation. For individuals charts (again assuming Normality), it is recommended that at least 50 individual measurements are acquired. Sample sizes would need to be much greater when dealing with non-Normal data. The data need to be acquired in a structured way (such as a nested ANOVA type of design) so that we can obtain estimates of sampling and measurement uncertainties, as well as of the intrinsic process variability (including within- and between-group variability).

3.11 Economic aspects of sampling designs

Sampling designs in the real world have to strike a balance between cost of sampling and measurement and the desire for high reliability. The costing can be built into the assessment of the design and several authors have discussed this in detail (Cochran, 1977, 1983; Thompson and Fearn, 1996; Thompson and Seber, 1996; Snedecor and Cochran, 1967).

Bibliography and References

Aubuchon, J.C. (1990) *Experimental Rank Regression (RREG) Command*, Draft Documentation, Minitab Inc., State College, PA.

Albert, J.H. (1996) *Bayesian Computation Using Minitab*, Duxbury, Inc., Belmont, CA.

Berry, D.A. (1996) *Statistics: A Bayesian Perspective*, Duxbury, Inc., Belmont, CA.

Carlin, B.P. and Louis, T.A. (1996) *Bayes and Empirical Bayes Methods for Data Analysis*, Chapman and Hall, London.

Carr-Brion, K.G. and Clarke, J.R.P. (1996) *Sampling Systems for Process Analysers*, 2nd edn, Butterworth-Heinemann, Oxford.

CITAC (Co-operation on International Traceability in Analytical Chemistry) (1995) *Guide 1- International Guide to Quality in Analytical Chemistry—An Aid to Accreditation*, English edn 1.0, H.M.S.O.

de Ruig, W.G. and van der Voet, H. (1994) Is there a tower in Ransdorp? Harmonization and optimization of the quality of analytical methods and inspection procedures, in *Reviews on Analytical Chemistry—Euroanalysis VIII*, Royal Society of Chemistry, Cambridge.

Cochran, W.G. (1977) *Sampling Techniques*, 3rd edn, Wiley, New York.

Cochran, W.G. (1983) *Planning and Analysis of Observational Studies* (eds L.E. Moses and F. Mosteller), Wiley, New York.

Corn, M. (1981) Strategies of air sampling, in *Recent Advances in Occupational Health* Number One, (ed. J.C. McDonald), Churchill Livingstone, Edinburgh, Chapter 18.

Crosby, N.T. and Patel, I. (1995) *General Principles of Good Sampling Practice*, Royal Society of Chemistry, Cambridge.

Davies, O.L. and Goldsmith, P.L. (1977) *Statistical Methods in Research and Production*, Longman, London.

Deming, W. and Edwards (1960) *Sample Design in Business Research*, Wiley, New York.

Desu, M.M. and Raghavarao, D. (1990) *Sample Size Methodology*, Academic Press, San Diego, CA.

Efron, B. and Tibshirani, R.J. (1993) *An Introduction to the Bootstrap*, Chapman and Hall, London.

Einax, J.W., Zwanziger, H.W. and Geiss, S. (1997) *Chemometrics in Environmental Analysis*, VCH, Weinheim.

Fraser, C.G. (1986) *Interpretation of Clinical Chemistry Laboratory Data*, Blackwell Scientific, Oxford.

Gardiner, W.P. (1997) *Statistical Analysis Methods for Chemists—A Software-based Approach*, Royal Society of Chemistry, Cambridge.

Gardiner, W.P. and Gettinby, G. (1998) *Experimental Design Technique in Statistical Practice— A Practical Software-based Approach*, Horwood Publishing, Chichester.

Good, P. (1994) *Permutation Tests—A Practical Guide to Resampling Methods for Testing Hypotheses*, Springer-Verlag, New York.

Good, P. (1998) *Resampling Methods—A Practical Guide to Data Analysis*, Birkhauser, Basel.

Gy, P.M. (1995) Sampling: are we interested in it at all? in *Quality Assurance and TQM for Analytical Laboratories* (ed. M. Parkany), Royal Society of Chemistry, Cambridge, pp 142-147.

Harvey, R.P. (1981) Statistical aspects and air sampling strategies, in *Detection and Measurement of Hazardous Gases* (eds. C.F. Cullis and J.G. Firth), Heinemann Educational, London, Chapter 6.

Henderson, A.M. (1995) Sampling: is it a weak link in total quality management? in *Quality Assurance and TQM for Analytical Laboratories* (ed. M. Parkany), Royal Society of Chemistry, Cambridge, pp 152-158.

Hettmansperger, T.P. and McKean, J.W. (1998) *Robust Nonparametric Statistical Methods*, Arnold, London.

Horwitz, W. and Albert, R. (1997) The concept of uncertainty as applied to chemical measurements. *Analyst*, **122** 615-617.

Johnson, N.L., Kotz, S. and Wu, X. (1991) *Inspection Errors for Attributes in Quality Control*, Chapman and Hall, London.

Keith, L.H. (ed.) (1996) *Principles of Environmental Sampling*, 2nd edn, American Chemical Society, Washington DC.

Keppel, G., Sautley, W.H. Jr and Tokunaga, H. (1992) *Introduction to Design and Analysis—A Student's Handbook*, 2nd edn, W.H. Freeman, New York.

Kestenbaum, D. (1997) Sand castles and cocktail nuts. *New Scientist*, 24 May, pp 25-28.

Kotz, S. and Lovelace, C.R. (1998) *Process Capability Indices in Theory and Practice*, Arnold, London.

Leidel, N.A., Busch, K.A. and Lynch, J.R. (1977) *Occupational Exposure Sampling Strategy Manual*, US DHEW (NIOSH) Publication No. 77-173, US Department of Health, Education and Welfare, Public Health Service, Center for Disease Control, National Institute for Occupational Safety and Health, Cincinnati, OH.

Mandel, J. (1964) *The Statistical Analysis of Experimental Data*, Dover, New York.

Montgomery, D.C. (1991) *Introduction to Statistical Quality Control*, 2nd edn, Wiley, New York.

Moreton, J. and Falla, N.A.R. (1980) *Analysis of Airborne Pollutants in Working Atmospheres: The Welding and Surface Coating Industries*, Royal Society of Chemistry, London.

Oakland, J.S. (1996) *Statistical Process Control—A Really Practical Guide*, 3rd edn, Butterworth-Heinemann, Oxford.

Olsen, E. and Nielsen, F. (1995) Sampling in the context of TQM—with special emphasis on log-normal distributions, in *Quality Assurance and TQM for Analytical Laboratories* (ed. M. Parkany), Royal Society of Chemistry, Cambridge, pp 131-141.

Parkany, M. (ed.) (1996) *The Use of Recovery Factors in Trace Analysis*, Royal Society of Chemistry, Cambridge.

Ramsey, M. and Argyraki, A. (1997) Estimation of measurement uncertainty from field sampling: implications for the classification of contaminated land. *The Science of the Total Environment*, **198** 243-257.

Ramsey, M.H., Argyraki, A. and Thompson, M. (1995) On the collaborative trial in sampling. *Analyst*, **120** 2309-2312.

Simon, J.L. (1997) *Resampling: The New Statistics*, 2nd edn, Resampling Stats., Inc., Arlington, VA.

Smith, R. and James, G.V. (1981) *The Sampling of Bulk Materials*, Royal Society of Chemistry, London.

Snedecor, G.W. and Cochran, W.G. (1967) *Statistical Methods*, 6th edn, Iowa State University Press, Ames, IA.

Sprent, P. (1998) *Data Driven Statistical Methods*, Chapman and Hall, London.

Staude, R.G. and Sheather, S.J. (1990) *Robust Estimation*, Wiley, New York.

Stuart, A. (1976) *Basic Ideas of Scientific Sampling*, Charles Griffin, London.

Thomas, C.L.P. and Schofield, H. (1995) *Sampling Source Book. An Indexed Bibliography of the Literature of Sampling*, Butterworth-Heinemann, Oxford.

Thompson, M. and Fearn, T. (1996) What exactly is fitness for purpose in analytical measurement? *Analyst*, **121** 275-278.

Thompson, M. and Ramsey, M.H. (1995) Quality concepts and practices applied to sampling—an exploratory study. *Analyst*, **120** 261-270.

Thompson, S.K. and Seber, G.A.F. (1996) *Adaptive Sampling*, Wiley, New York.

Thompson, J.M., Smith, S.C.H., Cramb, R. and Hutton, P. (1994) Clinical evaluation of sodium ISFETs for whole blood [Na^+] assay. *Annals of Clinical Biochemistry*, **31** 12-17.

Verdinelli, I. (1992) Advances in Bayesian experimental design (with discussion), in *Bayesian Statistics 4* (eds. J.M. Bernado, J.O. Berger, A.P. Dawid and A.F.M. Smith), Oxford University Press, Oxford, pp 467-481.

Wetherill, G.B. and Brown, D.W. (1991) *Statistical Process Control—Theory and Practice*, Chapman and Hall, London.

Wetherill, G.B. and Glazebrook, K.D. (1986) *Sequential Methods in Statistics*, Chapman and Hall, London.

Yano, H. (1991) *Metrological Control: Industrial Measurement Management*, Asian Productivity Organization, Tokyo.

4 Interpreting results
M. Gerson

4.1 Introduction

Having gathered data and assessed them for reasonableness and quality, we should process them in some way to extract the information it contains. Ultimately we shall make decisions based on the data. It is quite clear that, because of the variability that affects all measurements, the decision we make may not be a clear-cut one—it will not be a simple yes or no, accept or reject. There is always some uncertainty associated with it.

We could simply ignore the uncertainty and force all decisions to be a simple binary choice. There is no doubt that in many situations we shall make the correct decisions using this approach, but there is also no doubt that sometimes the wrong decision will be made. The problem is that we shall not know that it has!

A better approach is to acknowledge the existence of uncertainty and then to estimate the risks involved in making correct or wrong decisions. If the risks are unacceptably large, we know we have to get more data to refine our information and reduce the risks.

This chapter describes the quantitative techniques for calculating the risks we run when interpreting our data and making decisions. It is based around some simple examples so that the principles can be established clearly. However, what is described here is fundamental to all statistically based decision making, from these simple examples to the most complex multivariate examples found in other parts of the book.

4.2 Where to go

Information needed	Go to Section(s)
What is a confidence interval?	4.5.2 et seq.
What is the value of a confidence interval?	4.4.1, 4.5.3, Example 3
Are there any really important assumptions being made if I interpret data this way?	4.8
How many replicate runs do I need to get a reliable conclusion?	4.6

Information needed	Go to Section(s)
Replication is quite expensive for me. Are there any ways of making the experiment smaller or more efficient?	4.7, 4.9.2
Can I relate the design of my data collection experiment to my commercial requirements and practical constraints?	4.3, 4.7
I need to know that two formulations/ingredients/processes are equivalent. How should I do this?	4.9.1
I think I need to do a paired comparison. How do I do the analysis?	4.9.2
How do I work out a confidence interval for an estimate of variability?	4.5.4
Can I compare the variability of two different processes/people/test methods?	4.5.5
How can I separate out variability from different causes?	4.9.3

4.3 The objectives of experimentation and data collection

Experimentation or data collection is always carried out for a purpose. Clear objectives and good planning will make it much more likely that useful information and good decisions will result from the work done. That is not to say that we should, or can, know in advance what information we shall get from an experiment; indeed, if we did there might be little point in doing it. But some of the most valuable results arise in studies that are so well planned that exceptional events are quite clearly identifiable. Example 1 below is quite typical of situations where good planning and record keeping give not only results relating to the initial objectives but also very useful information about the unexpected.

> *Example 1.* Owing to the time required by a laboratory technician to weigh out ingredients, mix them and then clean all equipment thoroughly before preparing the next sample, it was possible to prepare only four monomer mixtures in a morning or afternoon. These samples were then included with the next production batch in a polymerisation cycle and it was believed the nature of the cycle or the oven that was used might have some effect on measured outcome. Although several factors were being investigated, the experiment was planned so that each morning's or afternoon's work contained a balanced set of combinations enabling important factors to be investigated within a polymerisation cycle rather between cycles. As

a result of this planning and of good record keeping, it was discovered that it was not, as had been thought, the nature of the cycle that mattered but the length of time samples stood before being polymerised. This unexpected discovery was extremely valuable and was made in addition to conclusions regarding the factors in the experiment.

The objective of the experimentation we are planning might be, for example, one of the following:

- To quantify the performance of a system.
- To discover which factors have an effect on the system.
- To estimate the size of any such effects.
- To compare the performance of two or more systems or sets of conditions.
- To 'pick the winner'.
- To show equivalence between two or more systems or sets of conditions.
- To model the response of a system to varying one or more of the inputs.
- To optimise the performance of a system (in terms of reducing process time or cost, minimising variability, maximising yield, etc.)

The way in which the data are gathered will obviously be related to the nature of the objective. But it should depend also on considerations concerning the importance and commercial value of the information and any decisions made as a result. Practical factors such as costs, available time and safety will play a part in the planning, as will the breadth of validity of the conclusions that can be drawn and the importance of any assumptions that are made. The balance of practicalities with the need to do sufficient experimentation to get useful information is illustrated in Example 2.

Example 2. A company needed to measure the initial activity of batches of a microorganism and to ensure that the batches had a satisfactory shelf-life. The QC release procedure involved taking an initial sample out of one bottle of the batch, diluting it down in a series of five 10-fold dilutions and then putting a standard quantity of the resulting material onto each of three agar plates, all of which were incubated together. The number of colonies per plate was used as a measure of the activity of the batch. Since each bottle was expensive, shelf-life tests involved taking another sample from the same bottle at the end of the shelf-life period and carrying out the same test procedure. After about 6 months of production concern was expressed that batches appeared to be very different from one another both in their initial activity and in the extent of their deterioration at the end of shelf-life. In addition, there were some customer complaints about the performance of the product, although these did not always relate to batches with poor activity in the QC release test.

Before embarking on a major investigation into the causes of batch-to-batch differences it was agreed that sampling and measurement variability should be investigated. To gain valuable information, the experimenters used five bottles from a single batch. Four samples were taken from each bottle, from the top, about one-third of the way down, two-thirds down and very near the bottom. Each sample was divided in two and a series of dilutions made for each half. Then each diluent was tested on two agar plates, with all 80 plates being incubated for the same time and at the same temperature.

The experiment was planned to use expensive material and people's time efficiently to gain as much information as possible about four sources of variability (bottles, samples, dilution series, plates). It resulted in the discovery that nonhomogeneity of batches was a very serious problem since the five bottles had radically different activities and even within a bottle the samples differed a lot. Further investigation and work based on this discovery provided a solution to all of the problems.

Whether the data are collected from trials on a full-scale production process, from experimental runs in the laboratory or from relatively poorly controlled use of a product by customers, the raw data will need analysing and interpreting in order that conclusions can be drawn, decisions made and actions taken. Some of this analysis will be graphical in form (Chapter 2) and this is often very powerful. But numerical analysis is also a powerful and often an essential tool. It is the numerical analysis associated with decision making that forms the basis of this chapter.

4.4 Setting up a known system on which to experiment

Usually we experiment on systems because we want to measure or understand them. We do not yet know their full underlying structure, but it is what we want to uncover. Clearly, using data from experiments where we do not know the structure makes it difficult to show how the techniques we want to use will behave. So we shall play 'god' by setting up a simple system we *do* understand, and then carry out the same experiment on it many times over. In this way we try to make the nature of experimentation transparent.

The system we are going to use is a computer simulation that uses the formula

$$y_{ij} = \mu + kT_i + \varepsilon_{ij}$$

to determine the output variable, y_{ij}, for a given setting of the control variable T_i. We might think of T_i as being the temperature at which we carry out some chemical process with y_{ij} as the measured yield. Of course, for the analogy to be a reasonable one, we might have to restrict our temperature range quite tightly so that nonlinearity does not become an issue.

4.4.1 Important assumptions about the ε_{ij} term

The ε_{ij} term is unpredictable and likely to be different each time, so that in our computer simulation, as in real life, we do not get exactly the same 'yield', y_{ij} each time we use a particular 'temperature', T_i. In real life it will result from a combination of all the factors affecting the actual yield (due to small differences in catalyst purity, reaction time, initial quantities of reagents, etc.) and all the factors affecting its measurement (sampling variability, sample preparation, measurement error, etc.). Although labelled as being related to the jth occasion on which we use temperature i, we have set up ε_{ij} in the computer simulation so that its actual size is random (see Section 4.10) and unrelated to either i or j. In real experiments we need to take great care to determine that the same is true, otherwise all the following analysis is invalid. Indeed, the neglect of this assumption probably leads to more incorrect conclusions than any other in the design and analysis of experiments and other data collection (see Chapters 2, 6, 7 and 8).

Whenever there are several small contributors to variability, with none of them dominating, a histogram of this variable contribution will tend to have the shape of a Normal distribution. In the computer simulation we shall also create ε_{ij} by taking numbers at random from a Normal distribution. This distribution will have a mean of zero and a standard deviation σ. In almost all real experimental situations we would not know μ or k or σ. Nor would we be absolutely certain that the relationship is a straight line. In this artificial example we can set these as we want; I have chosen

$$\mu = 60, \quad k = 1.2, \quad \sigma = 0.5$$

but you are at liberty to use other values and see what happens as a result.

In the rest of this chapter the computer simulation has been run for the stated number of times and the results have been recorded. There has been *no* 'doctoring' of the results or special choice of runs.

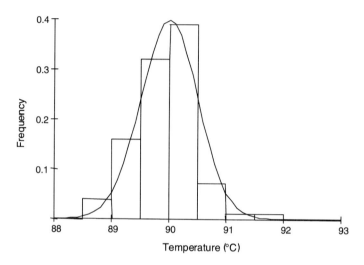

Figure 4.1 Histogram of 100 values of y when the temperaure is 25°C. The superimposed curve is the Normal distribution curve expected for the given values of μ, α, σ and T.

Here are the results of running the simple system 100 times over with a single value of T_i. I chose to use $T_i = 25$. I have calculated the average yield, $\bar{y}_{i\bullet}$, and the standard deviation, $s_{i\bullet}$, of these 100 individual values and drawn up a histogram of the individual values, Figure 4.1. The average and standard deviation are

$$\bar{y}_{i\bullet} = 89.925, \quad s_{i\bullet} = 0.491$$

The calculated value of $\bar{y}_{i\bullet}$ is not *exactly* the expected value of $\mu + kT_i = 60 + 1.2 \times 25 = 90$, nor is the value of $s_{i\bullet}$ *exactly* $\sigma = 0.5$, nor is the histogram *exactly* a Normal distribution. The Normal distribution frequency curve expected for the chosen values of μ, k and σ has been drawn over the histogram in Figure 4.1. You can see that the data distribution is reasonably close to Normal.

The use of the first person in the above section is deliberate. The results come from a particular set of runs that *I* carried out. If *you* try to do the same, but with a different starting point, you will get different *y*-values even if you use exactly the same algorithms for calculating ε_{ij}. However, your general conclusions will probably be the same.

4.5 Experimenting on the known system

4.5.1 Estimating the difference between two measurements

A great deal of experimentation involves comparing two or more controlled sets of conditions. For example, we might want to compare the activity of a catalyst from different suppliers, or to determine whether one laboratory technician tends to make more reproducible measurements than another. The same is also true if we are trying to model the properties of a polymer as a response to different curing times and temperatures, perhaps with the ultimate intention of optimising process conditions, increasing production capacity, etc.

The experiment I shall carry out on our known system is the simplest comparison. I shall run the simulation 8 times at a 'temperature' T_1 of 25 and another 8 times with 'temperature' T_2 of 30. Because of my temporary status as 'god' I know that, if it were not for the variable ε_{ij} terms,

$T_1 = 25$ would give yields of $\mu + k \times 25 = 60 + 1.2 \times 25 = 90$
$T_2 = 30$ would give yields of $\mu + k \times 30 = 60 + 1.2 \times 30 = 96$

The real *effect* of the change in temperature would be a difference of

$$96 - 90 = 6$$

in the yield. But because the ε_{ij} terms *are* there, in real life as in my model, that is not exactly what happens. In my first experiment I got

For $T_1 = 25$ and $n_1 = 8$, $\bar{y}_1 = 89.730$ and $s_1 = 0.426$

For $T_2 = 30$ and $n_2 = 8$, $\bar{y}_2 = 96.105$ and $s_2 = 0.449$

Temperature effect estimate, $E_1 = \bar{y}_2 - \bar{y}_1 = 6.376$

Combined standard deviation estimate, $s_c = \sqrt{\dfrac{s_1^2 + s_2^2}{2}} = 0.437$

(*Note:* s_1 and s_2 are both estimates of the variability of the ε terms. Neither one is better than the other, so s_c is an average of the two. For technical reasons we take the square root of the mean of the squares (RMS) rather than the simple average. The RMS is always slightly larger than the simple average of the standard deviations.)

The fourth such experiment gave me

For $T_1 = 25$ and $n_1 = 8$, $\bar{y}_1 = 90.202$ and $s_1 = 0.495$
For $T_2 = 30$ and $n_2 = 8$, $\bar{y}_2 = 96.198$ and $s_2 = 0.474$
Temperature effect estimate, $E_1 = \bar{y}_2 - \bar{y}_1 = 5.997$

Combined standard deviation estimate, $s_c = \sqrt{\dfrac{s_1^2 + s_2^2}{2}} = 0.485$

I ran this same experiment 100 times in total. Each time there was a different estimated effect and a different standard deviation estimate. I have collected them together in the two histograms shown in Figure 4.2.

The most striking thing about the histogram of estimated standard deviations (Figure 4.2b) is that the spread of estimates is very wide as a proportion of the real value, $\sigma = 0.5$. This means that even with many repeat measurements the estimate of standard deviation may be quite unreliable. Those who think they can get a reliable estimate of standard deviation from 3 or 4 measurements are really kidding themselves—see later in Section 4.5.3.

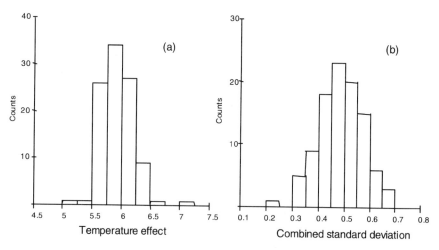

Figure 4.2 Histograms of (a) 100 values of E and (b) 100 values of s_c, calculated from the model of temperature effect.

The main things to notice about the histogram of the temperature effects (Figure 4.2a) are that:

- On average the histogram seems to be centred close to the value 6.0 that we calculated for the 'real' temperature effect (ignoring variability). In the jargon we say that 'the estimates are unbiased'.
- The histogram looks (roughly) like a Normal distribution.
- The spread of this histogram is only about half that of the histogram of individual values; Figure 4.1.

Do any of these things surprise you? Think about them.

4.5.2 A confidence interval for the effect of changing the temperature

The empirical findings about temperature can be backed up by statistical theory. This tells us that if, instead of 100 runs of the simple comparative experiment, I had done many thousands of runs, then:

- On average the histogram of temperature effects would be centred at 6.0.
- The histogram would look almost perfectly like a Normal distribution.
- The standard error of the histogram would be

$$\sqrt{(\sigma^2/n_1) + (\sigma^2/n_2)}$$

which becomes $\sigma\sqrt{2/n}$ when $n_1 + n_2 = n$.

The term *standard error* is used to differentiate the measure of spread for a derived statistic—in this case the difference between two averages—from the spread of variability of the individual values. It is partly dependent on the *design* of the experiment, i.e. the number of replicates of each temperature run, and partly on the number and size of the uncontrolled *causes of variability* that contribute to the size of σ.

One of the properties of all Normal distributions is that 95% of the shape lies within 1.96 standard deviations/standard errors of the mean. Putting all of this together, we can conclude that if we do large numbers of identical comparative experiments on the computer simulation, then

- 95% of the time the estimated temperature effect will be no more than 1.96 standard errors away from the 'true' effect.

Turning this statement round:

- We can be reasonably confident (95% confident) each time we do such an experiment that the 'true' temperature effect is no more than 1.96 standard errors away from the estimated effect.

This statement is generally shortened to saying that a 95% confidence interval for the temperature effect is

$$\text{Confidence interval} = \text{estimated effect} \pm 1.96 \text{ standard errors}$$
$$= \bar{y}_1 - \bar{y}_2 \pm 1.96 \, \sigma\sqrt{2/n}$$

So the 95% confidence interval for our estimate of temperature effect (using the results of the first run of 8 experiments), is

$$\text{Confidence interval} = 96.105 - 89.730 \pm 1.96 \times 0.5 \times \sqrt{(2/8)}$$
$$= 6.376 \pm 0.490$$
$$= 5.886 \text{ to } 6.866$$

At this stage we have chosen to use the interval that contains 95% of the values in a Normal distribution. There is no particular significance attached to 95% other than that it is a frequently chosen value. For a $100(1-\alpha)\%$ confidence interval the formula can be generalised to

$$\text{Confidence interval} = \bar{y}_1 - \bar{y}_2 \pm z_{\alpha/2} \, \sigma\sqrt{2/n}$$

$z_{\alpha/2}$ is the value of a standard Normal distribution ($\mu=0$, $\sigma=1$) that gives an upper tail area of $100\alpha/2\%$. Since the Normal distribution is symmetrical, this means that $100(1-\alpha)\%$ of it will be enclosed by the values $-z_{\alpha/2}$ to $+z_{\alpha/2}$.

Intervals of 90% and 99% are the most common alternative choices to 95%. Note that the 90% confidence interval is the narrowest of the three, but it is also the most likely to miss including the true parameter we are trying to estimate.

4.5.3 A revised confidence interval for temperature effect

We are nearly there but not quite. The trouble with the above formula is that we can only calculate the 95% confidence interval if we *know* the value of σ. In the case of our artificial example we do—it is 0.5. But usually we only have the estimate provided by s_c, and, as seen in Figure 4.2b, this estimate may not be very good. In particular, it may happen to be a considerable underestimate.

To remedy this problem—and at the cost of much mathematics—W. Gossett, a Guinness master brewer writing under the pseudonym Student, invented the *t*-distribution. This looks similar to the Normal distribution but has longer tails. Its exact shape depends on the amount of information (the number of degrees of freedom, ν) that has gone towards the estimate of σ.

When there are 8 replicate measurements of yield at a fixed temperature, there are $8-1=7$ independent 'gaps' between them (bits of information), so $\nu=7$. But putting together the estimates at each of the two temperatures, we have a total of $7+7$ bits of information, so $\nu=14$. Unfortunately, for more complicated experimental designs it becomes difficult to explain degrees of freedom in this intuitive way.

The modifications to the confidence interval formula that are required are to replace σ with its estimate s_c, and replace z_c with the appropriate t-value to give

$$\text{Confidence interval} = \bar{y}_1 - \bar{y}_2 \pm t_{\nu, \alpha/2}\, s_c \sqrt{2/n}$$

t compensates for the general underestimation of σ by s_c.

Table 4.1 gives a small selection of t-values for different degrees of freedom, ν, at a 95% confidence level. This table is a very short extract of the comprehensive tables available in many texts. Much modern software includes a function to calculate any t-value. Table 4.1 shows that with small ν, i.e. with very little information concerning the calculated standard deviation estimate, the t-value is large. This means that the 95% confidence range is also large.

With 'perfect' knowledge ($\nu = \infty$), Student's t-distribution becomes exactly the same as the Normal distribution. This is why the t-value is 1.96 for infinite degrees of freedom. The fit is already quite close for 15 or 20 degrees of freedom.

Table 4.1 Variation of the 95% confidence t-value with degrees of freedom

Degrees of freedom, ν	95% t-value, $t_{\nu,0.025}$
1	12.70
2	4.30
5	2.57
10	2.23
15	2.13
20	2.09
40	2.02
∞	1.96

We can now calculate the 95% confidence interval for our first estimate of temperature effect assuming that we do not know the true variability, σ, of the system and have to rely on its estimate s_c. The t-value for 14 degrees of freedom is 2.14, so

$$\text{Confidence interval} = 96.105 - 89.730 \pm 2.14 \times 0.437 \times \sqrt{(2/8)}$$
$$= 6.376 \pm 0.468$$
$$= 5.9 \text{ to } 6.8$$

Since the range of uncertainty is about 0.9 units wide, it is sensible to round the 95% confidence interval to 1 decimal place in this example and write it as 5.9 to 6.8.

In this way, a single experiment gives us a probable range for the 'real' effect of changing temperature by 5°C.

Example 3. A large chemical laboratory in Norway complained to the makers of a commercial assay kit for thyroid function. The laboratory routinely calculated the average patient measurement per week. They said there had been a change when a modified kit, claimed to be equivalent to the original, had been introduced. The weekly patient values they sent to the manufacturer are shown in Table 4.2. s_1 and s_2 have 6 and 4 degrees of freedom, respectively, so the combined estimate of variability is

$$s_c = \sqrt{\frac{6 \times 1.26^2 + 4 \times 0.92^2}{6+4}} = 1.136$$

and has $6+4 = 10$ degrees of freedom.

A 95% confidence interval for the change in weekly patient values is then given by (see Section 4.5.2 for unequal replicates)

$$\bar{y}_1 - \bar{y}_2 \pm t_{10} \cdot s_c \sqrt{\frac{1}{n_1} + \frac{1}{n_2}}$$
$$= 2.61 \pm 2.23 \times 1.136 \times 0.586$$
$$= 2.61 \pm 1.48$$

With appropriate rounding, this confidence interval is 1.1 to 4.1. While this interval is quite wide, it is entirely positive, indicating that there has almost certainly been a change in weekly patient values.

Because of the enormous commercial implications of such a conclusion, the marketing department of the company concerned referred the data to a statistician to check their calculations. Knowing quite a lot about thyroid function and a little about the climate of Norway, she queried the implied assumption that the ε_{ij} terms were random. She suggested it was possible that a major contributor to ε might be

INTERPRETING RESULTS 125

Table 4.2 Weekly patient thyroid function values before and after introducing a modified kit

	Before modification	After modification
	101.7	99.8
	100.3	97.4
	102.9	98.9
	100.9	98.5
	102.4	97.9
	100.1	
	99.5	
No. measurements, n	7	5
Average, $\bar{y}$	101.11	98.50
Standard deviation, s	1.26	0.92

the fact that people with high values of thyroid function have high metabolic rates, whereas hypothyroid people feel the cold. In Norway that year, spring had been a particularly sudden change from cold and snow to warmth and sunshine at a time that largely coincided with the change in assay kits. On persuading the laboratory to look back to previous years, it was seen that similar, if less dramatic, changes had occurred before.

This example emphasises the crucial importance of thinking carefully about each interpretation of results and not relying blindly on 'recipes'.

4.5.4 Confidence intervals for variability—Standard deviation

(It is strongly recommended that you skip straight to Section 4.5.6 when first reading this chapter so that you do not get too bogged down in detail before seeing the purpose and value of confidence intervals.)

For data like ours, where the individual values come from a reasonably Normal distribution, the square of an estimate of standard deviation, s^2, will have a distribution whose shape is that of χ_ν^2. This is the chi-squared distribution with ν degrees of freedom. An example of the distribution is given in Figure 4.3. The shaded areas in the plot indicate the proportion (probability) that an s^2 value of this magnitude will occur. Thus, the darker area to the right of the line marked $\chi_{\nu,0.025}^2$ shows that there is a probability of 0.025 (2.5%) that an s^2 value greater than 20.5 will occur, while the lighter and darker shaded areas together indicate a probability of 0.975 (97.5%) of an s^2 value being greater than 3.25. The lighter shaded area alone corresponds to a 0.95 (95.0%) probability of an s^2 value between 3.25 and 20.5.

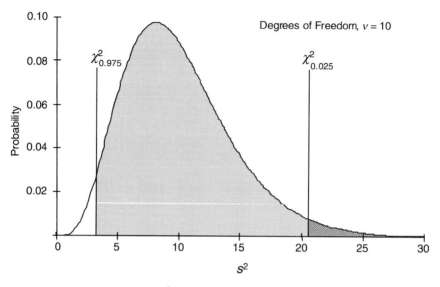

Figure 4.3 The distribution of s^2 when the individual measurements have a Normal distribution.

Of s^2 values obtained from identical experiments, 95% will lie within the range

$$\frac{\sigma^2 \chi^2_{\nu, 0.975}}{\nu} \quad \text{to} \quad \frac{\sigma^2 \chi^2_{\nu, 0.025}}{\nu}$$

Conversely, a 95% confidence interval for the value of σ^2 will be

$$\frac{s^2 \nu}{\chi^2_{\nu, 0.025}} \quad \text{to} \quad \frac{s^2 \nu}{\chi^2_{\nu, 0.975}}$$

By taking square roots of the above terms, a 95% confidence interval for the true value of σ is

$$s\sqrt{\nu/\chi^2_{\nu, 0.025}} \quad \text{to} \quad s\sqrt{\nu/\chi^2_{\nu, 0.975}}$$

Example 4. Repeat measurement of the radioactivity of a large radiation source has to be kept to a minimum, partly because it involves careful positioning of the source and partly because of the need to limit the exposure of the operator to radiation. To estimate the reproducibility of the measurement, a single source is positioned, measured and removed three times. This takes a technician a morning to carry out. The standard deviation of the three measurements is $s = 10$, and it has 2 degrees of freedom. A 95% confidence interval for the real measurement variability, σ, would be

$$10\sqrt{2/7.378} \quad \text{to} \quad 10\sqrt{2/0.0506}$$

which is

$$5.2 \quad \text{to} \quad 63$$

What does this tell you about relying on a single triplicate for an assessment of the reproducibility of a person, machine or system?

Anything that can build up the number of degrees of freedom helps create more reliable estimates, which is why combining estimates of variability as I did in Section 4.5.1 is so valuable.

4.5.5 Confidence intervals for variability—Ratio of two variances

We might also want to compare two sets of equipment, conditions or people in terms of their variability. For example, in choosing between two measurement systems, A and B, one of the criteria would be that of reproducibility.

To make this kind of comparison we could do n_A repeat measurements on system A and calculate s_A, and do n_B measurements on system B and calculate s_B. Then, if the confidence interval for the variability of one of these systems eliminates it from consideration (it is too large) and the other does not, our choice is made. Unfortunately, things are not usually as clear-cut, especially if the costs or other factors regarding the choice are unequal. Instead we can take the ratio

$$R_{AB}^2 = s_A^2/s_B^2$$

It does not matter which system is called A and which is called B, but in practice most people choose A to be the one with the higher estimate of standard deviation. The distribution for R_{AB}^2 is called the F-distribution, and so R^2 is often known as the F-ratio. It is characterised by the two degrees of freedom $\nu_A = n_A - 1$ and $\nu_B = n_B - 1$. You will also find F-ratios used extensively later in the analysis of variance (see Section 4.9.3 and Chapter 8). In these cases the degrees of freedom are not so simple to work out, but everything else is the same.

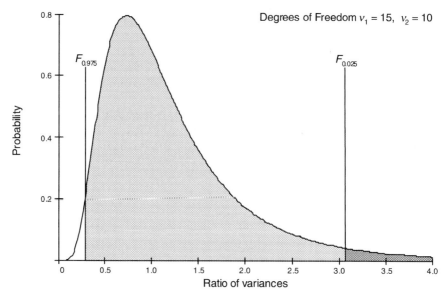

Figure 4.4 The distribution of ratios of variances—the F-distribution for 15 and 10 degrees of freedom.

The shape of the F-distribution is rather like that of the χ^2 distribution but more extreme. An example distribution is shown in Figure 4.4, where the interpretation of the shaded areas is very similar to that given for the χ^2 distribution in Figure 4.3.

The F-distribution has the interesting feature that $F_{\nu_A, \nu_B 0.975} = 1/F_{\nu_B, \nu_A 0.025}$. Consequently F-tables are printed only for the upper tails since you can get the lower tails by swapping over the degrees of freedom and taking the inverse. For example,

$$F_{3, 7, 0.975} = 1/F_{7, 3, 0.025} = 1/14.62 = 0.0684$$

The ratio of true variability for the two measurement systems A and B is σ_A/σ_B. A 95% confidence interval for this ratio is given by,

$$R_{AB}/\sqrt{F_{\nu_A, \nu_B, 0.025}} \quad \text{to} \quad R_{AB} \cdot \sqrt{F_{\nu_B, \nu_A, 0.025}}.$$

Take care with the change of order of degrees of freedom in looking up the F-values.

The use of confidence intervals for variance ratios is illustrated in Example 5, where it is claimed that one instrument has less variability than another. The validity of the claim can be assessed quantitatively.

Example 5. In our laboratory we have a rather out-dated, relatively laborious and somewhat unreliable method for measuring particle size. Call this method A. An equipment salesman has persuaded us to allow a demonstration by him of a new, more convenient and (of course) more expensive method, which we shall call method B. He claims, also, that it gives good reproducibility.

To compare the two methods, we take a sample of powder from a standard production batch and subdivide it. The salesman measures five of the subsamples while two of our laboratory technicians measure 8 subsamples between them. The results are shown in Table 4.3.

Table 4.3 Results from two methods of particle size determination

	Method A (current)	Method B (new)
	129.2	127.9
	127.8	128.3
	128.3	128.2
	127.5	128.6
	128.8	127.8
	127.9	
	128.5	
	129.0	
Mean	129.38	128.16
Standard deviation	0.61	0.32

To assess whether the methods are equivalent in the sense that on average they give the same result, we could construct a 95% confidence interval for the difference of the two means. But our main concern at present is to estimate the relative precision of the two methods.

The standard deviation estimates are $s_A = 0.61$ with 7 degrees of freedom, and $s_B = 0.32$ with 4 degrees of freedom. The estimated ratio is

$$R_{AB} = s_A/s_B = 1.9$$

This simple estimate alone would lead us to assume that the variability of method A is nearly twice that of method B, and that we should definitely invest in the new equipment. However, a 95% confidence interval for the true ratio, σ_A/σ_B, is

$$1.9/\sqrt{9.07} = 0.63 \quad \text{to} \quad 1.9\sqrt{5.52} = 4.5$$

This makes the decision much less clear-cut. The data from this trial show that it is possible that the current method A, might be less variable than method B, in which case we should certainly be making the wrong choice if we selected system B. It is possible also that method B is very much more precise than the current method. To clarify the situation, we need to get more degrees of freedom, especially for method B, so that we have better estimates of standard deviation and a much narrower confidence interval.

4.5.6 Who needs confidence intervals?

In industry, an experiment is carried out to gain the information and understanding on which to *make decisions* or to *take action*. It would have been nice to have done a single experiment on the computer simulation of Section 4.4 and to have come to the conclusion that

> The effect of changing the temperature from 25°C to 30°C is an increase in yield of 6.376%.

Indeed, that is the kind of statement most people would have made on the basis of the data of Section 4.5.1.

The trouble is that we now know that

- The conclusion is incorrect. The real answer is 6.0%.
- If we repeat the experiment we shall almost certainly get a different answer.

If that was all we knew we should still not be able to make an informed decision. Luckily the confidence interval is there to indicate the bounds of our ignorance. That first experiment in Section 4.5.1 gave us a confidence interval for the temperature effect (end of Section 4.5.3) of 5.9% to 6.8%.

It could be that even a 5.9% difference would be sufficient to justify the expense of putting in a temperature control system to avoid the problem. Or it could be that even differences as large as 6.8% are relatively unimportant in terms of the measurement that is being made. In either case, we can be (reasonably) confident that the decision taken is well supported.

A bigger problem arises if, say, a 6% difference would be unimportant but anything much greater would require expenditure on the control system. The knowledge that the initial estimate of 6.376% is so uncertain may suggest that it was not worth doing the experiment in the first place. And a simple-minded decision maker might prefer not to know about this uncertainty! But for the professional, who is used to working with risk and uncertainty, there is now an informed choice:

> Pay for extra experimentation (an increased number of runs, n, will reduce the width of the confidence interval) so that the final decision is based on better information.

or

> Pay to put in the temperature control system, even though there is a reasonable chance that it is not necessary, on the grounds that it is more likely that it is needed.

The choice between these two decisions will be dictated by factors such as the relative costs, how quickly the final decision must be taken, political or business pressures, how risk-averse the manager is, and so on. In this particular case the cost of a wrong decision on the hardware might be quite substantial, whereas more experimentation might be relatively cheap and quick. In other situations the reverse might be true.

4.5.7 Single-sided and double-sided intervals

(Once again you may wish to skip to Section 4.6 on first reading this chapter in order to maintain the flow of understanding.)

So far we have set up all our 95% confidence intervals so that there is $2\frac{1}{2}\%$ in each of the upper and lower tails. However, this symmetry is not essential and we might need to consider, in particular, whether the interval should be single-sided, with all 5% in one of the tails and nothing in the other. For a confidence interval involving the χ^2- or F-distribution, this will lead to the lower end of the interval being at zero or the upper end at infinity. For intervals involving the t-distribution, one or other end will be at infinity.

One situation where we would use a one-sided confidence interval is where there is a physical limit to a difference, ratio, etc. For example, if we measure each of a number of samples from a batch of material, the standard deviation of the samples, s_{sample}, is an estimate of the combination of sampling variability and measurement variability, $\sqrt{\sigma_{\text{sample}}^2 + \sigma_{\text{meas}}^2}$.

If we then take a single sample and measure it several times, the standard deviation of these measurements, s_{meas}, is an estimate of σ_{meas} only. The ratio

$$R_{AB}^2 = s_{\text{sample}}^2 / s_{\text{meas}}^2$$

is an estimate $(\sigma_{\text{sample}}^2 + \sigma_{\text{meas}}^2)/\sigma_{\text{meas}}^2$. This clearly cannot be less than 1, so a two-sided confidence interval would not be sensible.

Alternatively, we may have several measurements of the percentage impurity of a product and only be concerned with whether the true batch impurity level is below a specification limit or not. Again we might argue that a one-sided confidence interval was appropriate. However, in cases such as the latter, the decision to use a one-sided interval is much more contentious and should preferably be made *before* seeing the results. There is often an implication that only one kind of error is important. For example, it only matters that we detect (and presumably reject) this batch if there is anything more than a 5% chance that its impurity level might be even a little above the specification limit but that there are no important consequences to incorrectly failing a batch.

The t-value or F-value for a double-sided 95% confidence interval is numerically the same as the single-sided 97.5% confidence interval. This means that one set of tables can be used for single-sided and double-sided intervals simply by using an appropriate confidence level. However, it is important to know beforehand whether the tables are for single-sided or double-sided intervals, otherwise completely the wrong t-value or F-value will be chosen! For example, the two-sided 95% interval using a t-value with one degree of freedom will take the values ± 12.7, whereas a one-sided interval would use 6.3. The lower and upper values of a two-sided 95% interval for $F_{1,1}$ are $1/647.8$ and 647.8, whereas the one-sided upper value is 161.4.

4.6 Deciding on the size of a simple comparative experiment

4.6.1 A simple approach

For practical reasons it may not be easy or even possible to carry out more measurements when it is found that the confidence interval from the initial experiment is too wide to allow a sensible informed decision to be made. So the size of the experiment usually must be decided in advance, and Example 6 shows a simple approach to calculating the size.

> *Example 6.* An experiment was to be carried out to study the production factors that affect catalyst activity. A particular grade of catalyst was to be made in a 4-week production campaign. Samples collected during this trial were expected to take nearly 2 months to analyse, so it was clearly important to design the trial in the first place so that the confidence intervals would be small enough to allow decisions to be made without too much risk of being incorrect.
>
> It was decided from practical economics that a confidence interval of ± 3 units would be satisfactory for the main comparison. The 95% confidence interval for a comparison of n replicates under condition A with n replicates under condition B would be

INTERPRETING RESULTS

$$\bar{y}_A - \bar{y}_B \pm t_\nu s \sqrt{2/n}$$

So we need a value of n such that

$$t_\nu s \sqrt{2/n} \approx 3$$

Production variability was estimated from the standard deviation of past data as about 5.5, so this was used as the estimate of s. Putting this into the equation, and rearranging the equation to give a value for n, we get

$$n = t_\nu^2 \times 2 \times 5.5^2/3^2$$
$$n = t_\nu^2 \times 6.722$$

Now to get the estimate of n, we need the value t_ν. This depends on the number of degrees of freedom, ν, which depends on the value of n! Fortunately, the value of t_ν does not change greatly with ν unless ν is very small. So as an approximation we can use $t_\nu = 2.2$. This gives the value of n as

$$n = 2.2^2 \times 6.722 \approx 34$$

The rule of thumb is that for a confidence interval of *size* $\pm C$,

$$n \approx 10 \left(\frac{s}{C} \right)^2$$

In this example, the required amount of replication was regarded as much too costly in resources and time. To deal with this problem we found ways of reducing the overall variability of the process and this allowed us to reduce the required replication (see also Section 4.7). Although it took several weeks to plan and prepare for the trial, the result was extremely valuable and we could rely upon it knowing that the level of replication was appropriate to the amount of variability in the experiment.

4.6.2 *A more general method*

The simple method of estimating experiment size is based on controlling the risk that the true temperature effect (say) is within the 95% confidence interval while the measurement falsely indicates that it is outside the interval. This is a Type I error (see Chapter 1) and in our example there is a 5% chance that it will occur. This value is called α and in our example $\alpha = 0.05$. The complementary situation, that the true effect is outside the

95% confidence interval but the measurement indicates it to be inside the limit, also occurs. This is a Type II error and the probability that it occurs also depends on the size of the experiment. By convention, this error is called β and a frequently used value is 10%, so $\beta=0.10$. It would be useful to control both types of error at the same time. (The value $1-\beta$ or $100(1-\beta)\%$ is called the power—see Chapter 1.)

A small amount of mathematics generates the equations that allow this to be done. The equation required is

$$n = 2\left(\frac{(t_{\nu,\alpha/2} + t_{\nu,\beta}) \times s}{D}\right)^2$$

where $t_{\nu,\alpha/2}$ is the t-value for ν degrees of freedom and a $100(1-\alpha)\%$ confidence level, assuming a two-sided interval; $t_{\nu,\beta}$ is the one-sided t-value for ν degrees of freedom and a power of $100(1-\beta)$; D is the minimum difference to be detected; and s is our prior estimate of standard deviation of the measurements.

Again the problem with this equation is that the required t-values depend on n, and n depends on the t-values. Iterative procedures can be used to solve this problem, but we can make the approximation that, for reasonably large sample sizes, the sum of the t-values is 3.3.

Using the estimate of variability from Example 6, which was 5.5 units, but now requiring at least a 90% chance of detecting a difference, D, as large as 3 units, the formula for n gives

$$n = 2 \times \left(\frac{3.3 \times 5.5}{3}\right)^2$$

$$\approx 73$$

and the rule of thumb for such a situation is that

$$n \approx 20\left(\frac{s}{D}\right)^2$$

The approach to be taken to deciding on the size of the experiment depends on what kind of decision or action will be taken as a result of the experimental outcome and so whether Type II errors need to be controlled or not. But it must also be remembered that the value of n is only an *estimate* of the required replication and that it depends on how good the estimate of variability is, and even whether it is an appropriate estimate for the proposed experimentation. The fact that the estimate of replication is only approximate does not invalidate the need to obtain it but probably means that rounding of the figure to a close but more convenient one is perfectly acceptable.

4.7 Reducing the amount of work to be done

In the section above we found that when comparing the average outcome of two temperatures, two reagents, etc. the number of replicates is either

$$n \approx 10\left(\frac{s}{C}\right)^2 \quad \text{or} \quad n \approx 20\left(\frac{s}{D}\right)^2$$

If we find this level of replication unacceptable for the amount of information we get, there are four basic options:

- Increase the size of C (or D).
- Widen the temperature range so that the effect becomes larger, and hence C (or D) can be increased.
- Reduce σ, the true variability of the ε terms.
- Use the same experiment for investigating more than one factor.

Option 1 is by far the simplest but it means that by spending less on the experiment we get less information (or, equivalently, our conclusions are less certain).

Option 2 is excellent provided we can assume that the effect (of temperature, say) remains reasonably linear over the wider range. In practice this may be questionable or we may run into problems of practicality, safety, etc.

Option 3 is good if it can be managed. The technique of experimental blocks was invented by Fisher and has been developed to a fine art in some areas of science. If successful it has the potential to reduce the size of experiments dramatically, even though some ingenuity may be required. Chapters 6–8 discuss how such an experiment should be analysed to give the right confidence interval.

Option 4 may be an excellent way of investigating several factors for the price of just one. It can also allow us to look at possible interactions between these factors and so widen the range of validity of the conclusions. These very efficient and effective ways of experimenting were invented and developed by various British statisticians (Fisher, Cox, Plackett and Burman) and some have been popularised by the Japanese engineer, Taguchi. They are described in Chapters 6 and 7.

Choosing which option(s) to use is a skill addressed in much of this book. It depends on your practical circumstances and it can be built up with experience, scientific insight and imagination. It is, in my opinion,

4.8 Hypothesis testing and significance levels

In the first of our experiments to look at the effect of changing temperature from 25°C to 30°C, the 95% confidence interval was 5.9 to 6.8. In this example, both ends of the confidence interval, and so all parts of it, are positive. One of the values that is *not* in this interval is the value zero; for this to happen one end of the interval would have to be negative and the other positive. So, we can be reasonably sure that there *is* a real difference, as well as setting limits on how big it might be.

Some data interpretation, known as hypothesis testing, looks *only* at the question of whether there is likely to be a real difference or not, without giving the associated and very valuable information about how big the difference might be. Some of the reasons for doing this are better than others.

Bad reasons for doing hypothesis tests

- The analyst or the computer software has not learned how to work out confidence intervals.
- The decisionmaker wants life to be made 'easy' by having only a yes/no answer to whether there is an effect, even if the answer is wrong.
- The decisionmaker is not even aware that experiments can lead to incorrect conclusions.
- Giving confidence intervals does not suit the established reporting style of the company, journals, etc.

Acceptable reasons for doing hypothesis tests

- The experiment was planned in such a way that any confidence interval that contains the value zero has limits that are small enough to be of no practical importance, so that it is not essential to quote them.
- The experiment was carried out to investigate a more complex question such as

    Do 5 technicians all have the same precision of measurement?
    Is the relationship between yield and temperature a straight line effect or is there nonlinearity?

- There may be no difference, ratio or other parameter for which a confidence interval can be specified.

Even in the more complex situations, it may be possible to give a confidence interval for the ratio of the (apparently) best and poorest standard deviation, or for the difference between the average yield at an intermediate temperature compared with the average at two extreme temperatures.

In hypothesis testing there is a *Null Hypothesis* which, as its name implies, is usually of the form

H_0: there is no difference between these operators/temperatures/suppliers of chemicals.

This is frequently written in the form

$$H_0: \mu_1 = \mu_2$$

when we are comparing two averages, for example. The alternative to the Null Hypothesis is that there is a difference between the operators/temperatures/suppliers. This is called the *Alternative Hypothesis* and it is what has to be accepted if the Null Hypothesis is rejected as a result of a hypothesis test. It is usually written in the form

$$H_1: \mu_1 \neq \mu_2$$

We then calculate the probability, p, of getting *this* particular set of results from *this* experimental set-up *assuming* that the Null Hypothesis is true. We also often make other assumptions that strictly belong in the Null Hypothesis. We also made them when calculating confidence intervals, as discussed in Section 4.4.1.

If p is small this suggests either that we do, indeed, have an unusual set of results or that something was wrong with the Null Hypothesis. The conclusion is usually made that there is not no difference, i.e. that there is some difference. The Null Hypothesis is then rejected in favour of the Alternative Hypothesis. If, on the other hand, p is not small (usually taken to be $p > 0.05$) we do not reject the Null Hypothesis. Care should be taken *not* to assume that this proves that the Null Hypothesis is correct, any more than a 'not guilty' verdict proves that a man is innocent of a criminal offence. In either case it may simply be that the evidence against them is not very strong or that the case has not been well made. In fact, in most situations it is rather unlikely that the Null Hypothesis ($\mu_1 = \mu_2$) is *exactly* true (see Section 4.9.1 on equivalence).

Often a *p*-value of 0.05 (5%) or less will result in our saying that 'the difference is statistically significant at the 5% level' while a *p*-value of 0.01 or less will be called 'a difference that is highly statistically significant at the 1% level'. Unfortunately, the difficulty of saying or writing this technical jargon leads to most people (including me when I'm not careful!)

simply saying 'there is a significant/highly significant difference'. In ordinary English the word 'significant' means 'important'. So instead of expressing a sophisticated probabilistic statement about the likelihood of one set of results conditional on the experimental design and on all aspects of the Null Hypothesis, we appear to be making a definitive statement about both the existence *and* the importance of a real effect. This makes the listener/reader feel comfortable, but it hides still further the need for a confidence interval or the equivalent to remind us of the extent of our ignorance or lack of knowledge.

Hypothesis tests have different names, often relating to the probability distribution that is used. Thus, there are t-tests, F-tests, χ^2 goodness-of-fit tests, Mann–Witney tests and many others. In every case the Null Hypothesis is assumed to be correct in order to work out the probability, p, of getting an experimental result as extreme as the one actually observed. If p is small enough for us to conclude that the Null Hypothesis is incorrect, then p is also the probability that we have come to the wrong conclusion. But if p is not small it is usual for people to 'accept' that the Null Hypothesis is correct without questioning how likely it is that the conclusion is wrong. We should remember that we can never *prove* that the Null Hypothesis is true; we can only estimate the probability that it may be true. This imbalance is particularly important where experiments are being done to try to establish equivalence, for example, between a cheaper and a more expensive reagent in a paint formulation. There are other ways of getting round the problem but, wherever possible, the best solution is to calculate and quote confidence intervals rather than to do hypothesis testing.

4.9 Some more applications of confidence intervals

4.9.1 Equivalence studies

Quite often work is carried out to discover whether two procedures, raw materials, etc. are equivalent in their effects. Usually this is in order to replace the currently used material with one that is cheaper or has other advantages. For example:

- Can we use a modified starch instead of sugar without affecting taste?
- Will a new enzyme assay on average give the same measurement of activity as our current one?
- Will the final strength of an engineering plastic be affected if we use 20% less catalyst in making it?

- Do we really need to leave a mixture stirring for 40 minutes before going on to the next stage, or would 30 minutes do?

In each of these questions we are seeking to show that the two alternatives are equivalent to each other, not that they are identical. So far we have looked at examples where it is assumed that the alternatives are identical (i.e. that there is zero difference between the averages) and we have attempted to disprove this by showing that the confidence interval does not include a zero difference. On many occasions we shall fail even though the difference between the averages is very small. This is often due to the standard error estimate used in the t-test calculation being small, either because a large number of measurements have been made or because the set of measurements happens by chance to have a very narrow distribution.

One method to circumvent this problem is to define, at the time of planning the experiments, a maximum value for the difference for which there is no practical consequence. If the measured difference is less than this figure, then no statistical test is made. This approach is workable, but it does introduce a conceptual difficulty when the measured difference is larger than the no-practical-consequence range and yet is statistically not significant.

An alternative is to use an equivalence test (Hartmann *et al.*, 1995; Berger and Hsu, 1996). In this the 95% confidence interval of the measured difference between the averages is compared with a predefined range. If the confidence interval falls completely inside the defined range, then the two alternatives are said to be equivalent.

In terms of hypotheses, the equivalence test uses the Null Hypothesis that the true difference between the averages is greater than some predetermined value, θ. The Alternative Hypothesis is thus that the true difference between the averages is not greater than this value. Example 7 illustrates the use of the equivalence test.

Example 7. An analytical method has been implemented at two laboratories to measure the amount of colouring agent in a liquid hair lotion. Similar equipment has been supplied to the laboratories and the analysts at the two sites were trained in the method at the same course. The amount of colour is not critical to the effectiveness of the lotion, but large changes in colour could affect the customers' perception of the quality of the product and hence its sales. A range of ±1.5% in the amount of colour agent has been shown to have an acceptable range of perceived colour. It is therefore required to show that the laboratories produce equivalent assay results for the amount of colouring agent within this range.

Table 4.4 Assay results (% target value) for hair lotion colouring agent determined at two laboratories

Batch	Lab. 1	Lab. 2
1	101.9	100.6
2	101.6	101.0
3	100.9	100.2
4	101.3	100.6
5	101.6	100.6
6	101.8	101.1
7	102.0	100.7
8	101.6	100.9
9	101.4	100.5
10	101.6	101.1
Average	101.57	100.73
Standard deviation	0.316	0.291
Pooled SD	0.304	
$t_{18,\ 0.05}$	1.734	

A trial has been carried out in which samples from 10 batches of lotion were assayed at both laboratories. The assay results, expressed as a percentage of the target amount of colour agent, are shown in Table 4.4.

A 95% confidence interval for the difference of the averages for the sites is 0.604 to 1.076, which can be rounded to 0.6 to 1.1. As this interval does not include zero, the conclusion is that there is a difference between the laboratories. But in the equivalence test, we also look to see if the confidence interval is completely contained inside the range of acceptable difference. As 0.6 to 1.1 is completely inside the range -1.5 to $+1.5$ we conclude that the two sites are producing equivalent results.

In practice, it may be quite difficult to decide what the range of acceptable difference needs to be in a given situation. For example, what difference in taste will put customers off sufficiently that a significant number stop buying a product? Making the attempt to establish this, however, has the benefit of bringing into focus the potential costs of changing as well as the (usually) more obvious savings to be had. Without it, cost cutting quite often results in lower, not higher, profits.

4.9.2 Paired comparisons

In Section 4.5 we estimated the effect on yield of running a process at two different temperatures. To do this, replicate batches were made at each temperature. A critical assumption in the ensuing calculation of confidence interval was that the residual variability terms, ε_{ij}, behaved like random values from a distribution with constant standard deviation throughout the relevant period of time.

This assumption is frequently not true and sometimes this causes considerable difficulties and/or incorrect conclusions. But quite often, if we think about it, we can turn this to good effect and do less work to achieve the required precision for our experiment.

Example 8. Crude sugar is being converted into ethanol. It is known that impurities in the sugar can have an appreciable effect on the conversion efficiency but the purity of different batches is not under our control (although it might affect the price that is paid for them!). The comparison we want to make is between the conversion efficiency of two catalysts.

Instead of using one catalyst across one set of batches and then the other across a different set, the experiment is arranged so that both catalysts are used with each batch of sugar. Table 4.5 shows some data for this type of experiment. It can be seen that, although the efficiency varies quite widely from sugar to sugar, the difference between the catalysts is relatively consistent.

Table 4.5 Data for the catalyst conversion efficiency experiment on crude sugars

Crude sugar	Catalyst A	Catalyst B	Difference
1	23.3	23.6	−0.3
2	19.2	19.3	0.1
3	22.4	23.0	−0.6
4	27.3	28.1	−0.8
5	26.3	27.2	−0.9
6	25.6	25.8	0.2
7	19.9	20.5	−0.6
Average	23.429	23.929	−0.5000
Standard deviation	3.145	3.310	0.3055
Pooled SD		3.23	

A 95% confidence interval for the average difference in conversion efficiency is calculated as

$$\bar{d} \pm t_{\nu, 0.05} s_d / \sqrt{n}$$

where n is the number of paired comparisons; $\bar{d}$ is the average difference; s_d is the standard deviation of the differences; $\nu = n - 1$ is the degrees of freedom for s_d.

For the experiment in Example 8, the confidence interval works out to

$$-0.5 \pm 2.45 \times 0.3055/\sqrt{7}$$

which is -0.2 to -0.8.

To get the same precision from an unpaired experiment, the number of replicate runs required for each catalyst would have been about

$$n = 10\left(\frac{s}{C}\right)^2 = 10\left(\frac{3.23}{0.28}\right)^2 \approx 1330.$$

4.9.3 Analysis of variance for estimating different sources of variability

(You will need to read Sections 4.5.4 and 4.5.5 if you want to know how to set up the confidence intervals in this section.)

In the paired comparisons experiment of Example 8 we could analyse the data in a different way to look at the variability of the batches of crude sugar as well as the variability due the catalysts. The method used is the analysis of variance (ANOVA), which is described in detail in Chapter 8. Most statistics packages will do the calculations leading to the output of an ANOVA table such as that in Table 4.6.

Table 4.6 ANOVA table for the data in Table 4.5

Source of variation	Sum of squares	df	Mean squares	F	p-value	F_{crit}
Catalysts	0.8750	1	0.8750	18.74	0.005	5.99
Sugars	124.8286	6	20.8047	445.50	0.000	4.28
Residual[a]	0.2800	6	0.0467			
Total	125.9836	13				

[a] This term is sometimes denoted 'Error', which is unfortunate as it implies that mistakes are being made.

Much of this table can (and should) be ignored. But the column headed mean squares contains two important estimates of variability. The mean square (residual), 0.0467, is an estimate of σ_{resid}^2. σ_{resid} is the standard deviation of all sources of variability other than the crude sugars or the catalysts. It represents the variability that would result if we could repeat runs of a single sugar/catalyst combination over a short timescale. The estimate has a χ^2 distribution with $\nu=6$ degrees of freedom, so a 95% confidence interval for σ^2_{resid} (from Section 4.5.4) is

$$\frac{0.0467 \times 6}{\chi^2_{6,\,0.025}} \quad \text{to} \quad \frac{0.0467 \times 6}{\chi^2_{6,\,0.975}}$$

which is 0.0193 to 0.226. The corresponding 95% confidence interval for σ_{resid} is 0.14 to 0.48.

The mean square (sugars) is an estimate of the combined term,

$$\sigma^2_{resid} + 2\sigma^2_{sugar}$$

The coefficient 2 arises from the fact that each sugar is measured twice, albeit with different catalysts. σ_{sugar} is the standard deviation of real sugar–sugar differences. This mean square also has a χ^2 distribution, and in this example also has $\nu=6$ degrees of freedom. The estimate is completely independent of the estimate of σ^2_{resid} given by mean square (residual). Consequently the ratio

$$R^2 = \frac{\text{Mean square (sugars)}}{\text{Mean square (residual)}}$$

has an $F_{6,6}$ distribution and is an estimate of

$$\frac{\sigma^2_{resid} + 2 \times \sigma^2_{sugar}}{\sigma^2_{resid}} = 1 + 2 \times \frac{\sigma^2_{sugar}}{\sigma^2_{resid}}$$

Using the method in Section 4.5.5 we could calculate a 95% confidence interval for the ratio of the mean squares, and hence for $\sigma^2_{sugar}/\sigma^2_{resid}$. This is a case where we use a one-sided interval since $\sigma^2_{sugar}/\sigma^2_{resid}$ cannot be less than zero. But it is actually more likely that we would want a separate estimate of σ^2_{sugar} since it looks as though this is the most important cause of variability in the conversion process. We can get this estimate from

$$\frac{\text{Mean square (sugars)} - \text{Mean square (residual)}}{2} = \frac{20.847 - 0.0467}{2} = 10.40$$

Strictly speaking, we do not know the distribution for this, so we cannot get a 95% confidence interval. But there is an approximate formula given by Davies and Goldsmith (1976). If, as in this case, the mean square (residual) is very small compared with mean square (sugars), we can treat the estimate as if it were just a χ^2_6 distribution to give a 95% confidence interval for σ^2_{sugar} of 4.32 to 50.43 and so a 95% confidence interval for σ_{sugar} of 2.1 to 7.1.

4.10 Appendix

4.10.1 Calculation of approximately Normally distributed random numbers

The experiment simulation started in Section 4.4 requires random numbers drawn from a Normal distribution to simulate the effects of error in the experiments. A method is given here that is simple to implement if you do not have access to software that has a function to generate Normally distributed random numbers directly.

The random numbers generated in much computer software and in calculators come from a Uniform distribution, that is, each is equally likely to take any value between 0 and 1. We can create an approximately Normally distributed ε_{ij} by

- adding together 12 such numbers,
- subtracting 6 to make the mean of the ε_{ij} zero,
- multiplying the result by σ, the standard deviation we shall choose for this 'variable' or uncontrolled part of the system.

This algorithm for generating Normally distributed random numbers is now quite classical but still very useful. It makes use of the Central Limit Theorem, which can be stated in the following form:

> If the averages of groups of numbers are taken, then the distribution of the averages will tend towards a Normal distribution as the size of the group increases.

The approximation to the Normal distribution improves as the size of the group that is averaged is increased. A group of size 12 has been found to be a convenient compromise between a good approximation and the time it takes to compute the numbers.

References

Berger, R.L. and Hsu, J.C. (1996) Bioequivalence trials, intersection–union tests and equivalence confidence sets. *Statistical Science*, **11** 283-319.

Davies, O.L. and Goldsmith, P.L. (1976) *Statistical Methods in Research and Production*, Longman, London.

Hartmann, C., Smeyer-Verbeke, J., Penninckx, W., Heyden, Y.V., Vankeerberghen, P. and Massart, D.L. (1995) Reappraisal of hypothesis testing for method validation: detection of systematic error by comparing the means of two methods or of two laboratories. *Analytical Chemistry*, **67** 4491-4499.

5 Robust, resistant and nonparametric methods
J. Thompson

5.1 Introduction

The great strengths of conventional (parametric) statistics lie with the considerable base of theory that has been developed since the time when Laplace first published a treatise on probability in 1812. In that treatise, he rediscovered and further developed Bayes' theorem. However, that approach was rejected by mathematicians, who took over the development of statistics from the mid-nineteenth century, and who considered the approach as much too vague (Sivia, 1996). The Central Limit Theorem and other frequentist or classical theories (see, for example, Snedecor and Cochran, 1967) have been used to justify the overwhelming dominance of the Normal distribution in introductory statistics books. But even W.S. Gossett (known by the pseudonym 'Student', as inventor of the Student's *t*-test) recognised that the numerous chemical assays made under his supervision were never distributed in magical bell-shaped Normal or 'Gaussian' curves (see Student, 1927). The large amount of practical experience with real-life data has opened up a wider variety of approaches and encouraged scepticism among data analysts about the constraints of purely classical, Normal theory-based approaches, with which even Gauss was dissatisfied. In research and development, and in many other practical situations, we often have to deal with relatively small data sets, which we cannot definitely say originate from Normal distributions.

In the late 1930s, simpler methods of data analysis started to gain interest, marking a radical change from the conventional parametric approach. Conover (1980) describes the change:

> Thus approximate solutions to exact problems were found, as opposed to the exact solution to approximate problems furnished by parametric statistics. This new package of statistical procedures became known as 'nonparametric statistics'.

The simplicity of both calculation and theory of nonparametric methods makes them attractive to use by scientists and many methods are readily available in 'off the shelf', easily used statistics software. Sometimes, we need more measurements to achieve a similar strength of conclusion to that made by conventional parametric statistics, but with the great advantage of not making, or being limited by, assumptions about the data distribution. Because of this, they are also referred to as distribution-free methods.

Nonparametric methods protect us against the presence of outlier data that often wreak havoc when data sets are analysed using conventional parametric statistics. Methods that protect us against the untoward influence of outliers are referred to as *resistant*. Resistance comes from the insensitivity of such methods to localised misbehaviour, when compared with the behaviour of the bulk of the data. Although some resistant methods are nonparametric, many also come from modern robust and exploratory data analysis.

Robustness may be distinguished from resistance in that it implies 'an insensitivity to departures from an underlying probabilistic model' (Hoaglin *et al.*, 1983). A robust method generally works well for a family of closely related distributions and many such methods offer reasonable resistance to outliers as well.

5.2 Where to go

Information needed	Go to Section(s)
How can I check the distribution of my data?	5.4.1, 5.5.2
I want to see if there is a difference between two sampling methods but the data are pretty noisy.	5.4.2, 5.8, 5.9
Are there any quick graphical methods to check data distributions?	5.4.2, 5.9, 5.10, 5.11.2
How do I check for skewed data?	5.4, 5.5
I think there may be some outliers in the data, what do I do?	5.4, 5.5
I have done a paired experiment to check two sets of reaction conditions. The data do not look very Normal so can I still test for a difference?	5.8
What tools are there to check for effects of factors such as batch number, operator or reaction kit?	5.9, 5.11, 5.12
My spectrophotometer may be getting noisy, what test can I use to check for a change in noise level?	5.4.3
What confidence can I have in the results of nonparametric tests?	5.4.4, 5.4.5
There seem to be some 'funny' values in my calibration data, how can I reduce their effect on the calibration line?	5.13

5.3 Some simple and useful concepts

Several different types of data are encountered in scientific work. *Quantitative* data come in two broad classes:

> *Continuous* data may be obtained from measurements on an interval scale having an arbitrary zero (e.g. temperature in °C or °F) or on a ratio scale with a well-defined, fixed zero point (length, mass, etc.). It is meaningful to calculate ratios between measurements on a ratio scale (e.g. by saying there is 10% more mass in sample A than sample B) but it is not meaningful to say that the temperature of A is 10% more than that of B.
>
> *Discrete* data include measurements such as photon or nuclear particle counts.

Different kinds of *qualitative* data may also be collected:

> Categorical or *attribute* data consist of counts of items in different categories with no particular order to them (e.g. colours or flavours). Categorical data are important in quality control when one is concerned with inspection for defective versus nondefective items or when tracking noncompliance with standard operating procedures in good laboratory practice.
>
> *Ordinal* data consist of counts of items that can be put into specific rank-ordered categories (e.g. the scale of hardness for minerals or the strength of a flavour or of a smell).

Because of the constraints of space, this chapter will not deal with problems relating to discrete and qualitative data, but the reader interested in these areas will find useful starting points in the references at the end of the chapter.

The *location* of a set of data relates to its average position on a quantitative scale of measurement (continuous or discrete) or on a qualitative (ordinal) scale. The most well-known location parameter is the arithmetic mean, but the median is also often used in data summaries. These location parameters will be compared with others in this chapter in terms of their resistance, robustness and efficacy.

The *spread* of a data set is another important parameter used in a wide variety of ways. Commonly used measures of spread include the standard deviation and range, neither of which is resistant or robust. Spread measures are often used to construct confidence intervals around location parameters as well as in setting thresholds, such as action and warning limits, in determining whether a process is out of control or for compliance with environmental or occupational pollution control limits.

Ideas about distribution shapes are important to consider when looking at the capabilities of various robust and resistant and even nonparametric techniques. It is helpful to be able to describe shapes of distributions in a variety of qualitative and quantitative ways. Thus, we may wish to know about symmetry or lack of it, elongation of distribution tails, the number of modes (peaks) in a distribution, skewness, kurtosis (flat-toppedness or sharpness of shape), etc. Ways of analysing the shapes of data distributions using some important ranking methods developed within exploratory data analysis will be compared with traditional methods of assessing skewness and kurtosis and their applications will be illustrated. The chapter will also look at methods of comparing distributions and how these methods may be applied.

There are many different types of problem data that scientists involved in R&D may encounter and methods for highlighting and assessing them will be illustrated using robust/resistant and nonparametric methods.

5.4 Looking at continuous measurements variables

5.4.1 Some initial comments about the shapes of data distributions for continuous variables

In the majority of introductory texts on statistics or data analysis, there is still a focus on the so-called Normal or 'Gaussian' distribution. This is characterised by its symmetrical bell shape (see Figure 5.1a) and it is described mathematically by an equation derived by Gauss, the parameters of which are the mean and standard deviation. A skewed distribution, which can be 'transformed' to the bell shape of the Normal distribution by taking logarithms of the data, is the so-called log-normal

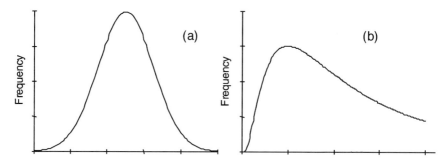

Figure 5.1 (a) A typical Gaussian or Normal distribution. (b) A typical log-normal distribution.

distribution (see Figure 5.1b). Regrettably, real data often do not conform to such ideal expectations and in describing a set of data we need to carefully explore its behaviour, to enable us to develop our strategy for analysis.

The exploration should include a variety of ways of plotting the data, including looking at the extremes as well as the behaviour of the bulk of the data, and various ways of summarising the shape of the data set both qualitatively and quantitatively. Ranking techniques now provide some simple but powerful tools to help us. We may also want to test the data for the presence of unusual measurements or outliers.

Data distributions may not necessarily conform to a particular shape, for a variety of reasons. The data may actually come from more than one distribution and these distributions may or may not overlap. If overlapping occurs, the presence of more than one distribution may not be obvious and the resultant distribution may not necessarily be the same basic type or shape as the underlying distributions.

These issues will be examined in more detail below in Section 5.5.2.

5.4.2 Estimating the location of a single set of data using arithmetic and geometric means

The data set in Table 5.1 is a set of 24 assays for cadmium in samples from a heterogeneous material. The arithmetic mean of these data is 60.08 and their geometric mean is 54.30 (estimated by calculating the arithmetic mean of the logarithms of the measurements and then taking the antilog of that mean; see Davies and Goldsmith, 1977, p. 34). *If* the data were found to be from a log-normal distribution (see below), the geometric mean would be the appropriate location estimate, but such an assumption may not be valid in this case. Let us see how the evidence unfolds as we explore.

This assumption of Normality or log-normality may depend on an understanding of the expected behaviour of the data (perhaps from previous experience), but with such a small set of data we cannot test whether it fits any specific distribution with any great confidence. It is

Table 5.1 Cadmium content (mg/kg) in a heterogeneous specimen of hazardous waste

34	45	45	50
39	42	75	39
160	88	87	57
44	43	46	59
46	43	94	51
140	40	43	32

worth noting that, when working with real data, the data distributions have often been found to have 'stretched-out tails' (Mosteller and Tukey, 1977, p. 23). Often such tails are not symmetrically stretched out.

Under such circumstances, the lack of resistance of the mean becomes all too obvious. Another example is shown in Table 5.2 for a sequential series of measurements of lead content of a specimen of an alkaline sludge from an industrial waste-water treatment process. There is just one unusual value in Table 5.2. With the unusual value retained in the analysis of the data, the mean is 191, without it the mean drops to 170. This unusual value is a genuine valid measurement, *not* a mistake. It stands out as very obviously different. Assays for other metals in this sludge specimen suggested that the sludge was very heterogeneous, despite valid and valiant attempts to mix it thoroughly prior to assay.

The data in Table 5.1 are more difficult to assess because there are several, possibly discrepant, data values, but that inference depends on assumptions about the underlying data distribution. Thus, the use of a location estimator that is more robust and resistant would be helpful in such cases.

Table 5.2 Sequential assay data for lead content of a lime sludge filter cake

210	190	180	800	160	190	180	180
150	180	160	150	160	150	170	200
170	180	120	190	120	140	140	190
200	190	210	160	150	160		

The mean belongs to the group of *L*-estimators, in which *L* stands for a linear function of the ordered values in the data set and has the general form

$$\mu^* = \sum_i w_i x_i$$

in which the x_i are the ordered data values and the w_i are weighting coefficients that are defined within the appropriate method. The *L*-estimators also include trimmed and Winsorised means (see below) and the median.

The median is simply the middle-ranked *x* value in an odd-numbered ordered data set or the average of the middle two values in an even-numbered ordered data set. It is much more resistant to the presence of outliers than the mean. Trimmed means are obtained by removing fixed percentages of values from the upper and lower ends of the ordered data set (percentages used include 5%, 10%, 20% and 25%). Staudte and Sheather (1990) provide examples of Minitab macros for the one-sample trimmed mean, an estimate of its asymptotic standard error and its associated 95% confidence interval. The 25% trimmed mean is the mean of the middle 50% of the ordered data set and is also known as the *midmean*. The median could be considered as a special case of the trimmed mean with trimming set at 50% (see Bell Krystinik and Morgenthaler, chapter 6 of Morgenthaler and Tukey, 1991). Winsorised means are obtained if k values in each tail of the ordered data set are downweighted (replaced) in value to those of x_{k+1} and x_{n-k}, respectively, before calculating the mean.

For the data in Table 5.1, the median is 45.5. The 25% trimmed mean (obtained by taking six measurements from each end of the data set and then calculating the average of the rest) is 47.67. The 25% Winsorised mean (adjusting the lowest six data values to the value of the seventh lowest ranked value and the highest six to the value of the seventh highest, and then calculating the mean of these 12 adjusted values, together with the 12 unadjusted values) is 49.33.

The median for the data in Table 5.2, where there is just one obvious outlier, is 175 and this is unaffected by changing the extreme value to any value above that of the median.

Thus, some of the *L*-estimators have some robustness and resistance and, for small data sets, the median and the trimmed means behave reasonably well. A detailed statistical evaluation of their performance has been done, resulting in a suggested simple rule for their use in small data sets (Rosenberger and Gasko, chapter 10 of Hoaglin et al., 1983):

For $n \leq 6$ use the median.
For $n = 7$ trim two observations from each tail.
For $n \geq 8$ trim 25% from each tail.

The class of *M*-estimators, which offer the greatest advantages in performance, flexibility and convenience, includes several robust location estimators (see Goodall, chapter 11 of Hoaglin et al., 1983), as well as the mean and median (by virtue of the general theoretical definition of this class). A considerable amount of research has been done on such estimators for both location and spread. *M*-estimators are iterative and include families derived by Tukey (the biweight) and Huber of the general form

$$x^*_{new} = \frac{\sum w_i(x^*_{old}) x_i}{\sum w_i(x^*_{old})}$$

where

$$w_i(x^*_{old}) = \frac{\psi(u_i(x^*_{old}))}{u_i(x^*_{old})}$$

and

$$u_i(x^*_{old}) = \frac{x_i - x^*_{old}}{c \cdot \text{MAD}}$$

MAD is the median of the absolute deviations from the median of the data set. It is a robust measure of spread (see below). The constant c is a so-called tuning constant.

For the Huber estimator of location,

$$\psi(u) = -1, \text{ if } u < -1, \quad \text{or } u \text{ if } |u| \leq 1 \quad \text{or } 1 \text{ if } u > 1$$

whereas, for the Tukey biweight location estimator,

$$\psi(u) = u(1 - u^2)^2, \text{ if } |u| \leq 1 \quad \text{or } 0 \text{ if } |u| > 1$$

If the iteration starts with a location estimator $x^*_{old} = \text{median}(x_i)$, then it can be used as a one-step estimator of x_{new} with reasonable success and no further iteration. For the one-step Huber estimator, a tuning constant of $c = 1.35$ works as well as the one-step Tukey biweight with $c = 7$ for samples of five observations from a variety of distributions. However, recent evidence suggests that the Tukey biweight, with a tuning constant $c = 6.5$, works better than the one-step Huber or trimmed means for samples of 10 or more observations (Bell Krystinik and Morgenthaler, chapter 6 in Morgenthaler and Tukey, 1991).

The one-step Huber estimate is 47.07 for the data in Table 5.1, higher than the median (45.5), whereas the one-step Tukey biweight estimate is 44.66, lower than the median but closer to it than the Huber. It is worth noting the differences in weighting that occur for the distribution of data values in Table 5.1, in particular the five highest values are downweighted to zero in the calculation of the one-step Tukey biweight but to nonzero values in the calculation of the one-step Huber. It is characteristic of the Huber estimator that the weighting values never become zero, however extreme the data values become, in marked contrast to the weighting for the Tukey biweight. Thus, in many respects, the latter is more suitable as a robust location estimator for data sets that are encountered experimentally or for environmental observations. Readers interested in Huber

5.4.3 Estimating the spread of a single set of continuous data

The standard deviation is the conventional spread or scale estimator but it is inherently lacking in resistance and robustness, performing even worse than the mean in this respect. This may readily be demonstrated using the data in Table 5.2, where we have one outlier. The standard deviation of this data set is 117.45 but, if we adjust the extreme value (800) to correspond to the next highest value of 210, the standard deviation drops dramatically to 24.88. We really need a spread measure that tells us something meaningful about the spread of the bulk of the data without being dramatically affected by extreme outlier data.

Several resistant and robust spread measures are worthy of assessment. Iglewicz (chapter 12 of Hoaglin *et al.*, 1983) reviewed various scale or spread measures, including the mean absolute deviation, the median absolute deviation and the fourth spread. The first two of these are defined as follows:

$$\text{Mean absolute deviation from the sample median} = \text{AD} = \frac{\sum_i |x_i - M|}{n}$$

$$\text{Median absolute deviation from the sample median} = \text{MAD} = \text{median}\{|x_i - M|\}$$

where M is the sample median of the data set x_i.

As with the previous discussion of resistance and robustness of location estimates, the reader may already begin to judge that the mean absolute deviation (AD) is less robust and resistant than the MAD. For the data in Table 5.1, the overall standard deviation is 32.648 and the AD is 23.625 which is already less influenced by the extremes than the standard deviation. However, the MAD = 6 and is therefore much less influenced by the extreme values in the data set. Thus, MAD is a useful simple, robust and resistant spread estimator that is calculated easily either by hand for small data sets or in a spreadsheet or in statistical software with spreadsheet capabilities.

Another series of spread measures, based on the ranked data, was described by Mosteller and Tukey (1977), who developed the concept of letter values (see also Hoaglin, chapter 2 in Hoaglin *et al.*, 1983). The median, denoted by the letter M, is the first of these letter values. Its depth into the ordered (ranked) data set is given generally by

$$\text{Depth of median} = \frac{1}{2}(1 + \text{batch size})$$

If we fold the ordered data from both ends so that they meet at the median (identify the two data values halfway between the ends of the ordered data and the median), the ranks at which the folds occur were termed *hinges* by Mosteller and Tukey (1977). These letter values were designated as the upper and lower hinges, or H values. The data values at the hinges are similar to the upper and lower quartiles, hence the other term given to them, *fourths*, with the alternative letter tag F (see Hoaglin, chapter 2 in Hoaglin *et al.*, 1983). Likewise, we can fold the ordered data over from the ends in to the fourths and so on. In doing this progressive folding, we move gradually into the tails of the ordered data set. The proportion of the data between each fourth and the nearest extreme is a quarter (25%). If we go to the next fold out, the proportion from there to the nearest extreme is now one-eighth ($12\frac{1}{2}$%), hence the letter tag E. The next fold gives us 1/16 (6.25%) in each of the tails and has the letter tag D, and so on. This progressive division of the ordered data set is letter value analysis and results in a letter value table. The letter value table for the data in Table 5.1 is shown in Table 5.3a.

The positions (depths inward from each end) in the ordered data set are calculated as follows:

$$\text{Depth of fourth} = ([\text{depth of median}] + 1)/2$$

$$\text{Depth of eighth} = ([\text{depth of fourth}] + 1)/2$$

or, more generally,

$$\text{Depth of Letter Value} = ([\text{depth of previous value}] + 1)/2$$

dropping any fraction from the depth of the median, fourth, etc. before adding 1.

Alternatives to a full letter value table are the 5-number summary (median, upper and lower fourths, maximum and minimum) and the 7-number summary (median, upper and lower fourths, upper and lower eighths, maximum and minimum). These are easily derived from the letter value table and in Table 5.3 they correspond to the values for M and H (or F) and for M, H (or F), and E respectively, as shown in Tables 5.3b and 5.3c, respectively.

Table 5.3
(a) Letter value display for data in Table 5.1

Letter	Depth	Lower letter value	Upper letter value	Mid-summary= (upper+lower)/2	Letter value spread= upper−lower
	$N = 24$				
M	$D_1 = (1+N)/2 = 12.5$		45.5	−	−
H or F	$D_2 = (1+D_1)/2 = 6.5$	42.5	67	54.75	24.5
E	$D_3 = (1+D_2)/2 = 3.5$	39	91	65	52
D	$D_4 = (1+D_3)/2 = 2.0$	34	140	87	106
Extremes	$D_5 = (1+D_4)/2 = 1$	32	160	96	128

(b) Five-number summary for data in Table 5.1

	Depth	Lower letter value	Upper letter value
	$N = 24$		
M	$D_1 = (1+N)/2 = 12.5$	45.5	
F	$D_2 = (1+D_1)/2 = 6.5$	42.5	67
Extremes	1	32	160

(c) Seven-number summary for data in Table 5.1

	Depth	Lower letter value	Upper letter value
	$N = 24$		
M	$D_1 = (1+N)/2 = 12.5$	45.5	
F	$D_2 = (1+D_1)/2 = 6.5$	42.5	67
E	$D_3 = (1+D_2)/2 = 3.5$	39	91
Extremes	1	32	160

As the letter value analysis proceeds, the next step is the calculation of the spreads between the upper and lower letter values giving us the fourth spread, etc. Thus, the fourth spread (often symbolised as d_F) tells us how spread out is the middle 50% of the data. The fourth spread for a Normal distribution is 1.349 × (standard deviation) with the upper and lower fourths at a distance of 0.6745 × (standard deviation) on either side of the mean. The fourth spread is a resistant measure of spread, being unaffected by the extremes. For the data in Table 5.1 it is 24.5.

If we divide the fourth spread by 1.349, we obtain the F-pseudosigma (=18.16 for Table 5.1 data) (Hoaglin, chapter 2 in Hoaglin et al., 1983). For a Normal data set, the F-pseudosigma should be approximately the same as the sample standard deviation but here it is clearly much less, again demonstrating the influence of the data in the tails on the standard deviation. We can calculate pseudosigmas for each of the letter value spreads by dividing the letter spread by the corresponding spread for a Normal distribution with a mean of zero and an standard deviation of 1.

If data is from a Gaussian distribution the sequence of pseudosigmas will be nearly constant in value but will increase systematically for elongated (heavy-tailed) distributions, and decrease systematically for distributions that have less elongated tails than the Gaussian.

We can also use the upper and lower fourths and the median to calculate another resistant measure of location, involving only simple arithmetic:

$$\text{Trimean} = \tfrac{1}{4}\,(\text{lower fourth}) + \tfrac{1}{2}(\text{median}) + \tfrac{1}{4}(\text{upper fourth})$$

For the data in Table 5.1, the trimean is 47.063.

It should be apparent by now that the concept of spread is rather arbitrary and somewhat more vague than the concept of location (Mosteller and Tukey, 1977, section 1F). Iglewicz (chapter 12 of Hoaglin et al., 1983) gives a general definition of a scale estimator as a nonnegative-valued function of the sample, $\mathbf{w}(x_1, x_2, \ldots, x_n)$, such that

$$\mathbf{w}(a + bx_1, \ldots, a + bx_n) = |b|\,\mathbf{w}(x_1, \ldots, x_n)$$

If we require a resistant scale or spread estimator, then the function $\mathbf{w}(\ )$ should not be greatly affected by a small number of outlying values, however else we might define our estimator. Iglewicz discusses the spread estimator related to Tukey's biweight location estimator, which gives zero weight to observations more than 4 standard deviations away from the median:

$$s_{\text{bi}} = \frac{n^{1/2}\bigl[\sum_{|u(i)|<1}(x_i - M)^2 \cdot (1 - u_{(i)}^2)^4\bigr]^{1/2}}{\bigl|\sum_{|u(i)|<1}(1 - u_{(i)}^2) \cdot (1 - 5u_{(i)}^2)\bigr|}$$

where

$$u_{(i)} = \frac{x_i - M}{c \cdot \text{MAD}}$$

and

$$c = 6$$

According to the studies of Lax (1975) reported by Iglewicz, this is a very good robust estimator of spread. Mosteller and Tukey (1977, pp. 208-9) recommend its use for $n \geq 8$. For the data in Table 5.1, the biweight spread estimate is 8.69 compared with MAD = 6.

Rousseeuw and Croux (1993) have developed two new robust spread estimators, (software for which is available direct from them), as alternatives to MAD. They claim that these are suitable for both symmetric and asymmetric distributions, because, unlike MAD, the estimators are not based on the absolute deviations from a central value. Interested readers may wish to try these estimators with their own data.

5.4.4 Estimating confidence intervals for location estimators for a single set of data

We may use the various spread estimators already discussed to determine confidence intervals. An example using a conventional Normal approach for a confidence interval around the mean involves a formula with the mean, standard deviation, the number of observations in the data set and the appropriate Student's t-value. Thus, for a two-sided (upper and lower) 95% confidence interval, we have (see Chapter 4),

$$\text{Confidence interval} = \text{Mean} \pm \frac{t_{(df,\alpha)} \cdot s}{\sqrt{n}}$$

where df is the number of degrees of freedom ($n-1$, not to be confused with the fourth spread d_F) and α is the significance level ($100(1-\alpha)=$ confidence level) and s is the sample standard deviation. This implies that if we repeated our set of observations, say 100 times, it is likely that 95 times out of 100 we will get the mean value within the range of the calculated confidence interval. It is assumed that the data are sampled from an underlying Normal distribution. For the data in Table 5.1, this gives 60.08 ± 13.79.

If it is unsafe to make the assumption of Normality, then confidence interval estimates can be derived for other location estimators using more resistant/robust spread estimators (see Iglewicz, chapter 12 of Hoaglin et al., 1983). The first of these is based upon the median and fourth spread and was suggested by Mosteller and Tukey, (1977, p. 209) as follows:

$$M \pm \frac{t_{n-1} d_F}{1.075 \sqrt{n}}$$

where M is the median, d_F is the fourth spread, n is the number of observations in the data set and t_{n-1} is the appropriate t value for the required level of confidence (e.g. 95%) with $(n-1)$ degrees of freedom. For the data in Table 5.1, the fourth spread is 24.5 (Table 5.3) and the appropriate t-value is 2.07, giving a 95% confidence interval of 45.5 ± 9.63.

A more robust interval, based on Tukey's biweight (T_{bi}) and its corresponding spread estimator (s_{bi}) is

$$T_{bi} \pm \frac{t_{0.7(n-1)} s_{bi}}{\sqrt{n}}$$

For the data in Table 5.1, this gives us a narrower interval of 44.69 ± 3.76.

We can also use nonparametric analogues of the one-sample t-test to estimate confidence intervals around the median. In statistical software, such as Minitab, the sign test and Wilcoxon's signed rank test (see Section 5.8) can be used in this way, giving us

$$\text{Sign test confidence interval} = 43.00 \text{ to } 57.35$$

$$\text{Wilcoxon signed rank confidence interval} = 44.0 \text{ to } 67.0$$

These confidence intervals contrast very sharply with the earlier ones because of their lack of symmetry about the median and it may be noted that they are also wider (and, hence, more conservative) than the interval derived from the fourth spread. The asymmetric confidence intervals derived from the sign and signed rank tests reflect the asymmetry seen in the boxplot and the stem and leaf plot of this data (see also Chapter 4).

5.4.5 Estimating confidence intervals around spread estimates for a single set of data

It should be made plain that, when we make an estimate of the spread of a set of data, it is *just* an estimate. So it is not unreasonable to try to put a confidence interval around this estimate, as was done with estimates for location (see also Chapter 4).

The conventional approach for calculating a confidence interval around the standard deviation is based on the use of the χ^2 statistic (see Snedecor and Cochran, 1967):

$$\frac{\sum (x_i - \mu)^2}{\chi^2_{0.025}} \leq \sigma^2 \leq \frac{\sum (x_i - \mu)^2}{\chi^2_{0.975}}$$

For the data in Table 5.1, this gives a 95% confidence interval for the standard deviation of 25.4 to 45.8 (around a standard deviation of 32.65). Note that this interval is not symmetrical. As usual, this approach assumes that the data is drawn from a Normal distribution.

A very different approach to estimation of location and spread and of confidence intervals around these estimates is the jackknife method.

Mosteller and Tukey (1997, pp. 133-45) illustrate its use in calculating both spread and confidence intervals for spread estimates such as standard deviation. The method will be shown here for the data of Table 5.1.

The method involves calculating the standard deviation for all the possible sets of data from Table 5.1 in which one data value has been removed. There are 24 observations in the full data set, so we calculate the overall standard deviation and the standard deviation values for each of 24 sets of 23 observations. We then calculate a set of 24 pseudo-values of the form

$$s_{*j} = 24 s_{\text{all}} - 23 s_{(j)}$$

where

$$s_{\text{all}} = \sqrt{\frac{\sum (x_i - \mu_{\text{all}})^2}{n-1}}$$

$$s_{(j)} = \sqrt{\frac{\sum (x_{i \neq j} - \mu_j)^2}{n-2}}$$

The jackknifed standard deviation is then the mean of the set of 24 s_{*j}, namely 33.76. The standard deviation of the 24 s_{*j} is $s_* = 42.07$, and this enables us to calculate a 95% confidence interval around the jackknifed standard deviation based on $t_{n-1} s_*$ giving us a rather disturbingly large upper confidence limit of 87.04. The lower limit in this case has no real meaning. Again this assumes that the data are from a Normal distribution.

A rather similar technique to the jackknife is the bootstrap, which is a data-based simulation method for statistical inference estimates (Simon, 1997; Efron and Tibshirani, 1993). Bootstrapping is a data-driven technique, involving taking random samples *with replacement* from a set of data. It makes no assumptions about the underlying distribution of which the data is a sample and in that sense is nonparametric. It is a very powerful and attractive method of data-driven analysis (see Sprent, 1998).

The use of the bootstrap may be illustrated by performing bootstrap estimates of some of the parameters for the data in Table 5.1 that have already been discussed, namely, the mean, median, standard deviation, AD and MAD. We may consider the data values in Table 5.1 as the equivalent of 24 positions on a roulette wheel at which a ball might settle after the wheel has stopped spinning. Each time we spin the wheel and toss the ball onto the wheel, allowing it to come to rest on the wheel, we are doing sampling with replacement. The series may be repeated any number of times as a random sampling exercise.

In the example given here, the bootstrap calculations were performed for 1000 series of sets of 24 randomly sampled observations with replacement from data in Table 5.1. These were performed using a very simple series of instructions in a low-cost, easy-to-use software package, *Resampling Stats* designed by Julian L. Simon (from Resampling Stats Inc.). Similar calculations could be done using macros written in Minitab (Staudte and Sheather, 1990, provide some useful examples of Minitab macros for both parametric and nonparametric bootstrapping that may be adapted for a variety of estimators of both location and spread) or other statistical software packages (some of which incorporate bootstrapping commands directly).

The results of these calculations are

Parameter	Estimate	95% bootstrap interval
Mean	60.08	48.96 to 74.27
Standard deviation	32.65	15.1 to 44.4
AD	23.63	10.1 to 36.7
Median	45.5	43 to 58
MAD	6.0	2 to 18

Notice that none of these confidence limits is symmetrical; their lack of symmetry results from the random sampling with replacement over a comparatively large series from a data set that is itself not symmetrical. We could have done a much larger series in a modest amount of computing time, as the calculations for the mean, standard deviation and AD only took 4.9 seconds and those for the median and MAD only took 4.3 seconds.

The rather bewildering array of spread estimates that we have seen is a consequence of the vagueness of the concept of spread and this has provoked a search for suitable estimators for a variety of contexts. Spread for the Normal distribution is defined from one of its parameters, the standard deviation, and, in a sense, the definition is rather tautological. As soon as we step outside of that ideal, and perhaps even utopian, world into the world of real and generally 'dirty' data, spread becomes more difficult to define and relate to the wide variety of possible theoretical distribution shapes or, much worse still, to real data distributions.

In many senses, the problems of location, spread and their respective confidence intervals relate to a desire to model the populations from which the data are sampled. The problem is worst for small-sized data sets, and only slightly better for medium-sized data sets because with such sets the information that we can reliably extract is limited. This is so whether reliance is placed on conventional Normal-based statistics or on the approaches described in this chapter.

5.5 Outlier tests for single sets of data

If we define an outlier as an extreme data value that does not seem to belong to the pattern exhibited by the rest of the data, perhaps we have a suitably vague concept again. R.A. Fisher (1922) cautioned that 'A point is never to be excluded on statistical grounds alone'. He was suggesting that it is appropriate to identify outliers but it is unwise to reject them without very good reasons. Outliers should only be rejected if they arise from genuine mistakes, such as transcription errors. Serendipitously, sometimes they have been the spur for major scientific discoveries. Otherwise, they may be useful markers of the unexpected, as in quality control for example. Hawkins (1980) has usefully and comprehensively reviewed many of the ideas relating to outlier identification up to the late 1970s, covering parametric, nonparametric and Bayesian approaches.

There is a variety of problems with classical parametric approaches to identifying outliers, many of which relate to the assumptions underlying such tests, and the interested reader is referred to Hawkins (1980) for a thorough discussion of these.

5.5.1 Exploratory data analysis methods for outlier detection based on the fourth spread

In developing the concepts of letter value analysis, Tukey (1977) invented some interesting distribution-free approaches to identifying unusual data. He started with upper and lower fourths and extended out from them by $1.5d_F$ in both directions. Observations beyond those boundaries Tukey termed *outside values*. For Normal data, only 0.7% of values are likely to be beyond each of these boundaries. For some distributions other than Normal, this would still not be appropriate for identifying outliers, as Hoaglin has shown (chapter 2 of Hoaglin *et al.*, 1983). Indeed, for heavy-tailed distributions, there will still be considerable proportions of the data beyond these boundaries. At $3d_F$ beyond the upper and lower fourths, Tukey labelled observations as *far-out*—beyond these boundaries they might well be considered outliers in a more general sense.

5.5.2 Exploring the shape of a single set of data

Graphical methods for shape investigation include histograms, stem and leaf plots, boxplots, suspended rootograms and probability plots (for defined distributions). The histogram is only a moderately useful tool for visualising distribution shape. Tukey (1977) proposed a variant of the histogram, termed the stem and leaf display, that shows more detail in a very useful way relating to the ordered data set. The data from Table 5.1 is shown in this way in Table 5.4.

Table 5.4 Stem and leaf table of the data in Table 5.1. $N = 24$, leaf unit $= 1.0$

Cumulative no. of observations counting from each end of ordered data	Stem	Leaf
	Lo	
2	3	24
4	3	99
10	4	023334
(4)	4	5566
10	5	01
8	5	79
6	6	
6	6	
6	7	7
5	8	
5	8	78
3	9	4
	Hi	140, 160

After sorting the data into increasing value order, the stem values and leaf unit size are determined. For the data in Table 5.1, most of the values are in the range 30–99 so it is appropriate to choose a leaf unit of 1.0. This means that values that differ by at least a value of 1.0 will be counted as distinct leaves. For these data, it is also appropriate to group leaves within a range of five units to the same stem. Thus, all values in the range 30–34 will be counted as leaves to a stem labelled 3 $(= 30/10)$. Data values in the range 35–39 will be counted as leaves to another part of the stem labelled as 3. In this way, all of the data values are attached to a stem label. If there are no leaves to attach to a stem, the stem value is still listed. The choice of stem range and leaf unit size is a compromise between having a simple and easily constructed table and having sufficient detail to show the major features in the data.

The columns called Stem and Leaf in Table 5.4 contain the stem labels and leaf values, respectively. The labels for the leaves are simply the unit value obtained by subtracting the corresponding stem values (multiplied by 10) from the data values. Thus, data value 32 has a stem label of 3 and a leaf label of 2. All leaves attached to the same stem are simply written in a line without spaces. Thus, the leaf 99 (second row in the table) indicates two values attached to stem 3 (i.e. there are two values of 39 in the data set).

The first column in the table is the cumulative number of values attached to stems. The counting is done both from the top of the table and from the bottom. The counts stop where they are the same. In this table, the counts stop at the value 10 (there are 10 leaves counting from the top of the table and 10 leaves counting from the bottom). The remaining four values in the centre are additional to both counts and are identified by the bracketed value. The median of the data set falls in the leaves of this stem.

Finally, this data set has two values (140 and 160) that are quite well separated numerically from the other values. Although the table could be extended to accommodate them, it would mean a long list of stem values with no leaves. Instead, the convention is to have two stems labelled Hi and Lo and to attach extreme values to these stems as appropriate.

Thus, the stem and leaf plot can be regarded as an alternative to a histogram. The distribution of values is shown by the numbers of leaves attached to stems, the location of the median is indicated where the cumulative sums become equal and possible outliers are shown by leaves attached to the Hi and Lo stems.

Tukey (1977) also proposed the boxplot (sometimes called the box and whisker plot) shown in Figure 5.2, in which the upper and lower edges of the box correspond to the fourths and a line, dot or asterisk between these indicates the position of the median. The whiskers have a maximum length (which Tukey termed a fence) corresponding to $1.5d_F$ beyond the upper and lower fourths, but are often drawn to positions within those boundaries corresponding to the most extreme data values inside these upper and lower boundaries. Any data values beyond those whiskers correspond to Tukey's outside values and are, thus, drawn to our attention by being plotted as individual data points.

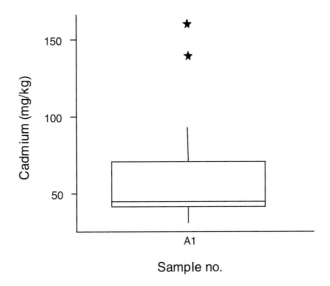

Figure 5.2 Boxplot of data from Table 5.1, assays for cadmium content in a heterogeneous specimen of hazardous waste.

Frigge, Hoaglin and Iglewicz (1989) have highlighted the problem that a number of statistical software packages produce boxplots according to different definitions of quartiles and fences. They offer recommendations for a single standard form based on the sixth of eight definitions, which describes standard fourths or hinges. For the lower quartile, defined in this way, we have,

$$Q_1 = (1-g)x_j + gx_{j+1}$$

where $[(n+3)/2] = j + g, j + g$ is an integer with $g = 0$ or $g = 1/2$ (chosen to make $j + g$ an integer), n is the number of observations.

The suspended rootogram is another kind of histogram with a Normal distribution fitted to it. As described in the Minitab manual:

> The fitting of the Normal distribution is based upon square roots of the counts in each bin [of the histogram] to stabilise variance.

The rootogram shows confidence intervals for the fit and so indicates how significant any departures from the Normal fit are (Velleman and Hoaglin, 1981).

Probability plots represent another way of graphically assessing closeness of the fit to a particular distribution. Four such plots for the data in Table 5.1 are shown in Figure 5.3, demonstrating a poor fit for each of them.

Traditionally, we try to quantify deviations from the Normal shape using skewness and kurtosis, which are 2.02 and 3.78, respectively, for the data from Table 5.1. The boxplot shows these distortions from Normal in a more easily digestible way but does not give any quantitation of this distortion.

An alternative approach to describing features of the shapes of data distributions in general involves some powerful techniques based on letter value analysis described in detail by Hoaglin (chapters 10 and 11 of Hoaglin et al., 1985). Some aspects of Hoaglin's approach are illustrated below with analyses of two sets of real data. We can take further steps in the letter value analysis of the data from Table 5.1, not only by looking at the spreads between successive pairs of upper and lower letter values but also at the mean of each pair, termed *mids* or *midsummaries*. If the data distribution is symmetrical, the *mids* will be all approximately the same, but, if the distribution is skewed to the right (e.g. as with a log-normal distribution) the *mids* will increase as we go down the letter value table. If the distribution is left-skewed the *mids* will decrease down the table. Table 5.3a shows the letter value analysis for data from Table 5.1, including

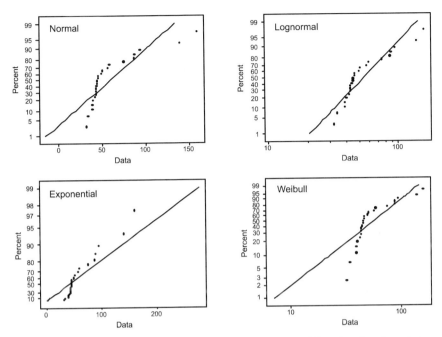

Figure 5.3 Normal, log-normal, exponential and Weibull probability plots for the cadmium data of Table 5.1.

mids and spreads, and the increasing *mids* hint at a right skewed set of data.

We really need a larger data set to look more carefully at distribution shape. Table 5.5 shows a set of 98 serum cholesterol measurements, obtained during a study of matrix and interference effects on the response of ion selective sensors. If the upper letter values are plotted against the corresponding lower letter values, they should lie more or less along a straight line with a negative slope (-1) if there is no skewness. Figure 5.4a shows the plot for the cholesterol data, demonstrating the asymmetry in the data set. An alternative plot that is diagnostic of skewness is the *mid-versus-spread* plot, which plots the *midsummaries* versus the letter value spread (including the median with zero spread). If the plot points downwards, this is diagnostic of left-skewness; an upward-pointing plot indicates right-skewness, as with the plot for the cholesterol data (shown in Figure 5.3b). If we have outliers in the data, they can string the plot over a wider range; so a plot of *mid*-versus-z^2 (where z is the standard Normal quantile) is better (Figure 5.3c).

Elongation, as distinct from skewness, can be diagnosed by plotting the pseudosigmas versus z^2. Where the elongation is symmetrical, a

Table 5.5 Serum cholesterol measurements

1.41	3.14	3.68	1.98	2.78	3.80	2.42	4.25	2.90
2.42	3.05	4.39	5.01	3.70	3.42	4.00	3.62	4.02
4.78	3.79	3.31	5.60	5.31	4.89	3.21	5.82	5.26
3.02	3.17	3.37	3.52	3.37	1.85	3.06	2.49	2.62
3.77	3.05	3.31	3.32	2.40	2.63	2.90	3.85	4.02
3.87	0.88	2.79	6.10	4.25	3.23	1.96	10.53	3.10
6.06	2.32	1.14	5.56	10.10	3.58	1.14	11.45	2.29
1.29	2.09	3.70	1.48	2.10	1.59	0.93	1.11	2.38
2.78	2.27	4.15	3.11	2.64	1.29	2.30	2.04	2.97
2.45	3.46	2.85	2.97	2.18	3.00	2.45	2.61	2.25
1.00	4.10	3.08	2.41	1.51	1.30	3.55	3.50	

horizontal line will be the result, clearly not the case for the cholesterol data (Figure 5.5a). If asymmetry is present, then we do better by examining the behaviour separately for the data each side of the median, using a pushback technique in which we calculate 'flattened' letter values.

First, it is necessary to estimate an appropriate scale (or spread), s, for the data set, which we can do by calculating the median of the pseudosigmas. If each pair of lower and upper letter values is represented by x_p and x_{1-p}, then the flattened letter values are $x_p - s \times z_p$ and $x_{1-p} - s \times z_{1-p}$ for each of the p upper and lower letter values. A flattened letter value plot versus the corresponding upper and lower z values for the cholesterol data (Figure 5.5b) indicates the asymmetrical elongation of this data set. The small number of very high cholesterol values produce the three elevated points on the right of this plot.

Further detailed analysis, along lines suggested by Hoaglin to fit data to the so-called g and h distributions, requires a larger number of observations than for the cholesterol data set used here.

5.6 Robust and resistant methods in evaluating data transformations and the value of transformation

There is sometimes a need to consider transforming raw data to aid us in making more effective displays or summaries or to help us make less complicated analysis. Some of the difficulties that are encountered with raw data include

- asymmetry
- many outliers in one tail
- parts of the data at different levels with different spreads
- large and systematic residuals when we try fitting a simple model to our raw data (Emerson and Stoto, chapter 4 in Hoaglin *et al.*, 1983).

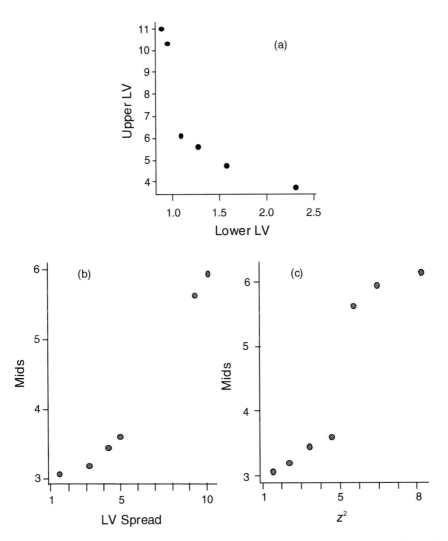

Figure 5.4 (a) Plot of upper letter values versus lower letter values for the serum cholesterol data in Table 5.4; (b) plot of midsummaries (from letter values) versus letter value spreads for the serum cholesterol data; (c) plot of midsummaries versus z^2 for the serum cholesterol data.

Simple but effective power transformations that assist us in dealing with such problems preserve the order of data in a batch, preserve letter values (except for minor differences resulting from interpolation) and are continuous, smooth, elementary functions. They have the general form

$$T_p(x) = \begin{matrix} c\,\log(x) + d, & p = 0 \\ ax^p + b & p \neq 0 \end{matrix}$$

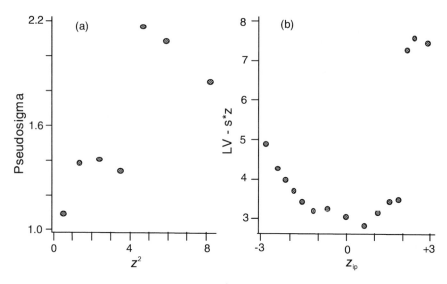

Figure 5.5 (a) Plot of pseudosigmas versus z^2 for the serum cholesterol data; (b) plot of flattened letter values versus upper and lower z values for the serum cholesterol data.

where a, b, c, d and p are real numbers and $a > 0$ for $p > 0$ and $a < 0$ for $p < 0$. The values of a, b, c and d are arbitrary, but the value of p matters and is chosen to help in the analysis of our data. For example, quadratic transformations may involve raising the observation values to the power of 2 or $\frac{1}{2}$ depending on the direction of the concavity of the data set.

Transformations may help us in various ways, in that they may facilitate more natural interpretation, promote symmetry, promote more stable spread among the various batches of our data, promote linear relationships or simplify structures of two-way or more-way tables enabling us to use simple additive models.

Many of the methods already discussed above in exploring shapes of raw data distributions are also valuable in assessing various kinds of data transformation. In addition to these, Emerson and Stoto (see above) and Emerson (chapter 8 of Hoaglin *et al.*, 1983) discuss in very helpful detail the various plots and strategies that can be employed in this quest.

5.7 Randomness in a data set

Demonstrating randomness in a series of consecutive measurements in a data set is useful in a variety of ways. For example, it may be useful in checking possible instrumental drift or shift in a process average as not being due to expected random sequences of values. Forensically, regulators auditing environmental measurements to test whether observers are cheating can use randomness as a marker of observer reliability. The idea of runs is familiar in the context of gambling or game playing, where we talk of a 'run of bad luck', and the term is used in the same sense here.

Although runs tests are not particularly powerful because of the general nature of the inferences and the weak assumptions, they are none the less uniquely useful, moderately versatile and have important practical applications, especially in quality control.

Nonparametric runs and runs-up-and-down tests are particularly helpful in this respect and Gibbons (1976) has a complete chapter on such tests with useful, detailed examples. Wetherill and Brown (1991) examine in detail the theory of runs in analysing control charts and Cusum charts, looking in particular at a Markov chain approach to assess the strengths and weaknesses of charts with action limits only versus those with warning limits included as well.

Robust/nonparametric control chart methods have been considered by Janacek and Meikle (1997).

5.8 Nonparametric methods for comparing locations of paired data sets

Although the paired t-test is a very commonly used test, it is based on an assumption of Normality in the data. It appears to cope only with relatively very modest departures of the data distribution from Normality and symmetry but has much higher Type I errors with skewed distributions.

A radically different approach, which appears to make huge sacrifices in the use of the data that we have taken great pains to collect, is to apply an old but well-tried and now well-understood test, the sign test. This is a very useful nonparametric paired sample test for inferences about the median of the differences between the paired measurements. It is simple to carry out and makes no assumptions about the underlying distribution. According to Conover (1980), it is the oldest nonparametric test, dating back to 1710. As with other paired sample tests, the data are reduced to a single set by considering differences in the paired measurements. In this case, we concern ourselves only with the signs of the differences, +, − or 0 (for tied results, i.e., paired measurements having the same value). It can be applied either as a two-tailed or as a one-tailed test and very helpful detailed discussions on its applicability are given in Gibbons (1976), Conover (1980) and Sprent (1993, 1998).

The test not only applies to continuous data but it also works as well with score data.

The test statistic is either of the following:

S_+ = number of positive signs observed among the differences

S_- = number of negative signs observed among the differences

Note that $S_+ + S_- = n$, the total number of signs (if there are tied results, this will be less than the total number of pairs of observations by an amount corresponding to the number of tied pairs).

The table for this test is that for the (symmetric) binomial distribution with $\theta = 0.5$, so the left-tail cumulative probability for S_+ is equivalent to the right-tail cumulative probability for S_-. We can use either of these. When using S_+ and testing for the one-sided Alternative Hypothesis that $p_+ > p_-$, use the right-tail probability for S_+. When testing for the other one-sided Alternative Hypothesis that $p_+ < p_-$, use the left-tail probability for S_+, and when testing for a two-sided hypothesis, use twice the smaller tail probability for S_+ (see Gibbons, 1976).

A variation of the sign test can be applied to testing hypotheses about any quantile, not just about the median (Conover, 1980; Sprent, 1993, 1998). A variation on the sign test was devised by Cox and Stuart to check for trends and is discussed by Sprent (1998). He suggests caution in its use, particularly with respect to the care needed when assumptions about independence of measurements in a time series do not hold, as characterised by seasonal or other periodic effects and serial correlations.

The Wilcoxon signed ranks test involves the ranking of the nonzero differences from a paired comparison and attaching the sign of the difference corresponding to a given rank to that rank. The sums of the positive ranks and the negative ranks are then calculated and used as test statistics. The signed ranks test works well with symmetric distributions but is less powerful than the sign test when asymmetry is significant (see Hettmansperger and McKean, 1998; Sprent, 1998). Cautionary advice on the pitfalls of these tests (which also apply to the paired t-test), especially in taking care with the experimental design and also with observational data, is offered by Sprent (1998) and Maritz (1995).

These tests may be illustrated and compared using the first 40 pairs of measurements from a set of 98 collected in a comparison of flame photometric and ion-selective electrode methods for assay of sodium in blood plasma (see Table 5.6). The table shows the raw data for the two types of assay and differences for each pair, the signed ranks and the calculation when a substantial number of nonzero differences are present (here, there is also the problem that many of these differences are tied, i.e.

Table 5.6 Paired assays of serum for sodium using both flame photometry (Flame) and ion selective electrode (ISE) measurements

No.	Flame	ISE	Flame−ISE	Sign	Signed rank	No.	Flame	ISE	Flame−ISE	Sign	Signed rank
1	158	145	+13	+	+30.5	21	139	136	+3	+	+14.5
2	128	130	−2	−	−9.5	22	138	135	+3	+	+14.5
3	141	137	+4	+	+21	23	139	137	+2	+	+9.5
4	147	141	+6	+	+26	24	142	142	0		
5	137	137	0			25	145	142	+3	+	+14.5
6	133	132	+1	+	+4	26	142	140	+2	+	+9.5
7	146	146	0			27	135	135	0		
8	144	137	+7	+	+27.5	28	133	133	0		
9	152	138	+14	+	+32	29	129	128	+1	+	+4
10	142	146	−4	−	−21	30	132	133	−1	−	−4
11	142	138	+4	+	+21	31	141	141	0		
12	142	135	+7	+	+27.5	32	132	136	−4	−	−21
13	140	141	−1	−	−4	33	135	135	0		
14	143	130	+13	+	+30.5	34	138	138	0		
15	140	138	+2	+	+9.5	35	128	132	−4	−	−21
16	140	136	+4	+	+21	36	151	152	−1	−	−4
17	140	143	−3	−	−14.5	37	135	140	−5	−	−25
18	142	132	+10	+	+29	38	154	155	−1	−	−4
19	143	139	+4	+	+21	39	134	137	−3	−	−14.5
20	137	130	+3	+	+14.5	40	133	134	−1	−	−4

The total of positive ranks, T_+, is 381.5.

We can find an asymptotic approximate p-value for the 32 nonzero pairs using the following formula, if there are no ties, which links T_+ with its standardised normal value:

$$z_{+,R} = [T_+ - 0.5 - n(n+1)/4]/\{\sqrt{[n(n+1)(2n+1)/4]}\} = 2.188$$

However, here we have many ties, so we must adjust the denominator to correct for this by replacing it with

$$\sqrt{\{[n(n+1)(2n+1)/24] - [(\sum u^3 - \sum u)/48]\}}$$

where u is the number of absolute differences that are tied for a given nonzero rank and the sum is over all sets of tied ranks (Gibbons, 1976).

This gives us a z-value of 3.03 and a corresponding p-value of 0.0012, thus signifying a definite bias between the methods.

having the same value). The mean difference between the pairs is 2.000 (mmol/l) with a standard deviation of 4.701. The median difference is 0.500.

The conventional paired t-test gives a test t-value of 2.69, suggesting rejection of the Null Hypothesis of a mean difference of zero (no bias) with a p-value of 0.01.

The test value for the sign test may be calculated using either the 12 negative or the 20 positive signs of the differences, and choosing the critical value for the test appropriately. Here, the positive signs are used. There are 8 zero differences that are not used in the test.

Most tables of critical values for the sign test only include values of n for which $n = S_+ + S_- \leq 20$. For larger values of n, we calculate the right-tail standardised variable corresponding to n as

$$z_{+,R} = \frac{S_+ - 0.5(1+n)}{0.5\sqrt{n}} = \frac{20 - 0.5(1+32)}{0.5\sqrt{32}} = 1.24$$

The p-value corresponding to $z_{+,R} = 1.24$ is obtained from a table of the standardised Normal variate, and is 0.1075. The Null Hypothesis is that the median difference is zero, and as p is greater than 0.05 we cannot reject the Null Hypothesis at the 95% confidence level.

Here the number of nonzero differences is not sufficient to allow us to be confident that the median difference is significantly different from zero. This is an illustration of the lower power of the sign test. The Wilcoxon signed rank test gives a p-value of 0.0012 (adjusted for ties), suggesting rejection of the Null Hypothesis, as with the paired t-test.

If we examine the histogram of the differences between the pairs (Figure 5.6), the asymmetry of the plot is evident and it can be seen that the small number of extreme differences will distort the t-test behaviour, pulling the confidence interval away from zero (0.496 to 3.504). Note that the 95% sign and signed rank confidence intervals (0.000 to 3.000) both include zero. On that basis, we can assume that the estimated bias between the tests is not significant for this number of pairs of observations. We really need more pairs to be sure of the bias.

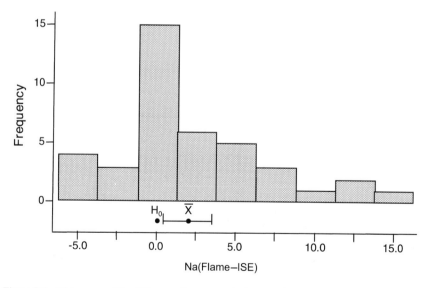

Figure 5.6 Histogram of the differences between paired measurements of serum sodium assayed by flame photometry (Flame) and ion selective electrode (ISE) methods.

5.9 Nonparametric methods for comparison of two sets of unpaired data

Pairing of measurements is an efficient approach to comparison but sometimes it is not possible. In this situation we need an unpaired comparison test.

The Wilcoxon–Mann–Whitney (WMW) test for differences in medians is such a test (it is also known as the Mann–Whitney, the Mann–Whitney–Wilcoxon, the Wilcoxon rank-sum or simply the Wilcoxon test). Table 5.7 shows two sets of unpaired measurements that we start to analyse by combining these sets into a single ordered set to which we then assign upward ranks as shown in the table. Note that the numbers of measurements in each set (m and n) need not be the same.

Table 5.7 Unpaired measurements of cadmium content (mg/kg) in a hazardous waste specimen by two independent analysts together with the Wilcoxon–Mann–Whitney test results

	Set a		Set b	
	Value	Rank	Value	Rank
	34	2	45	11.5
	39	3.5	75	19
	160	24	87	20
	44	10	46	13.5
	46	13.5	94	22
	140	23	43	8
	45	11.5	50	15
	42	6	39	3.5
	88	21	57	17
	43	8	59	18
	43	8	51	16
	40	5	32	1
Medians	43.5		50.5	
Sum of ranks		135.5		174.5

The 95.4% confidence interval for the difference in population medians is −16.00 to 10.99, p-value for the test = 0.418 (adjusted for ties in ranks), hence, the Null Hypothesis (no difference in medians) is accepted.

The Wilcoxon version of the test proceeds by obtaining the sums of the ranks (S_m and S_n) for the two sets of data that were combined. The test statistic may be either of these rank sums, which may be compared with table values published in some textbooks (Gibbons, 1976; Conover, 1980; Sprent, 1993). An alternative version known as the Mann and Whitney U-test, for samples of m and n observations, respectively, is given by

$$U_m = S_m - \frac{m(m+1)}{2} \quad \text{and} \quad U_n = S_n - \frac{n(n+1)}{2}$$

Either *U*-value may be used as the test statistic and tables for these are given in the textbooks already cited. Many statistical software packages offer the test, calculate the appropriate table value and compare it with the test statistic.

In recent years, with the increasing power of personal computers, the use of permutation tests based on ranks has considerably extended the capabilities of nonparametric statistical methods such as the WMW test (Sprent, 1998; Good, 1994). As with bootstrap methods, permutation tests use the available data to make estimates and appropriate software is now commercially available (StatXact and LogXact from Cytel Software Corp., 675 Massachusetts Avenue, Cambridge, MA 02139-9571, USA). It is also possible to write macros in statistical software, such as Minitab and S-plus. These methods have many useful properties and are expected to become major tools as appropriate software becomes more widely available. No further discussion of permutation methods is offered in this chapter and the interested reader should consult the texts referred to earlier in this paragraph for details. Staudte and Sheather (1990) provide Minitab macros for an unpaired trimmed means equivalent of the unpaired *t*-test, which also provides estimates of the asymptotic standard errors of the trimmed means and of their differences and a 95% confidence interval for the difference in the trimmed means.

The Kolmogorov–Smirnov two-sample test is a more general distribution shape comparison test that enables us to assess whether there is *any* significant difference between two sets of data. The approach is very simple. The cumulative frequency functions of the two data sets to be compared are calculated and may be plotted as cumulative histograms. The test statistic is the largest absolute difference between the two histograms. The test does not depend on the two distributions having the same shape, spread or location.

The squared ranks test for differences in spread (Conover, 1980) is very useful in situations in which we have little information about distribution shape. As with the Wilcoxon–Mann–Whitney test, Conover proposed squaring the ranks of the squared deviations $(x_i - \mu_x)^2$ and $(y_j - \mu_y)^2$ in the combined sample. Sprent (1998) discusses a permutation version of this test available in the StatXact software.

5.10 Nonparametric goodness-of-fit tests to specific distributions

The Kolmogorov–Smirnov test may be used to evaluate goodness of fit to specified distributions with the same location and spread as that of the data under test. This test looks at the significance of any major deviation between the actual data distribution and that with which it is being compared.

5.11 Nonparametric comparisons of the effects of one factor on more than two sets of data

Extending the approach of unpaired comparisons to more than two sets of data involves the use of analysis of variance (ANOVA). As with many other least squares methods, ANOVA tests are not robust, although they are widely used. This author has had the experience of working with many data sets where the least squares methods fail, even if transformation of the data has been attempted. The WMW method for unpaired comparisons was extended to provide a powerful and robust, distribution-free approach by Kruskal and Wallis (see Gibbons, 1976; Conover, 1980; Sprent, 1993, 1998; Hettmansperger and McKean, 1998).

5.11.1 Kruskal–Wallis one-way analysis of variance (ANOVA) by ranks, including multiple comparisons methods

As with the WMW method, the data for the combined groups is ranked and the rank sums for each group are used to compute the test statistic:

$$H = \frac{12}{N(N+1)} \sum_{j=1}^{k} \frac{[R_j - n_j(N+1)/2]^2}{n_j}$$

where k = number of samples (i.e. groups of data); R_j = sum of the ranks in the jth sample; n_j = the number of observations in the jth sample; N = the total number of observations.

When ties occur amongst the ranks, a correction is made to H:

$$H_{corr} = \frac{H}{1 - \frac{\sum u^3 - \sum u}{N(N^2-1)}}$$

where u is the number of observations in all samples combined tied for any rank. The test statistic, H, is compared with χ^2 with $k-1$ degrees of freedom.

The data, as laid out in four columns in Table 5.1, is actually from four sets of replicate assays and Kruskal–Wallis ANOVA gives a test statistic value of 2.08 ($p = 0.557$, with 3 degrees of freedom). The medians for the four columns are, respectively, 45, 43, 60.5 and 50.5.

If the test suggests a significant difference among the samples or groups, then they may be compared pairwise using various multiple comparisons tests. Dunn's multiple comparisons test is suitably conservative (described by Gibbons, 1976) and involves testing the absolute pairwise differences between the average ranks.

Gardiner (1997) describes the application of the Kruskal–Wallis ANOVA test to the analysis of two-factor factorial designs. This approach can be used with more complex designs (Hettmansperger and McKean, 1998).

5.11.2 Exploratory one-way ANOVA

If boxplots for several groups of data are plotted side by side (as in Figure 5.7), we may compare them in a form of graphical one-way ANOVA.

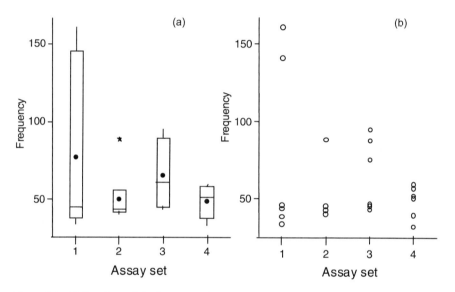

Figure 5.7 (a) Boxplots of the four sets of replicate cadmium assays in Table 5.1. (Note that dots within boxes are means, lines within boxes are medians and the asterisk outside the box is an outlier beyond the upper fourth plus 1.5 times the fourth spread. (b) Dotplots of the four sets of replicate cadmium assays in Table 5.1.

If the boxplots are notched, indicating confidence intervals around the corresponding medians, then we can examine the plot to see whether the notches overlap. The confidence interval around the median could be calculated using the sign or signed rank tests. If they do, then it is reasonably likely that the medians are not significantly different. If there is no overlap, then there is a possibility that the medians are significantly different. We have to be somewhat guarded about this conclusion, in the absence of a multiple comparisons test, but it is a useful exploratory indicator showing us where to focus our attention.

Sometimes, we have data that may require transformation, as when spread differs from group to group or lacks symmetry. Then spread-versus-level plots for evaluating data transformation strategies may be useful for one-way ANOVA (see Figure 5.8). For further details see Emerson and Stoto (chapter 4 and Emerson (chapter 8) in Hoaglin *et al.* (1983).

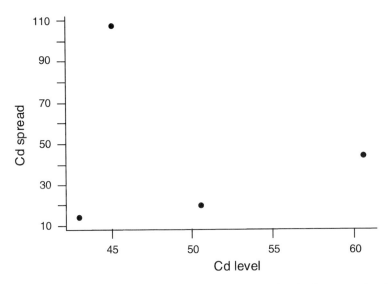

Figure 5.8 Spread (fourth spread) versus level (median) plot for the four sets of replicate cadmium assays in Table 5.1.

5.11.3 Cross-classified ANOVA designs—nonparametric and resistant/robust methods for two-way ANOVA and more complex designs

When analysing randomised complete block designs, an extension of the sign test may be used in which block differences are removed by replacing observations with their ranks, separately within each block (see Conover, 1980; Gibbons, 1976; Sprent, 1993, 1998).

This may be illustrated with a subset of data from a trial of five different commercial, ion-selective, near-patient analysers, with which plasma specimens were assayed for sodium, potassium and calcium ion concentrations. Thus, each block in the two-way analysis relates to sodium ion assays of a specimen of blood plasma from a patient, either in intensive care or undergoing liver transplant surgery. 'Treatments' for each block are the assays performed by each analyser. The data for this subset is in Table 5.8, which also shows the rankings used in the Friedman ANOVA.

178   DESIGN AND ANALYSIS IN CHEMICAL RESEARCH

Table 5.8 Two-way design for comparison of 5 ion-selective, near-patient analysers for assay of plasma sodium. The data values are given in the rows labelled with the specimen number. The ranks for the Friedman test are given in italics below the corresponding value

	Near-patient ion-selective analyser				
Specimen no.	Ektachem	AVL 984	Corning 654	Nova Stat	Radiometer KNA2
1	119	115.2	116	119.8	120
	3	*1*	*2*	*4*	*5*
2	134	140.1	131	134.3	134
	2.5	*5*	*1*	*4*	*2.5*
3	139	145.7	138	138.4	137
	4	*5*	*2*	*3*	*1*
4	142	149.9	142	141.9	145
	2.5	*5*	*2.5*	*1*	*4*
5	139	151.3	143	143.1	145
	1	*5*	*2*	*3*	*4*
6	135	146.1	139	139.2	140
	1	*5*	*2*	*3*	*4*
7	144	142.4	136	139.2	140
	5	*4*	*1*	*2*	*3*
8	146	139.9	137	140.5	141
	5	*2*	*1*	*3*	*4*
9	143	127.3	136	138.7	139
	5	*1*	*2*	*3*	*4*
10	130	133.9	132	133.9	136
	1	*3.5*	*2*	*3.5*	*5*
11	136	143.9	135	135.4	137
	3	*5*	*1*	*2*	*4*
12	133	127.8	128	132.7	133
	4.5	*1*	*2*	*3*	*4.5*

The usual form of the test statistic calculation for the Friedman analysis (see Sprent, 1993) is

$$T = \frac{b(t-1)(S_t^2 - C)}{S_r^2 - C}$$

where

$$S_r^2 = \sum_{ij} r_{ij}^2, \quad S_t^2 = \frac{\sum_i s_i^2}{b}, \quad C = \frac{bt(t+1)^2}{4}$$

where we have t treatments in b blocks and the ranks r_{ij} of individual responses x_{ij} to the i treatments. This is compared with the χ^2 table value for $t-1$ degrees of freedom with the appropriate p-value.

A variation on the above test statistic (see Conover, 1980; Sprent, 1993, 1998) tests significance using a comparison with table values of the F-distribution with degrees of freedom $(k-1)$ and $(b-1)(k-1)$.

Table 5.9 Results of Friedman two-way ANOVA and paired comparisons between analysers of data from Table 5.8. $S = 12.47$, $df = 4$, $p = 0.015$ (unadjusted), $p = 0.014$ (adjusted for ties in ranks)

Analyser	No. of assays	Estimated median	Sums of ranks R_j^{**}	Analyser pair	$\|R_j - R_i\|$	>12.69
A	12	143.26	42.5	A–C	|42.5–20.5| = 22.5	yes
C	12	136.47	20.5	A–E	|42.5–37.5| = 5.0	no
E	12	138.62	37.5	A–N	|42.5–34.5| = 8.0	no
N	12	138.67	34.5	A–R	|42.5–45.0| = 2.5	no
R	12	139.38	45.0	C–E	|20.5–37.5| = 17.0	yes
				C–N	|20.5–34.5| = 14.0	yes
				C–R	|20.5–45.0| = 24.5	yes
				E–N	|37.5–34.5| = 3.0	no
				E–R	|37.5–45.0| = 7.5	no
				N–R	|34.5–45.0| = 10.5	no
Grand median = 139.28						

E = Ektachem, A = AVL 984, C = Corning 654, N = Nova Stat, K = Radiometer KNA2.

The sums of ranks, R_j, in Table 5.9 are calculated using

$$R_j = \sum_{i=1}^{b} R(x_{ij})$$

The results of applying the Friedman two-way ANOVA by ranks to these data are displayed in Table 5.9, along with those for conventional least squares two-way ANOVA (see Table 5.11). It is interesting to note that the conventional approach fails to pick out the differences between analysers, unlike the Friedman test, which suggests at least one significant difference.

A multiple comparisons test can be done to evaluate pairwise differences between analysers (see Conover, 1980) and the results of this are also shown in Table 5.9. First, using the individual ranks, $R(x_{ij})$, from the b blocks (specimens) and k treatments (analysers), we calculate the following term:

$$A_2 = \sum_{i=1}^{b} \sum_{j=1}^{k} (R(x_{ij}))^2 = 643$$

Then, using the sums of ranks, R_j, for individual treatments, we calculate the term

$$B_2 = \frac{\sum_{j=1}^{k} R_j^2}{b} = 570.67$$

We can now use these terms to calculate the threshold above which the absolute difference between pairs of treatment rank sums may be regarded as significant at the chosen probability threshold, α, for rejection of the ANOVA Null Hypothesis:

$$|R_j - R_i| > t_{1-(\alpha/2)}\sqrt{\frac{2b(A_2 - B_2)}{(b-1)(k-1)}} = 12.69$$

where R_i and R_j are treatment rank sums, A_2 and B_2 are as calculated above, and $t_{1-(\alpha/2)}$ is the $1-(\alpha/2)$ quantile of the t-distribution with $(b-1)(k-1)$ degrees of freedom.

The multiple comparisons of the individual differences between rank sums calculated for the Friedman test in Table 5.8 are given also in Table 5.9. Thus, 4 out of the 10 possible pairs of analysers have been identified as having significant between-analyser bias at the 5% level.

Another useful approach to the analysis of two-way or more-way tables is median polish. This is a powerful, iterative method for resistant analysis. It offers the possibility of deriving additive, multiplicative or more complex additive-plus-multiplicative models to fit our data (see Emerson and Hoaglin, chapter 6 in Hoaglin et al., 1983; and Emerson and Wong (chapter 3) and Cook (chapter 4) in Hoaglin et al., 1985).

An additive model of a two-way table describes the relationship between the response variable, y, and the two factors whose influence we wish to investigate and is of the form

$$y_{ij} = \mu + \alpha_i + \beta_j + \varepsilon_{ij}$$

where the y_{ij} are the individual measured responses in the two-way table, μ is an overall typical value for the whole table (in the case of median polish, we would use medians), sometimes called the 'common value'. Incremental contributions for level i of the row factor and level j of the column factor relative to the overall value are referred to as the row effect, α_i, and the column effect, β_j, respectively, and the random departures (residuals) from this model are ε_{ij}.

Median polish achieves an additive fit iteratively by finding and subtracting row medians and column medians. Starting with the rows, for each row we find the median and subtract it from each observation in that row. Then we do the same thing with the columns, in principle continuing alternately 'polishing' rows and columns by subtracting medians until all rows and columns have zero medians. For small tables, the whole process can be done by hand, as Emerson and Hoaglin (in chapter 6 in Hoaglin et al., 1983) neatly illustrate. Minitab incorporates an algorithm for median polish developed by Velleman and Hoaglin (1981). That algorithm could easily be incorporated into spreadsheets (e.g. Excel) or any program utilising Basic or Fortran, as the authors have published program listings in both languages in their 1981 book (along with many other useful program listings for resistant exploratory data analysis algorithms, all of which are incorporated into Minitab). A similar process to that involved in median polish, in which we take the data apart to estimate additive or more complex models, may be done using means instead of medians (see Tukey *et al.* (chapter 1), Schmid (chapter 5) and Halvorsen (chapter 6) in Hoaglin *et al.*, 1991). Obviously, we must be aware when polishing with means to estimate the components of a model that this method may lack resistance and robustness. So the use of a resistant exploratory technique such as median polish is advisable as a way of checking the validity of the means-derived model.

The data analysed by Friedman's two-way ANOVA in Table 5.8 have also been analysed by median polish (using Minitab), starting the polishing process with columns and using four iterations. Table 5.10 shows the common value and the column and row effects from this calculation for the simple additive model.

This type of data analysis can be extended to three or more way tables using the algorithm in Minitab (Cook, chapter 4 in Hoaglin *et al.*, 1985) and to more complex multiplicative or additive-plus-multiplicative models (Emerson and Hoaglin, chapter 3 in Hoaglin *et al.*, 1985).

Table 5.10 Median polish ANOVA of ion-selective analyser results of Table 5.8

Common effect	Column effects		Row effects	
	Analyser	Effect	Specimen no.	Effect
	Ektachem	−0.550	1	−19.413
	AVL 984	4.400	2	−4.488
138.787	Corning 654	−2.250	3	0.763
	Nova Stat	0.000	4	5.463
	Radiometer KNA2	0.625	5	5.588
			6	0.588
			7	0.413
			8	1.588
			9	−0.413
			10	−4.888
			11	−2.238
			12	−6.413

Four iterations (assuming an additive linear model) were performed in Minitab 12, starting the polish with columns.

A conventional least squares ANOVA (see also Chapter 8) on the same data has been made for comparison with the Friedman and median polish methods. The results are shown in Table 5.11. Unlike the Friedman ANOVA, the least squares method suggests no significant difference between analysers at the 5% level. This arises from outlier problems in this data set that distort the least squares ANOVA results.

Table 5.11 Least squares two-way ANOVA of the data of Table 5.8

Source	df	Sum of squares	Mean square	F	p
Analyser	4	111.5	27.9	2.38	0.066
Specimen	11	2813.6	255.8	21.8	0.000
Error	44	516.1	11.7		
Total	59	3441.3			

It is possible to perform such polishing using trimmed means or Tukey's biweight as more robust alternatives (Emerson and Hoaglin, chapter 6 in Hoaglin et al., 1983). It is also worth noting that median polish can cope well with missing values but the number of iterations may need to be increased considerably to obtain convergence. An alternative method better able to cope with this difficulty with many fewer iterations and with reasonable resistance is the square combining table (Godfrey, chapter 2 in Hoaglin et al., 1985). It can cope with up to 20% of data being wild and the presence of many holes in the table. If more than 20% of the data are wild, it is preferable to use median polish even though this may require large numbers of iterations.

5.12 Estimating functional relationships or making paired comparisons with robust and nonparametric regression methods for two variables

Regression is a powerful tool much used in exploring quantitative relationships between response variables and independent variables (see also Chapters 11 and 12). In this chapter, the emphasis is on robust, resistant and nonparametric regression methods. The data set in Table 5.6 is used to illustrate some of these techniques. It is a subset from a performance validation of serum sodium measurements, here listing 40 paired comparisons of flame photometric sodium concentration measurements versus assays using a commercially available sodium ion-specific glass electrode system (Radiometer KNA1). The former measures concentrations of sodium in the total volume of the serum, whereas the latter measures ion activity in the aqueous portion of the serum. Ion activity estimates may be affected by the selectivity of the electrode system as well as other matrix and interference effects, which would be expected to be different from those affecting the flame photometry.

ROBUST AND NONPARAMETRIC METHODS

Table 5.12 Results of regression analysis of data from Table 5.6 of flame photometric assays versus ion-selective electrode assays, using various robust, resistant and nonparametric regression methods compared with least squares approach

Regression method	Slope	Standard error of slope	Intercept	Standard error of intercept	Half-slope ratio
Rank	0.900	0.138	15.3	19.06	
Least squares	0.859	0.132	21.4	18.14	
Tukey's*	0.9474		7.711		0.300
Theil's**	0.9286		10.286		

*Tukey's three group resistant line.
**Jheil's median of pairwise slopes.

As actual serum specimens (from patients in a critical-care unit and from others undergoing liver transplant surgery) were used, these matrix and interference effects vary widely, hence the scatter in the plot of the data in Figure 5.9. Table 5.12 shows the regression estimates for this data set, assessed using some of the methods discussed below.

No regression method is perfect, even if it is robust and/or resistant. Thus, it is sensible to evaluate how well different regression approaches deal with your own data sets. When doing such evaluation, some reasonable agreement between methods is desirable. If we examine the slope and intercept estimates and look at the residuals plots, this should guide us as to whether a linear relationship is reasonable. We may also find out from these clues whether there are other problems that might require transformation (or perhaps some form of nonlinear regression) or the use of additional explanatory variables (with stepwise multiple regression).

5.12.1 Tukey's three-group resistant line regression

This is a relatively simple, iterative technique (see Emerson and Hoaglin, chapter 5 in Hoaglin *et al.*, 1983). It first involves dividing the data points into three roughly equally sized groups. The median *x* and *y* values for each group are calculated and the two outer sets of medians are used to calculate an estimate of the slope and the three sets of medians are used in calculating the intercept. The residuals from this initial estimate are then used to refine the estimate and this process continues until convergence criteria are satisfied. The method is available in Minitab as the RLINE command in the exploratory data analysis section. It can be used for stepwise multiple regression by selectively stripping the contribution of each variable out of the remaining residuals (see Emerson and Hoaglin,

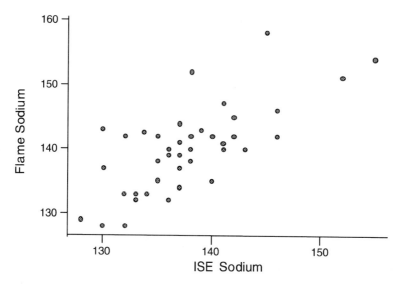

Figure 5.9 Scatter plot of flame photometric versus ion selective assays of plasma sodium data of Table 5.6.

as above, for a description of the way this may be done). This method has a breakdown bound of about 12½% of the data being wild (possible outliers) before it is seriously affected. The output from Minitab can include the slope, intercept (termed as level) and the half-slope ratio (the ratio of the slope between the first and second groups within the data to that between the second and third groups). The latter is a useful exploratory indicator of possible nonlinearity, which may be checked against plots of the residuals.

5.12.2 Theil–Kendall regression using the median of pairwise slopes

In this method, all the possible slopes between all the pairs of points from the data set are calculated (in Minitab, this is done using the WSLOPES command) and the median of these is the slope. The intercept is estimated as the median of the intercepts obtained by drawing lines of the estimated slope through all the points. According to Sprent (1993, 1998), this is a very robust method and it has a breakdown bound of 29%. Sprent (1998) gives details of how to calculate confidence intervals for the slope and intercept.

5.12.3 Hettmansperger's rank regression methods

Regression methods using either rank-based or signed rank-based scores have been developed by Hettmansperger and his colleagues (see Hettmansperger and McKean, 1998) and are available in Minitab in the undocumented RREG command, details of which can obtained in a technical report from Minitab.

McKean has put various data sets and rank-based analyses from the above text, based on use of S-plus routines, online at the Web site http://www.stat.wmich.edu/home.html.

The rank-based scores approach is preferable because of its ability to deal with asymmetry in residuals distributions, whereas the signed rank scores-based approach works better when there is symmetry in the residuals distributions. This seems to be a very promising method, which may be used also for multiple regression problems and also in ANOVA.

5.12.4 Rousseeuw's least median of squares (LMS) and least trimmed squares (LTS) regression methods

LMS and LTS regression methods are described in Rousseeuw and Leroy (1987), which acts as a manual for the software for this method (available direct from Rousseeuw—last updated in January 1996—or as a procedure within S-Plus). Methods used for identification of multivariate outliers and leverage points have been described by Rousseeuw and van Zomeren (1990) and are incorporated in the 1996 version of the software. Both LMS and LTS regression methods are iterative and work by minimizing a robust measure of the scatter of the residuals and have a breakdown of 50% (the maximum possible). They are suitable for both bivariate and multiple regression problems.

5.12.5 Tukey's biweight regression method

Tukey's biweight regression method uses the same approach as described earlier for the biweight estimate of location to calculate weights from the residuals. These are used to weight the various components of the regression calculation, in an iterative approach to progressively more refined estimates of the slope and intercept (see Mosteller and Tukey, 1977). This biweight regression is incorporated into the 'rreg' command in STATA software in an iteratively reweighted least squares procedure (Hamilton, 1998). It begins with an OLS regression; there then follows a step in which observations with large residuals are downweighted using Huber functions for weighted least squares regression. After several such iterations, the weighting function used is switched to Tukey's biweight, as suggested by Li (chapter 8 in Hoaglin et al., 1985).

STATA also offers other robust regression procedures, including 'qreg' (quantile or least absolute values regression) and 'bsqreg', which uses bootstrapping, with the quantile regression, to estimate standard errors when dealing with situations involving outliers of high influence or leverage (for further details see http://www.stata.com). Staudte and Sheather (1990) give a useful series of Minitab macros for iterative Welsch robust regression estimates.

References

Conover, W.J. (1980) *Practical Nonparametric Statistics*, 2nd edn, Wiley, New York.
Davies, O.L. and Goldsmith, P.L. (1977) *Statistical Methods in Research and Production*, 4th edn, Longman, London.
Efron, B. and Tibshirani, R.J. (1993) *An Introduction to the Bootstrap*, Chapman and Hall, London.
Fisher, R.A. (1922) On the mathematical foundations of theoretical statistics. *Philosophical Transactions of the Royal Society*, **222A** 322.
Frigge, M., Hoaglin, D.C. and Iglewicz, B. (1989) Some implementations of the boxplot. *American Statistician*, **43** 50-54.
Gardiner, W.P. (1997) *Statistical Analysis Methods for Chemists. A Software-based Approach*, Royal Society of Chemistry, Cambridge.
Gibbons, J.D. (1976) *Nonparametric Methods for Quantitative Analysis*, Holt, Rinehart and Winston, New York.
Good, P. (1994) *Permutation Tests. A Practical Guide to Resampling Methods for Testing Hypotheses*. Springer, New York.
Good, P. (1999) *Resampling Methods–A Practical Guide to Data Analysis*, Birkhäuser, Boston, Basel and Berlin.
Hamilton, L.C. (1998) *Statistics with Stata® 5*, Duxbury Press, Pacific Grove, CA.
Hawkins, D.M. (1980) *Identification of Outliers*, Chapman and Hall, London.
Hettmansperger, T.P. and McKean, J.W. (1998) *Robust Nonparametric Statistical Methods*, Arnold, London.
Hoaglin, D.C., Mosteller, F. and Tukey, J.W. (eds.) (1983) *Understanding Robust and Exploratory Data Analysis*, Wiley, New York.
Hoaglin, D.C., Mosteller, F. and Tukey, J.W. (eds.) (1985) *Exploring Data Tables, Trends and Shapes*, Wiley, New York.
Hoaglin, D.C., Mosteller, F. and Tukey, J.W. (eds.) (1991) *Fundamentals of Exploratory Analysis of Variance*, Wiley, New York.
Janacek, G.J. and Meikle, S.E. (1997) Control charts based on medians. *The Statistician*, **46** (1), 19-31.
Maritz, J.S. (1995) *Distribution-free Statistical Methods*, 2nd edn, Chapman and Hall, London.
Morgenthaler, S. and Tukey, J.W. (eds.) (1991) *Configural Polysampling*, Wiley, New York.
Mosteller, F. and Tukey, J.W. (1977) *Data Analysis and Regression. A Second Course in Statistics*, Addison-Wesley, Reading, MA.
Rousseeuw, P.J. and Croux, C. (1993) Alternatives to the median absolute deviation. *Journal of the American Statistical Association*, **88** 1273-1283.
Rousseeuw, P.J. and Leroy, A.M. (1987) *Robust Regression and Outlier Detection*, Wiley, New York.
Rousseeuw, P.J. and van Zomeren, B.C. (1990) Unmasking multivariate outliers and leverage points. *Journal of the American Statistical Association*, **85** 633-639.
Simon, J.L. (1997) *Resampling: The New Statistics*, 2nd edn, Resampling Stats, Inc., Arlington, VA.
Sivia, D.S. (1996) *Data Analysis. A Bayesian Tutorial*, Clarendon Press, Oxford.
Snedecor, G.W. and Cochran, G. (1967) *Statistical Methods*, 6th edn, Iowa State University Press, Ames, IA.

Sprent, P. (1993) *Applied Nonparametric Statistical Methods*, 2nd edn, Chapman and Hall, London.
Sprent, P. (1998) *Data Driven Statistical Methods*, Chapman and Hall, London.
Staudte, R.G. and Sheather, S.J. (1990) *Robust Estimation and Testing*, Wiley, New York.
Student (1927) Errors of routine analysis. *Biometrika*, **6** 151-164
Tukey, J.W. (1977) *Exploratory Data Analysis*, Addison-Wesley, Reading, MA.
Velleman, P.F. and Hoaglin, D.C. (1981) *Algorithms, Basics and Computing of Exploratory Data Analysis*, Duxbury Press, Pacific Grove, CA.
Wetherill, G.B. and Brown, D.W. (1991) *Statistical Process Control. Theory and Practice*, Chapman and Hall, London.

6 Experiment design—Identifying factors that affect responses
S. Godbert

6.1 What is design of experiments?

The development chemist within, for example, the pharmaceutical industry is concerned with the process of taking a drug candidate right the way from discovery to an established secure supply by manufacture on a production scale. The route of manufacture involves not only the preparation of the intermediates and final drug substance, but also the isolation of the product and the measurement of its quality. There are many benefits to making this, often lengthy, process more efficient to reduce the time it takes to bring a drug to market. Experimental design is an established and proven methodology for product and process improvement in all industries. Design of experiments (DOE) is an invaluable tool for identifying critical parameters, optimising chemical processes and identifying robust operating regions for our processes.

DOE is an active approach to product and process improvement that involves making a series of controlled interventions (experimental runs) according to a prespecified plan. This is a very efficient way of collecting information to achieve greater understanding and control over products and processes. Many sets of experiments involve the study of effects of two or more parameters. In general, factorial designs are most efficient for this type of experiment.

A well-designed experiment uses the smallest number of experimental runs to get the greatest amount of information; that is, it is small and efficient. In addition, a well-designed experiment should allow us to clearly separate the effects of the parameters from the noise in the system so that the data can be used to identify the important parameters.

A design for identifying important parameters is known as a screening design, i.e. it screens out the important few parameters from the trivial many. The objective of a screening experiment is to proceed from a stage where very little is known with certainty about the roles played by the experimental parameters to a stage where the relative importance of the parameters can be assessed. This entails the identification of those parameters that have a significant influence on, for example, yield and selectivity in synthetic chemistry and peak separation in analytical chemistry.

DOE is not a substitute for creative chemistry. These techniques work best in conjunction with a chemist's own knowledge of the underlying chemistry. Although DOE will not salvage a flawed multistage synthesis, for example, it may identify why it has happened. It can also show how to get the best out of a flawed stage, where no other options are possible.

It is important to carry out a DOE study on an appropriate scale, using appropriate technology. If the objective is to make 50 g lots of a material then laboratory-scale flasks, heaters, filters, etc. are appropriate, but if the objective is to make 50 tonne lots then production scale equipment is needed. However, there may be compelling financial and logistical reasons why the use of large-scale equipment is wholly impractical. In reality, at the preliminary stages of exploration, amounts of starting material are very limited. It makes good sense to carry out the initial designs easily and quickly on a small scale in the laboratory. These reactions should provide a basis for scale-up to large-scale laboratory and ultimately industrial-scale production levels. It is important to verify this as early as possible using jacketed vessels, reaction calorimeters, small plant reactors, etc. However, some processes such as crystallisation, exothermic reactions and phase transfer reactions can be very scale dependent, owing to bulk transfer and heat transfer effects. In these cases, it is more appropriate to use small-scale experimentation as a guide to identify the important factors and then to work on a larger scale to obtain the more predictive model. Most scale-up issues can be anticipated but many organisations do a very poor job of capturing those issues and sharing them between project teams. Experimental design provides a framework for capturing and sharing that information in a learning organisation.

If you don't understand the critical factors at the current scale then scaling up is not going to fix it!

DOE is one of the most powerful tools available for the design, characterisation, and improvement of products and processes. It is a group of techniques used to organise and evaluate testing so that it provides the most valuable data and makes efficient use of resources. With this technique, you will be able to decide whether a change in your process is worth the added cost, how to improve the yield or other quality characteristic, and how to reduce the variability of your product or process. DOE highlights the presence of background noise that is rarely appreciated (though of course still present) when using a traditional approach. As a result, attention is drawn to the importance of process control. And, finally, an experimental design approach is also totally consistent with the need of the regulators of many industries to see evidence of a structured approach.

190   DESIGN AND ANALYSIS IN CHEMICAL RESEARCH

6.2 Where to go

Information needed	Go to Section(s)
So what is experimental design?	6.1
What does all the jargon mean?	6.3
How much resource will I need?	6.4, 6.7
I can't do all the experiments in one day.	6.5.2
I have far too many factors that may affect the responses.	6.5.3
I have a lot of factors (>15).	6.6.3
Some important factors are outside my control during manufacturing but I could look at them in the lab.	6.6.4
I have already set specifications for all my input factors, is there anything DOE can add?	6.6.6
I'm not sure that all my factors have a linear relationship with all the responses over the ranges chosen.	6.6.7
Some of my factors have more than two settings (levels).	6.6.9
What can go wrong?	6.9, 6.10
Show me what the sequence of experiments might look like.	6.11
I need to know how robust my process is.	6.6.6, Chapter 7

6.3 Terminology

Factors. These are the controllable parameters used as inputs to the products and processes under evaluation. As they are varied they may be expected to change the output of the response variable. Also referred to as treatments, independent variables, predictors, x-variables, and input variables. Examples of factors are temperature, time, rate of addition, amount of reagent or solvent in a chemical reaction; amount of solvent, temperature, pH, flow rate and wavelength for an analytical method.

Continuous and discrete factors. Factors can be quantitative or qualitative. Quantitative factors are those that can take numerical values on a continuous scale (e.g. temperature, time). Qualitative factors are usually assigned names and are discrete in nature (e.g. column supplier, analyst, site).

Responses. These are the outcomes of interest. Also referred to as output variables, dependent variables, *y*-variables. Examples of responses are the yield of product, the level of impurities from a chemical reaction or the resolution and percentage area of peaks from an analytical method.

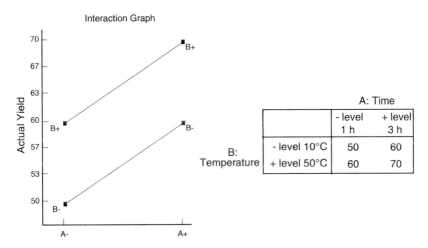

Figure 6.1 Time (A) and temperature (B) interaction plot when no interaction exists.

Levels. These are the particular values or settings of a factor. For example, a temperature may be set at either 10°C or 30°C and the column may be from Supplier A or Supplier B.

Factor effect. The change in the response caused by varying the level of the factor.

Interaction. The measured change in response as a result of the combined effect of two or more factors. A two-factor interaction indicates how the effect of one factor changes as the level of the second factor is varied. For example, Figure 6.1 shows an interaction plot when there is no interaction between temperature and time. Figure 6.2a is an interaction plot when an interaction between temperature and time exists. This is a typical relationship between temperature and time, e.g. if your reaction is run at a high temperature then a shorter time is generally required and vice versa. Figure 6.2b is also an interaction plot when an interaction between temperature and time exists. Higher-order interactions, involving three or more factors, are also possible in which the effect of one factor depends upon the levels of two or more factors. Fortunately, these are less common in real life.

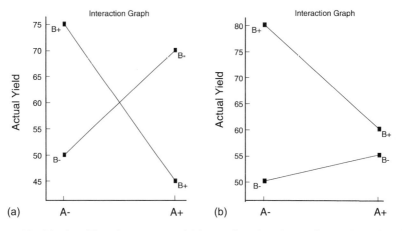

Figure 6.2 (a) Time (A) and temperature (B) interaction plot when an interaction exists. (b) Time (A) and temperature (B) interaction plot with a different interaction.

Centre points or controls. Replicated centre points or controls are run at levels that are the mid-points between the extreme levels of each factor. We can use the replicated centre points as our estimate of background variability.

Pilot/sizing experiment. A small experimental set performed before embarking on the chosen design to check either that the extreme combinations of factor levels will work or to determine the repeatability of the system.

Experimental run. An experiment with a specified combination of levels for each factor.

Experimental design. A systematic series of experimental runs.

Replication. A replicate is the complete repetition of an experimental run. Any differences between replicates are not a result of changes in the factors but are simply a reflection of background variability. Changes in the responses as the factor levels are changed can then be compared to this background variability.

Duplication. Duplication is not the same as replication. Duplication occurs when several measurements are made on the same experimental run at the same time. Duplication gives information only on measurement error and/or product uniformity. It almost always underestimates experimental error. Repeated sampling from the same reaction while considering time as a factor will also underestimate the experimental error.

Experimental error. There are three main categories of experimental error:

- Background variability or noise due to unassignable causes; quantify through replication.
- Bias or systematic error due to assignable causes; minimise through blocking and randomisation.
- Blunders due to mistakes in experimental practice; avoid through careful experiment practice and well-defined worksheets for DOE.

If only one replicate for each experimental run is used, there is no direct estimate of the background variability. One approach to the analysis of such a design is to assume that certain high-order interactions are negligible and combine them to estimate the background variability. This is an appeal to the sparsity of effects principle; that is, most systems are dominated by a few of the main effects and low-order interactions, and most high-order interactions are negligible (Table 6.1).

Table 6.1 Relative sizes of main effects and interactions

Main effects	Often important
2-Factor interactions	Often important, especially if the main effect involved is important
3-Factor interactions	Sometimes important but small compared to 2-factor interactions
4-Factor interactions	Rarely important
n-Factor interactions	If you get to here you have something really unusual!

Randomisation. Randomisation is an experimental technique used to remove the effect of potential bias errors. There is always a risk that the experimental result may be influenced by nonrandom, often time-dependent errors. Such risk may be counteracted by randomisation. This means that in any situation where the experimenter has a choice about the order of doing things, then the choice of random order should be made. Examples are the order of executing the experimental runs and the order of analysing samples drawn from a reaction (particularly if several samples are analysed on the same occasion). Running the experiment in random order makes the interpretation of factor effects more straightforward.

Blocking. Blocking is useful when there is a known factor that may influence experimental results but the factor is not itself of interest. For example, it may not be possible to conduct all the experimental runs using a single batch of material. Different batches would then be used as blocks in the experimental design. The experimental runs are divided into blocks so that the factor effects of interest can still be estimated independently of the changes in the batches of material.

6.4 Getting started

The first step in any experimental work should be a clear definition of the objectives of the experiment and the reasons for undertaking it. The objective should be translated into precise questions that the experiment can be expected to answer. It is wise to assemble a team that possesses all the relevant knowledge about the product or process you are about to investigate. Experts from other groups may be able to bring a new perspective and it is often useful to include someone who is not familiar with the process to provide some out-of-the-box views (a statistician can often fulfil this role and also provide help with the design and analysis).

A good tool for identifying the possible factors affecting the performance of a system is a cause and effect or fishbone diagram. Its purpose is to stimulate thinking about possible factors affecting process performance. The skeleton of a cause-and-effect diagram is given in Figure 6.3. Use the headings (materials, methods, etc.) to identify possible factors that may affect this aspect of the product or process (see Section 6.11 for more examples). In a DOE context, the diagram is helpful in identifying factors to include in an experimental design. A useful rule is to include in the design any borderline factors, since to eliminate them early in the process is much cheaper than to identify and introduce them further down the line.

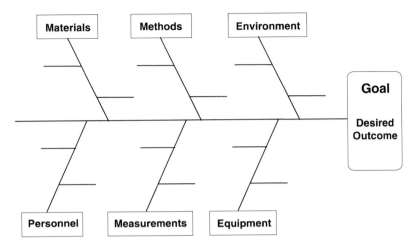

Figure 6.3 Skeleton of a cause-and-effect diagram.

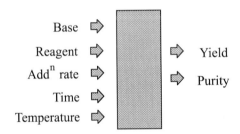

Figure 6.4 Format for defining factors and responses for experimentation.

When the factors to be investigated have been determined and the outcomes of interest have been defined, a summary of the factors required to investigate all the responses of interest can then be displayed as in Figure 6.4. The next step is to draw sketch plots of the assumed relationships between the responses and each of the factors. This will help to determine the experimental design you require and can be used after the completion of the work to show the information gained as a result of performing the experiment. An example is given in Figure 6.5.

The choice of levels for each factor is usually well defined if factors are qualitative (site, supplier, batch). If factors are quantitative, we can choose two or more levels, the choice being dependent on the size of the experiment (number of runs and the amount of replication) and the

Factors	Settings	Relationship for Yield	Relationship for Purity
Base	400-600 µl	 ⌢	 ↘
Reagent	1000-1600 µl	 ↗	 →
Addition Rate	0.5 to 1.5 ml/min	 ↗	 ↗
Time	40-120 min	 ⌢	 ⌢
Temperature	0-30 °C	 ↗	 ⌢

Figure 6.5 Characterisation of relationship between factor and response.

nature of the anticipated response. If the relationship between a response and factor is known to be linear, then two levels of the factor will be sufficient. If the relationship is curved, then at least three levels are needed to characterise the response. In screening experiments, where the objective is to identify important factors, two levels of each factor are usually sufficient. This chapter will deal solely with designs that have all factors at two levels. This is not a necessity of DOE and designs with more than two levels will be discussed in Chapter 7. A useful rule of thumb to decide on the actual settings to use as the two levels of the factor is to divide extreme ranges of the factor into four equal parts and take the one-fourth and three-fourths values as the choice of levels. Thus, if the minimum and maximum values are 20 and 40, respectively (the temperature range that can be tolerated by a solvent in a reaction, for example), choose 25 and 35 as the low and high levels.

Scientific learning is iterative, the results of each stage of investigation generating questions to be answered during the next. It may even turn out that a different quality characteristic, not initially suspected of being important, will dominate future investigations. Therefore, a general rule of thumb is never to spend more than 25% of your total experimental resources on the initial set of experiments.

Although the mathematics behind DOE is very easy and all the calculations can be performed in your favourite spreadsheet, a purpose-built DOE software package can make the design and analysis much more fun. There are many excellent packages on the market. Throughout this chapter, Design Expert has been used (Stat-Ease Corporation, 2021 East Hennepin Avenue, Minneapolis, MN 55413-9827, USA).

EXPERIMENT DESIGN

6.5 Two-level designs

To set up a design, the experimental domain or design space must be specified. For a two-level design this means assigning a low level and a high level to each factor. These low and high levels are often coded as −1 and +1 or simply as − and +. The design space for two factors will be a square and for three factors a cube (Figure 6.6).

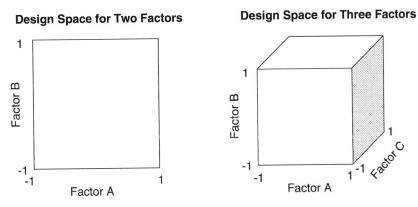

Figure 6.6 Design space in two and three factors.

6.5.1 Full factorial

A full factorial design with k factors, each with two levels, is denoted by 2^k. These designs are efficient for investigating the k factors and all the relationships (interactions) between these factors. Because there are only two levels for each factor, we must assume that the response is approximately linear over the range of the quantitative factor levels chosen.

A full factorial design consists of all possible combinations of the levels of each of the factors. This is best illustrated using an example.

6.5.2 Example: Determining the best storage conditions using a full factorial

An experiment was performed to determine the best storage conditions for a product. The response of interest is the level of impurity, which we wish to minimise. Four factors were chosen for investigation. The factors and their low and high levels are shown in Table 6.2.

Table 6.2 Factors and levels to determine best storage conditions

Factors	Low level (−)	High level (+)
NaCl (mg/ml)	10	90
Acid	HCl	H_3PO_4
Vial size (ml)	10	15
Stopper	W	D

Four factors will give 16 (2^4) different combinations of the factor settings as shown in Table 6.3. Table 6.4 shows the same design in coded form, using − and + for the low and high levels. It is much easier to see the pattern of the highs and lows of each factor in this form.

Table 6.3 Study design and results for the experiment to determine the best storage conditions

NaCl (mg/ml)	Acid	Vial Size (ml)	Stopper	Total Impurity (wt%)
10.00	HCl	10.00	W	8.71
90.00	HCl	10.00	W	13.33
10.00	H_3PO_4	10.00	W	93.51
90.00	H_3PO_4	10.00	W	71.53
10.00	HCl	15.00	W	7.46
90.00	HCl	15.00	W	12.23
10.00	H_3PO_4	15.00	W	85.91
90.00	H_3PO_4	15.00	W	46.82
10.00	HCl	10.00	D	4.95
90.00	HCl	10.00	D	3.63
10.00	H_3PO_4	10.00	D	43.53
90.00	H_3PO_4	10.00	D	6.37
10.00	HCl	15.00	D	4.04
90.00	HCl	15.00	D	4.43
10.00	H_3PO_4	15.00	D	57.88
90.00	H_3PO_4	15.00	D	20.35

Notice that for the factor in the first column the settings alternate between the − and +; for the second column they alternate in blocks of two, −−++; for column three they alternate in blocks of four, −−−−++++; and for the last factor they alternate in two blocks of eight. This is known as *standard order*. The design should not be run in this order, instead the run order should be randomised. However, seeing the design in this format helps to show the pattern of the full factorial.

The coded table can be used to calculate the factor effects. The main effect of NaCl is determined by the average of the response when the factor is at its high level (+) minus the average of the response when the factor is at its low level (−), i.e.

Table 6.4 Coded format of the study design and results for the experiment to determine the best storage conditions

NaCl (mg/ml)	Acid	Vial Size (ml)	Stopper	Total Impurity (wt%)
−	−	−	−	8.7
+	−	−	−	13.3
−	+	−	−	93.5
+	+	−	−	71.5
−	−	+	−	7.5
+	−	+	−	12.2
−	+	+	−	85.9
+	+	+	−	46.8
−	−	−	+	5.0
+	−	−	+	3.6
−	+	−	+	43.5
+	+	−	+	6.4
−	−	+	+	4.0
+	−	+	+	4.4
−	+	+	+	57.9
+	+	+	+	20.4

$$\frac{\text{Sum of responses for '+' signs}}{\text{Number of '+' signs}} - \frac{\text{Sum of responses for '−' signs}}{\text{Number of '−' signs}}$$

$$= \frac{13.3 + 71.5 + 12.2 + 46.8 + 3.6 + 6.4 + 4.4 + 20.4}{8} -$$

$$\frac{8.7 + 93.5 + 7.5 + 85.9 + 5.0 + 43.5 + 4.0 + 57.9}{8}$$

$$= 22.325 - 38.25$$

$$= -15.925$$

This gives a factor effect for NaCl of −15.925. That is, as NaCl moves from its low setting of 10 to its high setting of 90, there is on average a 15.9 wt% decrease in the impurity.

Interaction effects can also be determined using this method. A column for the interaction of interest is calculated by multiplying the columns of the pluses and minuses of the two factors together. We then use this new column of pluses and minuses in the same way as for one of the original factors. For example, the interaction between NaCl and Acid is determined as follows:

Produce the column for the interaction giving the sequence of pluses and minuses as + − − + + − − + + − − + + − − +.

Calculate the average responses for the pluses and for the minuses.

Calculate the difference of the average responses.

The interaction effect of NaCl and Acid is −18.

The method can be used to determine the effect due to any factor or interaction and is described more fully in Box et al. (1978).

The factor effects for all the main effects and interactions for this example are given in Table 6.5.

Table 6.5 Factor effects for study to determine best storage condition

Factor	Effect
A: NaCl	−15.925
B: Acid	45.900
C: Vial	−0.800
D: Stopper	−24.275
AB	−18.000
AC	−1.950
AD	−2.975
BC	−0.175
BD	−18.100
CD	7.850
ABC	−2.425
ABD	−0.400
ACD	2.300
BCD	7.325
ABCD	1.875

A good way of displaying this information is in a Pareto chart. This is basically a bar chart of the effects in decreasing order of magnitude, ignoring the direction of the effect. Figure 6.7 shows the Pareto chart for this example. From this we see that the largest effects are due to the NaCl, Acid and Stopper main effects and the NaCl × Acid and the Acid × Stopper interactions.

The main effect of a factor should be individually interpreted only if there is no evidence that the factor interacts with other factors. When there is evidence of one or more such interaction effects, the interacting factors should be considered jointly. So, for example, there is little point in looking at the effect of Acid on its own since we know that it exerts its effect in conjunction with NaCl. Plots of the interaction effects are shown in Figures 6.8 (a) and (b). In these we see that by choosing HCl as the acid

(B−) there is a much smaller effect caused by changing the levels of NaCl and Stopper.

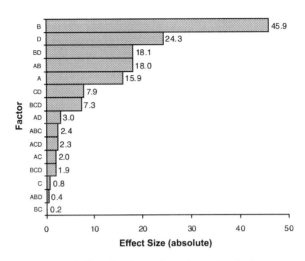

Figure 6.7 Pareto chart of effects for the study to determine the best storage conditions.

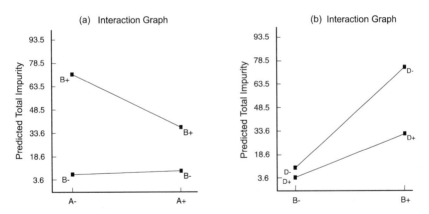

Figure 6.8 (a) Interaction plot for A: NaCl and B: Acid. (b) Interaction plot for B: Acid and D: Stopper.

The effect estimates can also be calculated using analysis of variance (ANOVA), see Chapter 8. As we have no replicates in this design, we do not have a direct estimate of the background variability. However, we can use the high-order interactions to estimate it, as mentioned earlier. Therefore, by estimating only the main effects and two-factor interactions, we can use the remaining data to give us an estimate of background variability.

The results of the ANOVA are given in Table 6.6. The coefficient estimate is half the effect estimated previously. This is because the effect estimate is the difference in the means at the different levels, whereas the coefficient estimate is the rate of change in response to one unit change in the factor (the levels are calculated in coded terms from -1 to $+1$, which is two units). The ANOVA table shows that the important coefficients correspond to the four largest effects picked out earlier.

Table 6.6 Analysis of variance table for factorial model with main effect and two-factor interactions to study the best storage conditions

	Root MSE	7.40		R-squared	0.9817
	Dep. Mean	30.29		Adj. R-squared	0.9451
	CV	24.44		Pred. R-squared	0.8127

Source	Sum of squares	df	Mean square	F-value	Prob. > F
Model	14704.99	10	1470.50	26.83	0.0010
Residual	274.01	5	54.80		
Corr. total	14979.00	15			

Factor	Coefficient estimate	df	Standard error	t for H_0, Coeff. $= 0$	Prob. > |t|
Intercept	30.29	1	1.85		
A:NaCl	-7.96	1	1.85	-4.30	0.0077
B:Acid	22.95	1	1.85	12.40	<0.0001
C:Vial Size	-0.40	1	1.85	-0.22	0.8374
D:Stopper	-12.14	1	1.85	-6.56	0.0012
AB	-9.00	1	1.85	-4.86	0.0046
AC	-0.98	1	1.85	-0.53	0.6208
AD	-1.49	1	1.85	-0.80	0.4580
BC	-0.087	1	1.85	-0.047	0.9641
BD	-9.05	1	1.85	-4.89	0.0045
CD	3.92	1	1.85	2.12	0.0874

For even a moderate number of factors, the total number of treatment combinations in a 2^k factorial design is large. Since resources are usually limited, the number of replicates that the experimenter can employ may be restricted. Frequently, available resources allow only a single run of the design to be made, unless the experimenter is willing to omit some of the original factors. A single run of a 2^k design is sometimes called an unreplicated factorial. With only one replicate, there is no direct estimate of background variability.

One approach to the analysis of an unreplicated factorial is to assume that certain high-order interactions are negligible and to combine them to estimate the variability, as we have done in the above example.

6.5.3 Blocking

It may be impractical to perform a complete full factorial experiment under exactly the same experimental conditions. This can be the case for many reasons—perhaps only a few experimental runs can be performed in a day, or the raw materials for the evaluation may come in small batches so that several batches are required to run the whole experiment. By grouping the experimental runs into blocks in which the experimental conditions are homogeneous, differences between the experimental units can be accommodated. The choice of which runs to put into which block relies on the technique of confounding. Blocks do not contain all the factor combinations.

The problem is how best to divide, say, a 16-run full 2^4 factorial into two blocks of 8. Table 6.7 shows the coded table for the four-factor full factorial including columns for all the interaction terms. To split the 16 runs into two blocks of 8, we have to sacrifice one of the interaction terms, i.e. we will not be able to estimate it. We generally choose the highest-order interaction, in this case the ABCD interaction. To split the design into two we simply choose the runs where the ABCD interaction is -1 and put them into one block, and put the runs where it is $+1$ in the other block. We now have the design split into two blocks of size 8. In

Table 6.7 Coded table for four-factor full factorial with all interactions

A	B	C	D	AB	AC	AD	BC	BD	CD	ABC	ABD	ACD	BCD	ABCD
−1	−1	−1	−1	1	1	1	1	1	1	−1	−1	−1	−1	1
−1	−1	−1	1	1	1	−1	1	−1	−1	−1	1	1	1	−1
−1	−1	1	−1	1	−1	1	−1	1	−1	1	−1	1	1	−1
−1	−1	1	1	1	−1	−1	−1	−1	1	1	1	−1	−1	1
−1	1	−1	−1	−1	1	1	−1	−1	1	1	1	−1	1	−1
−1	1	−1	1	−1	1	−1	−1	1	−1	1	−1	1	−1	1
−1	1	1	−1	−1	−1	1	1	−1	−1	−1	1	1	−1	1
−1	1	1	1	−1	−1	−1	1	1	1	−1	−1	−1	1	−1
1	−1	−1	−1	−1	−1	−1	1	1	1	1	1	1	−1	−1
1	−1	−1	1	−1	−1	1	1	−1	−1	1	−1	−1	1	1
1	−1	1	−1	−1	1	−1	−1	1	−1	−1	1	−1	1	1
1	−1	1	1	−1	1	1	−1	−1	1	−1	−1	1	−1	−1
1	1	−1	−1	1	−1	−1	−1	−1	1	−1	−1	1	1	1
1	1	−1	1	1	−1	1	−1	1	−1	−1	1	−1	−1	−1
1	1	1	−1	1	1	−1	1	−1	−1	1	−1	−1	−1	−1
1	1	1	1	1	1	1	1	1	1	1	1	1	1	1

doing this we will no longer be able to estimate the ABCD interaction, but high-order interactions are unlikely to be important anyway. We say that the ABCD interaction is confounded with block. We will return to confounding later.

What happens to our estimates of factor effects if there is a large difference between the two blocks? Because there are equal numbers of +1s and −1s in each of the blocks for each of the effects, we will be able to estimate them cleanly (i.e. not confounded with anything else). To illustrate this we will use the four-factor full factorial example above. The 16 runs are split as described and 100 is added to each of the results in block 1 to simulate a block effect.

Table 6.8 Study design for four-factor full factorial in two blocks

Block	NaCl	Acid	Vial Size	Stopper	Total Impurity
1	90.00	HCl	10.00	W	113.3
1	10.00	H_3PO_4	10.00	W	193.5
1	10.00	HCl	15.00	W	107.5
1	90.00	H_3PO_4	15.00	W	146.8
1	10.00	HCl	10.00	D	105.0
1	90.00	H_3PO_4	10.00	D	106.4
1	90.00	HCl	15.00	D	104.4
1	10.00	H_3PO_4	15.00	D	157.9
2	10.00	HCl	10.00	W	8.7
2	90.00	H_3PO_4	10.00	W	71.5
2	90.00	HCl	15.00	W	12.2
2	10.00	H_3PO_4	15.00	W	85.9
2	90.00	HCl	10.00	D	3.6
2	10.00	H_3PO_4	10.00	D	43.5
2	10.00	HCl	15.00	D	4.0
2	90.00	H_3PO_4	15.00	D	20.4

The main effect of NaCl is calculated in the same way as before as the differences between the average response for the '+' signs and the average response for the '−' signs:

$$\frac{113.3 + 146.8 + 106.4 + 104.4 + 71.5 + 12.2 + 3.6 + 20.4}{8}$$
$$- \frac{193.5 + 107.5 + 105.0 + 157.9 + 8.7 + 85.8 + 43.5 + 4.0}{8}$$
$$= 72.325 - 88.25$$
$$= -15.925$$

This result is identical with that obtained earlier without any block effect.

This 'new' design is shown in Table 6.8 and the complete set of coefficients is given in Table 6.9. We see that the coefficient estimates for the factor effects remain the same as in Table 6.6 and therefore so do the conclusions. The difference appears in the intercept, which is the overall mean. This is hardly surprising, since we have added 100 to half the results; we would therefore expect the overall mean to increase by 50, which it has. The other difference is that we now have estimates for the two blocks. The difference in the blocks is $(49.06-(-49.06)=98.12)$, which provides a good estimate of the size of the block effect we introduced.

Table 6.9 Analysis of variance table for factorial model with blocks

	Root MSE	8.06		R-squared	0.9826
	Dep. Mean	80.29		Adj. R-squared	0.9392
	CV	10.04		Pred. R-squared	0.7221

Source	Sum of squares	df	Mean square	F-Value	Prob. $> F$
Block	38514.06	1	38514.06		
Model	14704.99	10	1470.50	22.63	0.0043
Residual	259.94	4	64.99		
Corr. total	53479.00	15			

Factor	Coefficient estimate	df	Standard error	t for H_0, Coeff. $= 0$	Prob. $> \|t\|$
Intercept	80.29	1	2.02		
Block 1	49.06	1			
Block 2	−49.06				
A:NaCl	−7.96	1	2.02	−3.95	0.0168
B:Acid	22.95	1	2.02	11.39	0.0003
C:Vial Size	−0.40	1	2.02	−0.20	0.8524
D:Stopper	−12.14	1	2.02	−6.02	0.0038
AB	−9.00	1	2.02	−4.47	0.0111
AC	−0.97	1	2.02	−0.48	0.6538
AD	−1.49	1	2.02	−0.74	0.5014
BC	−0.088	1	2.02	−0.043	0.9675
BD	−9.05	1	2.02	−4.49	0.0109
CD	3.92	1	2.02	1.95	0.1233

What if we require more than two blocks? To divide the 16 runs into four groups of 4 runs each we use a quarter fraction of the full factorial. However, it is necessary to confound more terms with blocks, and possibly with each other. In fact, some of the two-factor interactions can not be estimated. Depending on the requirements of your design, this may or may not be reasonable. When blocking a design, always look to see which terms will be confounded with the blocks.

6.5.4 Fractional factorial

The number of runs in a full factorial increases dramatically as the number of factors increases, as shown in Table 6.10. These designs allow us to estimate a large number of main effects, two-factor interactions and higher-order interactions. This is illustrated for an experiment with seven factors in Table 6.11. In practice, main effects tend to be larger than two-factor interaction effects, two-factor interaction effects are larger than three-factor interaction effects, and so on.

Table 6.10 Number of experiments required to perform full factorial

Number of factors	2	3	4	5	6	7	8
Number of experiments	4	8	16	32	64	128	256

Table 6.11 Number of main effects and interactions in a seven-factor full factorial

Average	Main effects	Interactions						
		2-factor	3-factor	4-factor	5-factor	6-factor	7-factor	Total
1	7	21	35	35	21	7	1	$128 = 2^7$

Since higher-order interactions are rarely important, we can save experimental runs by not estimating them. There tends to be redundancy in a 2^k design if k is reasonably large—redundancy in terms of excess number of interactions that can be estimated. Fractional factorial designs exploit this redundancy. We can choose a fraction of the full factorial that allows us to estimate, say, the main effects and two-factor interactions but not the three-factor and higher-order interactions.

These fractional factorials are among the most widely used designs for product and process investigation and for process troubleshooting. A major use of fractional factorials is in screening experiments, where many factors are considered with the purpose of identifying those factors (if any) that have large effects and eliminating factors that have little or no effect on the response. Screening experiments are usually performed in the early stages of a project. The factors identified as important are then investigated more thoroughly in subsequent experiments.

Screening experiments are used to identify important factors and to suggest changes in their settings to improve process performance. The designs we use most frequently for screening experiments are two-level designs; that is, each factor is evaluated at a 'low' setting and a 'high' setting.

When there are several factors to be evaluated, the system being studied is most likely driven by some subset of main effects and possibly a few low-order interactions. If further information is needed, additional experimental runs can be added to a fractional factorial to convert it into a full factorial experiment. Since these extra runs will be performed after the first set, the full factorial is divided into two blocks.

Suppose we take our four-factor full factorial example, divide it into two blocks as described previously, and analyse only one of the blocks, say the one where the ABCD interaction is +. What we have is a half fraction of a 2^4 design. The coefficient estimates we obtained from the full factorial and from half the experiment are given in Table 6.12. The estimates of the main effects are of similar magnitude, but what has happened to our estimates for the two-factor interactions?

Table 6.12 Coefficient estimates for four-factor full factorial and the half fraction

Factor	Full factorial	Half fraction
A:NaCl	−7.96	−4.30
B:Acid	22.95	24.10
C:Vial Size	−0.40	−0.60
D:Stopper	−12.14	−13.35
AB	−9.00	−5.08
AC	−0.98	−10.03
AD	−1.49	−1.57
BC	−0.087	−1.57
BD	−9.05	−10.03
CD	3.92	−5.08
ABC	−1.21	
ABD	−0.20	
ACD	1.15	
BCD	3.66	
ABCD	0.94	

Table 6.13 Coded table for the half fraction of 2^4 with interactions

A	B	C	D	AB	AC	AD	BC	BD	CD	ABC	ABD	ACD	BCD	ABCD
−1	−1	1	1	1	−1	−1	−1	−1	1	1	1	−1	−1	1
−1	−1	−1	−1	1	1	1	1	1	1	−1	−1	−1	−1	1
−1	1	1	−1	−1	−1	1	1	−1	−1	−1	1	1	−1	1
1	1	1	1	1	1	1	1	1	1	1	1	1	1	1
1	−1	−1	1	−1	−1	1	1	−1	−1	1	−1	−1	1	1
1	1	−1	−1	1	−1	−1	−1	−1	1	−1	−1	1	1	1
1	−1	1	−1	−1	1	−1	−1	1	−1	−1	1	−1	1	1
−1	1	−1	1	−1	1	−1	−1	1	−1	1	−1	1	−1	1

To investigate what has happened, let us look at the coded table for the half fraction (Table 6.13). Can you see any patterns? Some things that should be noted are:

ABCD is the same as the average	ABCD = Average
A has the same pattern as BCD	A = BCD
B has the same pattern as ACD	B = ACD
C has the same pattern as ABD	C = ABD
D has the same pattern as ABC	D = ABC
AB has the same pattern as CD	AB = CD
AC has the same pattern as BD	AC = BD
AD has the same pattern as BC	AD = BC

Therefore, when we estimate the effect of A, what we actually have is an estimate of A + BCD ($-4.30 = -7.96 + 3.66$). We assume that the three-factor interaction, BCD, is unlikely to be important and therefore the effect is most likely due to the factor A. When we estimate the effect of AC, or BD, what we have is an estimate of AC + BD ($-10.03 = -0.98 + -9.05$). This is why the estimates for AC and BD are equal in Table 6.12. We say that AC is confounded with BD and that A is confounded with BCD. Thus, with this design we know that one or both of AC and BD are important but we cannot distinguish which.

A common convention, and one followed here, is that the effect measured by a fractional design is written inside brackets. So, [A] is the effect estimated for factor A by a fractional design. As A is confounded with BCD in the fractional design we are using in the example, we write [A] = A + BCD. The unbracketed effects are those obtained from the full factorial design.

We may be able to use our scientific judgement to determine which of the two-factor interactions is most likely to be important. It is worth noting that generally if the two-factor interaction is important then the main effects will be important. In this case we might attribute the effect to the [BD] interaction. When we try to determine which of the two-factor interactions are important in these ways we are making certain assumptions. If we need to verify which are important, then we can 'fold-over' the design. This means performing the other half fraction. This will then allow us to determine the effects of the two-factor interactions cleanly.

Note that, having performed the first eight experiments, we do have some information about important effects and we have learnt something about the product or process. The knowledge gained from the set of runs can help us to determine what our next set of experiments should be. Remember that experimentation is an iterative process. Suppose we thought that the two-factor interactions were likely to be insignificant and we proceeded with the eight-run design. Having looked at the results, we would have a clear indication that our original assumption was not true. We can revise our design to investigate the two-factor interactions in more detail.

In this example we have fractionated the design and confounded the two-factor interactions. This is not always the case. The confounding pattern will depend on the number of factors and the fraction you use, e.g. half, quarter. For example, with five factors a full factorial would require 32 runs and the half fraction 16 runs. The half fraction will still allow us to estimate the main effects and two-factor interactions cleanly.

Further details on fractional factorials and blocking may be found in Box *et al.* (1978).

6.6 Confounding

Two-level fractional factorials are useful in screening large numbers of factors during the early stages of investigation, when it is not yet known which of these factors have important effects on the response. We are looking for large main effects and, with screening designs, we can do this in relatively few runs. The factors that are found to be important during the screening stage can be studied in more detail later, if necessary.

The observed effect in a fractional factorial design is the sum of the two effects that are confounded. In practice we usually assume that it is the lower-order interaction from among the confounded terms that is responsible—this may not always be a safe assumption. The degree of confounding is referred to as the resolution of the design.

6.6.1 Resolution

The resolution of a design is often used in a traffic light analogy as in Figure 6.9. It is important to know the risk you are taking with the fractional factorial you choose.

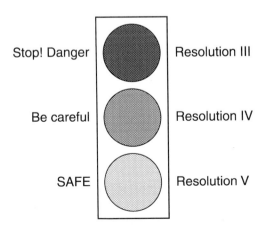

Figure 6.9 Traffic light representation of the resolution of a design.

- A design of resolution V does not confound main effects and two-factor interactions with each other, but does confound two-factor interactions with three-factor interactions. The easy way to remember this is that the two-factor interactions are confounded with three-factor interactions, two plus three is five giving a resolution V design.
- A design of resolution IV does not confound main effects and two-factor interactions with each other but does confound two-factor interactions with other two-factor interactions. Using the same rule as before, two plus two is four giving a resolution IV design.
- A design of resolution III does not confound main effects with each other but does confound main effects with two-factor interactions; one plus two is three giving a resolution III design.

Table 6.14 shows the resolution chart in Design Expert for two-level factorial designs up to 10 factors. The darker shadings correspond to the low-resolution designs.

Table 6.14 Table of resolutions of designs for up to ten factors

No. of Runs	No. of Factors								
	2	3	4	5	6	7	8	9	10
4	Full	1/2 Fract							
8		Full	1/2 Fract	1/4 Fract	1/8 Fract	1/16 Fract			
16			Full	1/2 Fract	1/4 Fract	1/8 Fract	1/16 Fract	1/32 Fract	1/64 Fract
32				Full	1/2 Fract	1/4 Fract	1/8 Fract	1/16 Fract	1/32 Fract
64					Full	1/2 Fract	1/4 Fract	1/8 Fract	1/16 Fract

6.6.2 Example: Crystallisation study to determine the factors affecting particle size uniformity using a resolution III screening design

Sometimes, owing to lack of resources, a resolution III design is all you can afford to do. In crystallisation studies this is often the case because to control all possible factors in the experiments carefully is time consuming. It is desirable to have uniformity of crystal size from the crystallisation stage of a product.

Suppose there are five factors that may affect the uniformity of crystal size. Owing to the length of time required to perform each experiment, an

Table 6.15 Study design and results for crystallisation experiment for a product

Temperature (°C)	Antisolvent, vol	Seeds, wt	Origin of seeds	Cooling Period (min)	Uniformity, D value
40	0.5	0.005	M	10	23.4
60	7.5	0.001	M	10	16.4
60	0.5	0.005	S	120	55.2
50	4.0	0.003	S	65	41.7
40	7.5	0.001	S	120	66.2
40	0.5	0.001	M	120	61.3
50	4.0	0.003	S	65	46.5
60	0.5	0.001	S	10	16.3
40	7.5	0.005	S	10	23.1
60	7.5	0.005	M	120	50.4
50	4.0	0.003	M	65	42.5
50	4.0	0.003	M	65	43.2

eight-run fractional factorial, resolution III, design was used. The design is shown in Table 6.15 and the confounding pattern in Table 6.16. The coefficient estimates are given in Table 6.17 and the Pareto plot in Figure 6.10. It is clear that the cooling period has the greatest effect on the uniformity of crystal sizes for product, U.

Table 6.16 Confounding pattern for crystallisation experiment

[A] = A + BD + CE
[B] = B + AD + CDE
[C] = C + AE + BDE
[D] = D + AB - ABD - ACE + BCE
[E] = E + AC + BCD
[BC] = BC + DE + ABE + ACD
[BE] = BE + CD + ABC + ADE

Table 6.17 Coefficient estimates for the crystallisation experiment

Factors	Coefficient estimate
A: Temperature	−4.47
B: Antisolvent	0.0025
C: Seeds	−1.01
D: Origin	−1.17
E: Cooling Period	19.23
BC	−1.26
BE	0.035

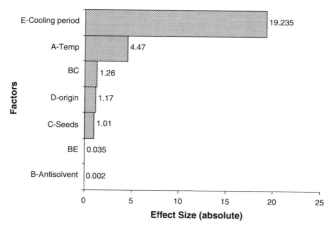

Figure 6.10 Pareto chart of effects for the crystallisation experiment.

Temperature also has some effect, although considerably less than the cooling period. Effect plots for cooling period and temperature are given in Figure 6.11. Increasing the cooling period from 10 to 120 minutes increases the uniformity by almost 40 units. Increasing the temperature decreases the uniformity by 10 units. Therefore, low temperature and long cooling period appear to be the crucial settings for the uniformity of crystal sizes for this crystallisation process.

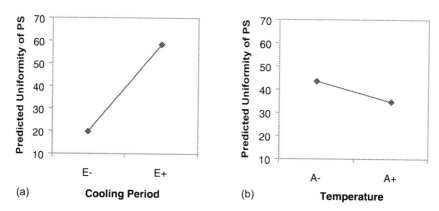

Figure 6.11 (a) Effect plot of cooling period on particle size uniformity. (b) Effect plot of temperature on particle size uniformity.

These experiments were performed on a small scale. The factors found to be important at this scale should be investigated further on a larger scale.

6.6.3 Irregular fraction designs

The fractions of the factorials we have looked at so far have been half, quarters, etc. and all have numbers of runs that are powers of 2, e.g. 8, 16, 32. There are other fractions that can provide a reasonable design with the number of runs not in powers of 2. For example, suppose we have four factors to investigate. What if we are not able to perform 16 runs but think that all two-factor interactions might be important? The half fraction in eight runs would leave ambiguities. However, we could perform a three-quarters fraction of the 2^4, that is 12 runs, which is a resolution V design. This would enable us to estimate the main effects and two-factor interactions without confounding. For more information, see John (1971).

6.6.4 Plackett–Burman designs

Plackett and Burman developed a set of screening designs in which the number of runs is a multiple of 4. For example, we can screen 11 factors with just 12 runs in a Plackett–Burman (PB) design. PB designs have an interesting confounding pattern, with each main effect confounded with various fractions of main effects and two- and three-factor interactions. For example, in the 12-run, 11-factor PB design the two-factor interaction AB is confounded with

$$AB = -\tfrac{1}{3}C - \tfrac{1}{3}D - \tfrac{1}{3}E + \tfrac{1}{3}F - \tfrac{1}{3}G - \tfrac{1}{3}H + \tfrac{1}{3}I + \tfrac{1}{3}J - \tfrac{1}{3}K$$

Like fractional factorial designs, the PB design can be folded over to disentangle the main effects from these two-factor interactions. However, the resulting design is a resolution IV design in which the two-factor interactions are partially confounded with other two-factor interactions. When the number of runs is a power of 2, the PB designs are identical to the equivalent fractional factorial. The most useful PB designs are the 12-, 20-, 24-, 28- and 36-run experiments.

PB designs are generally resolution III designs and therefore can be dangerous since they are designed to estimate main effects only. However, when you have to consider a large number of factors, say more than 15, they may be the best you can do to screen out the important effects. The design is very efficient providing there are no interactions and no outliers. For more information, see Carlson (1992).

6.6.5 Taguchi designs

Taguchi designs cover three types of design, one similar to fractional factorials, one similar to Plackett–Burman and the other a new type that will be described briefly below.

We can distinguish between factors over which we have control (Taguchi describes these as inner array or control factors) and factors over which we may have control in R & D but over which we will have little control in manufacturing (Taguchi's outer array or noise factors). We need to identify a combination of the control factors that is insensitive to changes in the noise factors but which optimises the response and minimises cost. By building the noise factors into the design, we can identify interactions between the control factors and the noise factors that will indicate control factors sensitive to noise variability.

A Taguchi design is obtained by combining a design for the inner array with a design for the outer array. The inner array design can be any screening or response surface design (see Chapter 7). More often than not, a highly reduced fractional factorial is used for the inner array (resolution III) owing to the large numbers of experiments involved. The outer array design is constrained to be a full factorial design. The inner array consists of the control factors and the outer array is made up of noise factors. Since every inner array point is repeated at each level of the outer array, Taguchi experiments get big quite quickly. An example of a Taguchi design is given in Table 6.18.

Table 6.18 Example of a Taguchi design

Inner Array		Outer Array				
		−	+	−	+	: Temp.
		−	−	+	+	: Time
Amount of Reagent	Addition Rate					
−	−					
+	−					
−	+					
+	+					

The objective of a Taguchi design is to maximise some 'performance statistic'. These performance statistics are a function of the mean response and of the variability. The thinking is that to ensure a robust process we need to choose a function that reflects process variability in addition to the process mean. At its simplest the mean and the standard deviation at each inner array design point can be used as performance statistics. We can produce models of the mean response and for the standard deviation and, say, maximise the mean subject to a constraint on the standard deviation or minimise the standard deviation subject to a constraint on the mean. Alternatively, we can combine measures of the mean and variability in more complex performance statistics such as the signal-to-noise ratio for the process:

$$S/N = -10 \log\left(\frac{\text{mean}^2}{\text{variance}}\right)$$

There are criticisms of this approach such as:

- The performance statistics often have undesirable properties.
- There is a tendency to ignore interactions and assume a linear model.

For more information on Taguchi designs, see Montgomery (1991), Frigon and Mathews (1997) or Logothetis and Wynn (1989).

6.6.6 D-Optimal designs

Despite the wide range of standard designs, there are numerous occasions when the tabulated designs do not meet all the requirements of the experimenter. Some situations where the standard designs may not be appropriate are:

- When it is essential to exclude certain combinations of factor settings, e.g. high temperature and high pressure could cause an explosion!
- When a standard design exists but it involves a specified number of runs that does not match the number available to the experimenter.

The frequency of such situations makes it clear that there is a need for factorial-type designs that are tailored to a researcher's specific requirements.

In selecting a D-optimal design you need to specify the model you believe to be important, i.e. according to which main effects and two-factor interactions, you wish to estimate. A subset of points is chosen from a candidate design, usually a full factorial, so that you are able to estimate the effects of the terms included in the model using the minimum number of runs. The main drawback of these designs is that they are highly dependent on the model you specify. If there are unexpected two-factor interactions, the design will not highlight them because you have not asked it to look for them. If one of the runs from these designs cannot be carried out for some reason, then you may not be able to estimate all the effects in your chosen model. For further information on optimality, see Chapter 7 and Carlson (1992).

6.6.7 Robustness designs

For robustness designs, the goal becomes one of defining an operating region for a product or process where the response will be close to optimal, robust to variation and inexpensive to produce. Having used a response surface design (see Chapter 7) to define a region you believe to be robust, i.e. a relatively flat response surface or a region within which all values are acceptable, a highly fractionated design (resolution III) can be used to confirm appropriate limits for the region.

6.6.8 Centre points

A potential concern in the use of two-level factorial designs is the assumption of linearity in the factor effects. Of course, perfect linearity is unnecessary and the designs will work quite well when the linearity assumption holds only approximately. Centre points are run at a level that is at the mid-point of the + and − levels of each factor. In coded terms, they are at the zero point level. One important reason for adding the replicate runs at the design centre is that the centre points do not affect the usual effect estimates in the two-level factorial. When centre points are added to the design, we get three levels of each factor and so we can test for evidence of nonlinear response to all factors. If, as is usual, the factorial points in the design are unreplicated, we can use the replicated centre points as our estimate of background variability.

6.6.9 Example: Determining robustness using a fractional factorial

This example shows the use of experimental design to demonstrate the effect of relatively small changes in chromatographic parameters on the performance of an HPLC assay.

The actual experiment that was carried out looked at the separation between a number of pairs of peaks and at the retention time. Here we concentrate on the separation between a single pair of peaks, A and B for simplicity. A chromatographic resolution greater than 2 is deemed reasonable.

Table 6.19 Factors, with levels, for investigation to determine robustness of analytical method

Factor	Nominal	Low level	High level
Flow (ml/min)	0.8	0.6	1.0
Temperature (°C)	30	25	35
TFA (% v/v)	0.05	0.045	0.055
%MeOH at time = 0 min (% v/v)	5	4	6
%MeOH at time = 20 min (% v/v)	30	28	32
%MeOH at time = 35 min (% v/v)	90	88	92

Brainstorming identified six possible factors that could affect the responses of interest and the normal operating range for each of these factors (Table 6.19). A quarter fraction of a 2^6 factorial was chosen as the design because the 16 runs required were judged to be a reasonable amount of effort to expend on this problem while still allowing us to estimate main effects clear of two-factor interactions. It was also decided to include four centre points to provide an independent estimate of background variability and to confirm the suitability of the nominal settings chosen.

Diagnostic plots from a preliminary analysis of the response showed very clearly that a logarithmic transformation of the data was advisable (see Section 6.7 for more detail). The experimental design and results are shown in Table 6.20.

The Pareto chart of effects is given in Figure 6.12; note that the effect sizes are not in the original units as a log transformation has been used (take the exponential of the effect sizes given to obtain the values in the original units). The main effect of B (Temperature) and C (%TFA) are clearly important. There also appears to be quite a strong effect due to the interaction of factors A (Flow Rate) and E (Intermediate MeOH).

Although two-factor interactions can be important when neither of the corresponding main effects is, it is comparatively rare. Let us look more closely at the aliasing structure for this design (Table 6.21). Here we see that the AE interaction is confounded with the BC interaction. The latter seems to be a much more reasonable candidate for inclusion in the model since both the B and C main effects were important. If there were serious doubts about which of the interaction terms to include, then a fold-over of the design would resolve the issue. Here the analytical chemists, using their scientific judgement, are content to assume that it is the BC interaction that is the important effect.

The statistical analysis output is given in Table 6.22. The R-Squared values show that the model with the two main effects plus their interaction explains about 99% of the variation in the data. Since we have centre points in the design, we can look at lack of fit for the model (which is not significant with a p-value of about 0.6) and whether there is evidence of any curvature in the response (which is highly significant with a p-value less than 0.0001). This latter result is not entirely unexpected since the centre point represents the normal running conditions of this assay, and we would expect these conditions to have been chosen to maximise the peak separation.

A square plot (Figure 6.13) is useful to investigate the actual effect of the important factors on the response. Here we see the predicted response (in the original units) for each of the combinations of high and low levels of the two important factors. The best predicted resolution is 7.68 at low

Table 6.20 Study design and results for HPLC assay robustness experiment

Run	A:Flow (ml/min)	B:Temp. (°C)	C:%TFA (%)	D:%MeOH at 0 min	E:%MeOH at 20 min	F:%MeOH at 35 min	A to B resolution
1	1.0	35	0.045	6	28	88	2.49
2	0.6	35	0.055	4	28	88	5.66
3	1.0	25	0.055	4	28	92	7.41
4	0.8	30	0.050	5	30	90	5.71
5	0.6	25	0.045	4	28	88	4.25
6	1.0	25	0.045	4	32	88	4.11
7	0.6	35	0.055	6	28	92	5.81
8	0.6	35	0.045	4	32	92	2.49
9	0.6	35	0.045	6	32	88	2.55
10	0.8	30	0.050	5	30	90	5.65
11	0.8	30	0.050	5	30	90	5.73
12	1.0	35	0.055	4	32	88	5.33
13	1.0	25	0.045	6	32	92	4.25
14	1.0	35	0.055	6	32	92	5.45
15	1.0	25	0.055	6	28	88	7.53
16	0.6	25	0.045	6	28	92	4.38
17	0.6	25	0.055	6	32	88	7.03
18	1.0	35	0.045	4	28	92	2.56
19	0.8	30	0.050	5	30	90	6.10
20	0.6	25	0.055	4	32	92	6.81

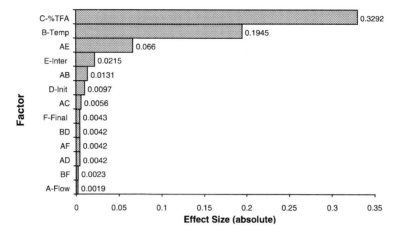

Figure 6.12 Pareto chart for HPLC assay.

Table 6.21 Confounding pattern for HPLC assay robustness experiment

[A] = A + BCE + DEF
[B] = B + ACE + CDF
[C] = C + ABE + BDF
[D] = D + AEF + BCF
[E] = E + ABC + ADF
[F] = F + ADE + BCD
[AB] = AB + CE
[AC] = AC + BE
[AD] = AD + EF
[AE] = AE + BC + DF
[AF] = AF + DE
[BD] = BD + CF
[BF] = BF + CD

temperature and high %TFA and the worst is 2.69 at high temperature and low %TFA. The software can also be used to give a predicted response at the centre of the design assuming a linear response; in this case we get a prediction of 4.55. The actual responses at the design centre were 5.65, 5.71, 5.73 and 6.10, which are all considerably higher than predicted, indicating curvature in the response surface. Note that in this example better separation is achieved in one corner of the design space than at the routine operation setting. Remember, though, that this setting may well be a compromise to achieve acceptable results across all the responses. We have now established that the chromatographic resolution

Table 6.22 Analysis of variance table for the factorial model for HPLC assay robustness experiment

Source	Sum of squares	df	Mean square	F-value	Prob. > F
Model	2.41	3	0.80	698.17	<0.0001
Curvature	0.19	1	0.19	163.14	<0.0001
Residual	0.017	15	1.150×10^{-3}		
Lack of fit	0.014	12	1.136×10^{-3}	0.94	0.5985
Pure error	3.619×10^{-3}	3	1.206×10^{-3}		
Corr. total	2.61	19			
	Root MSE	0.034		R-squared	0.9929
	Dep. Mean	1.56		Adj. R-squared	0.9915
	CV	2.17		Pred. R-squared	0.9883
	PRESS	0.031		Adeq. precision	61.770

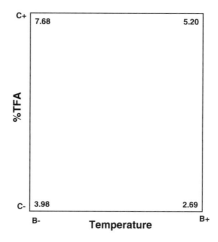

Figure 6.13 Square plot for the effect of temperature and %TFA on the resolution of peaks A and B.

between peaks A and B is not affected by four of the six initial factors, as either main effects or through interactions.

Two factors have been shown to affect the peak separation and, over the ranges of values studied for these factors, the peak separation is greater than the minimum acceptable value of 2. However, the analyst might feel that the amount of change in the peak separation over the whole factor range is too big for a robust assay. If it is, then the ranges of Temperature and %TFA must be reduced to ensure that the assay is operated in a region of the design space to achieve acceptable peak separation—but at least the analyst can now do this in an informed way.

6.6.10 Example: Assessing process deviation ranges using a highly fractionated factorial design

A process for the production of an ester has been developed and passed to manufacturing with an indication of the process deviation ranges (PDRs, i.e. the limits within which the factors are allowed to vary during normal production) and the specification limits on the outputs. No indication of the robustness of the method has been indicated. The production team taking on the process wish to verify that if the process remains within the PDRs the output ester is within all of the specification limits.

Table 6.23 Factor and responses to determine robustness of the production of an ester

Factors	PDR
A: Acetyl Chloride Batching (kg)	1.23–1.27
B: Starting Material (kg)	51.5–58.5
C: Reaction Temperature (°C)	49–56
D: Reaction Time (hr)	3–6
E: Sodium Acetate Batching (kg)	1.31–1.37
F: Water Batching (l)	8.1–8.5
G: Drying Temperature (°C)	30–35
Quality	**Spec. limits**
Starting Material (%w/w)	< 2.5
Output Water (KF) (%w/w)	5.0–5.6
By-Product (%w/w)	< 3.0
Yield (%w/w)	–

There is no specification for yield. The expected value is 86% w/w.

The factors under investigation and the outputs are in Table 6.23. There are seven factors of interest and the design chosen has 16 runs (a resolution IV design). The design is given in Table 6.24 and the results in Table 6.25.

The confounding pattern is given in Table 6.26. Clearly, we have values for the ester that are outside the specification limits. The coefficient estimates for each of the responses are given in Table 6.26. These are displayed in the Pareto charts in Figures 6.14 to 6.15. From these we see that Water Batching is the important factor for Starting Material, Output Water and By-Product levels. Figure 6.16 shows the effect plots for the Water Batching factor on each of these outcomes. We see that Water Batching at the high level is better for the Residual Water and Starting Material outcomes and Water Batching at the low level is better for the By-Product outcome. However, the Residual Water is outside the specification limits for the low level of Water Batching. This would

Table 6.24 Study design for the robustness study of production of an ester

Experiment no.	Acetyl Chloride (kg)	Starting Material (kg)	Reaction Temp. (°C)	Reaction Time (h)	Sodium Acetate (kg)	Water Batching (L)	Drying Temp. (°C)
1	1.23	51.5	49	6	1.31	8.5	35
2	1.27	58.5	56	3	1.37	8.1	30
3	1.23	51.5	49	3	1.31	8.1	30
4	1.27	58.5	56	6	1.37	8.5	35
5	1.23	58.5	49	6	1.37	8.1	35
6	1.27	51.5	56	3	1.31	8.5	30
7	1.23	58.5	49	3	1.37	8.5	30
8	1.27	51.5	56	6	1.31	8.1	35
9	1.27	58.5	49	6	1.31	8.1	30
10	1.23	51.5	56	3	1.37	8.5	35
11	1.27	58.5	49	3	1.31	8.5	35
12	1.23	51.5	56	6	1.37	8.1	30
13	1.23	58.5	56	3	1.31	8.1	35
14	1.23	58.5	56	6	1.31	8.5	30
15	1.27	51.5	49	6	1.37	8.5	30
16	1.27	51.5	49	3	1.37	8.1	35

Table 6.25 Results for the robustness experiment for ester production

Experiment no.	Starting Material (%w/w)	Output Water (KF) (%w/w)	By-Product (%w/w)	Yield (%w/w)
1	1.8	5.5	2.1	85.3
2	2.1	5.6	1.8	83.6
3	3.8	5.7	2.1	93.9
4	2.6	5.4	2.2	83.9
5	2.4	5.8	1.9	90.1
6	2.6	5.6	2.1	89.1
7	2.3	5.5	2.0	88.9
8	2.8	5.6	2.1	86.4
9	6.2	5.8	1.8	84.6
10	2.8	5.6	1.9	88.5
11	2.6	5.5	2.2	91.5
12	2.0	5.6	2.0	83.9
13	5.8	5.7	1.4	76.0
14	2.0	5.5	1.9	81.8
15	1.8	5.5	2.1	85.8
16	6.2	5.5	1.7	92.6

appear to be a critical factor. Either the PDR range for Water Batching or the specification limits for the Residual Water needs to be altered.

Looking at the results for the experiments where Water Batching was at the high level we do still have problems with the level of Starting Material remaining. There is some evidence of an interaction for the

Table 6.26 Coefficient estimates and confounding pattern for the ester production experiment

Factor	Confounding pattern	Starting Material	Residual Water	By-Product	Yield
A: Acetyl Chloride	A + BCE + BFG + CDG + DEF	0.25	−0.025	0.044	0.57
B: Starting Material	B + ACE + AFG + CDF + DEG	0.14	0.012	−0.056	−1.57
C: Temperature	C + ABE + ADG + BDF + EFG	−0.28	−0.013	−0.031	−2.47
D: Time	D + ACG + AEF + BCF + BEG	−0.41	0.000	0.056	−1.39
E: Sodium Acetate	E + ABC + ADF + BDG + CFG	−0.34	−0.025	−0.00625	0.54
F: Water	F + ABG + ADE + BCD + CEG	−0.80	−0.075	0.11	0.23
G: Drying Temperature	G + ABF + ACD + BDE + CEF	0.26	−0.013	−0.019	0.17
AB	AB + CE + FG	−0.12	0.000	0.056	0.28
AC	AC + BE + DG	−0.56	0.000	0.081	1.03
AD	AD + CG + EF	0.40	0.013	−0.00625	−0.62
AE	AE + BC + DF	0.15	−0.037	−0.044	−1.26
AF	AF + BG + DE	−0.16	0.013	0.044	0.16
AG	AG + BF + CD	−0.075	−0.050	0.069	1.24
BD	BD + CF + EG	0.46	0.025	−0.000625	1.44
ABD	ABD + ACF + AEG + BCG + BEF + CDE + DFG				

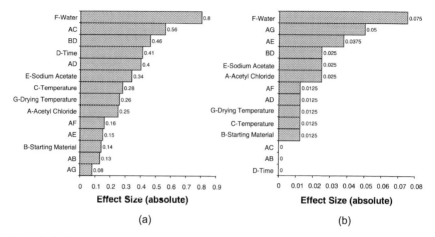

Figure 6.14 (a) Pareto chart for Starting Material Content. (b) Pareto chart for Water Context.

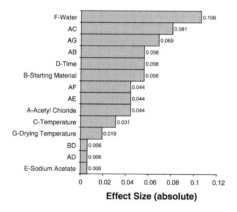

Figure 6.15 Pareto chart for By-Product Content.

Starting Material outcome. This is labelled as the interaction between Acetyl chloride and the Reaction Temperature. We must be careful though, since there are other two-factor interactions that are confounded with this one. Further work would be required to determine which of the interactions is important.

Obviously this process is not robust. The majority of the experimental runs were outside the specification limits for at least one of the outcomes. Further work will be required to provide an adequate process. There is no doubt that these problems could have been avoided if a designed

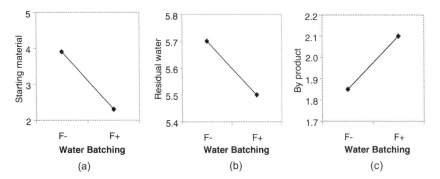

Figure 6.16 Effect plots of (a) Water on Starting Material content. (b) Water on Residual Water Content. (c) Water on By-Product Content.

approach had been used before introduction to manufacturing. It is worth noting that the investigation of the process at such a late stage is also very expensive. It is much cheaper to investigate earlier during the development of the product/process.

6.6.11 Other designs

Experiments need not be limited to factors at two levels, although the use of two levels is often necessary to keep the experiment at a manageable size. Where factors are quantitative, experiments at more than two levels may be desirable when curvature of the response is anticipated. As the number of levels increases, the size of the experiment increases rapidly and fractional designs are recommended. Multilevel designs are discussed in Chapter 7.

6.6.12 Mixed-level designs

There are many situations where there is a need for an experiment that incorporates two-, three- and four- level factors at the same time. These are possible but again the number of runs required will increase. For more details on mixed-level designs, see Montgomery (1991).

6.7 Data analysis and interpretation

Multiple regression analysis is a powerful technique for fitting a linear model to a series of data values. Chapter 11 gives more details on model fitting and model checking. Below, we look at the essential model checks and some of the useful information obtained from the model fitting from the experiment design point of view.

Before using the results from the analysis there are various diagnostic checks that should be performed. One of these is a plot of the residuals (the difference between the measured response and that predicted by the model equation for the same factor levels) versus the predicted values. This will determine whether a transformation of the response is required. The points on this plot should be randomly scattered. Figure 6.17a shows a plot where there is a random scatter of points, while Figure 6.17b shows a definite pattern often described as a funnel. The latter pattern indicates that a logarithmic transformation should be applied to the data before performing the analysis. A plot of the residuals versus the run order can also be useful in determining whether there has been a trend or step change in the results over time. Such an effect could be caused by some uncontrolled external variable such as room temperature.

Once the appropriate model has been established, the R-squared value and the variability estimate can be used to evaluate the model. The R-squared value is the proportion of the variation in the data that has been explained by the model. That is, how much of the variability in the observed response values can be explained by the design factors and their interactions. The R-squared value is always between 0 and 1. A value of 1 indicates that the statistical model explains all of the variability in the data. A value of 0 indicates that none of the variability in the response can be explained by the experimental factors. The acceptable R-squared value will depend on your application. For HPLC the R-squared value is seldom below 0.9 but, generally, an R-squared value above 0.75 is reasonable. The estimate of the variability (often called the mean square

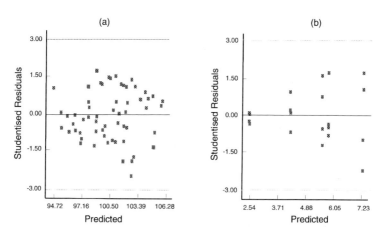

Figure 6.17 (a) Plot of residuals versus predicted showing a random pattern. (b) Plot of residuals versus predicted showing a 'funnel' pattern.

error in ANOVA) can also be compared to the variability that you generally see between replicates in similar products and processes.

We have already discussed the coefficient estimates from the model and their graphical display using the Pareto chart. The effects can be displayed also using the main effect plots, interaction plots and square plots. These graphical techniques provide a useful way of displaying the results. However, a statistical test (usually a t-test) on the coefficient estimates can be used to determine whether the factor effect is significant compared to the noise in the system (see Chapter 4).

The best validation of a model is, of course, that it provides good predictions, and these can be confirmed by comparing the prediction with the results from new experiments.

6.8 Choosing a design

It is important to tailor your design so that it is 'fit for purpose', i.e. that it can meet the objectives for the current stage in the project life-cycle. Use small pilot studies to evaluate appropriate levels of factors and the reproducibility at the current settings. Apply screening designs (such as fractional factorials) as early as possible, to identify the correct choice of discrete factors (such as solvent and reagent) and the most important continuous factors (such as time and temperature).

Before embarking on the design, it is usually preferable to look at the whole process first (e.g. reaction and work-up) to establish the overall issues. Then carry out a study to establish the best reaction conditions—the purer the process stream, the easier the work-up. Sometimes work-up and reaction are inextricably linked, in which case the study needs to cover both. As a result, the design will inevitably be harder to control and more time-consuming to implement. In general, it is better to split the process into independent subsets whenever possible.

There is much to be gained by running experiments sequentially (Figure 6.18), beginning with fractional factorial designs to screen large numbers of factors and to refine other experimental parameters. You can use a resolution III or IV design at this stage. If more information is required about confounded effects then the design can be augmented using a fold-over. This will enable the interesting effects to be cleanly estimated. Centre points can then be added to the design to determine whether the linear model sufficiently describes the relationship between the factors and the responses. If there is significant evidence of curvature, then axial or star points can be added to the design to make it into a response surface design (see Chapter 7).

At any stage of the sequence, the experimenter can withdraw if sufficient information has been obtained. By using the sequential approach,

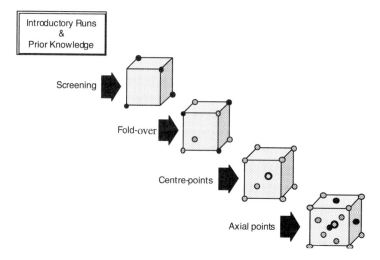

Figure 6.18 Schematic of the sequential nature of experimentation.

knowledge from previous steps and experiments can be used to govern the design for future experimentation.

The steps involved in performing an experiment are described below.

Assemble the right team. Does your team possess all the relevant knowledge about the product or process you are about to investigate? Are there experts in other groups who could bring a new perspective? It is often useful to include someone who is not familiar with the process to provide some out-of-the-box views. A statistician can often fulfil this role as well as provide help with design and analysis.

Recognition and statement of the problem. What are the objectives of the experiment and the reasons for undertaking it? Have you translated these objectives into precise questions that the experiment can be expected to answer?

Brainstorm the factors. The cause and effect diagram is useful for this. Include any borderline factors from the cause-and-effect diagram in your initial design since eliminating them early in the investigation process is much cheaper than having to introduce them later on.

The choice of factors to be included in the design should be considered carefully. What range should be investigated for each factor? What interactions might be important? Are the factors continuous or discrete?

Those factors not relevant to the experiment but which could influence the result should be carefully controlled or kept constant. If the factor settings cannot be controlled within reasonably tight limits, then treating them as fixed at their nominal settings will introduce extra variability into the analysis. The methods for dealing with these, in order of preference, are:

- Control them so they do not vary.
- Balance them in blocks so that their effect can be removed and not cause bias in the stipulated effects.
- Measure them and remove their effect if necessary using analysis of covariance.
- Randomise their effects over all of the runs in the experiment.

Can all combinations of factor levels be achieved or are there constraints on the design space? Do an initial ranging study to answer some of these questions if necessary.

Brainstorm the responses. In selecting the response variable, the experimenter should be certain that the variable really provides useful information about the process under study. Multiple responses are not unusual and can be handled, as shown in the ester example. The measurement error in the responses is also important. If the measurement error is large, then only relatively large factor effects will be detected by the experiment unless additional replication is included. Is there a cheap surrogate response that will be adequate to demonstrate the effects of the factors?

Choosing the design. Resolution V designs are the safest, although, if you have a large number of factors, it may be worth doing resolution IV or III rather than dropping factors. The choice will depend on the number of runs you can afford.

How many experimental runs can you perform under relatively homogenous conditions? If the number of runs is less than that needed for the complete experiment, you should consider blocking the design. How homogenous is your experimental material? If material is not homogenous you should again consider blocking your experiment.

Do you have any preliminary estimates of the precision likely to be achieved by the experiment? If not, do you want to run a pilot experiment first to estimate variability?

Can the investigation be divided into several small experiments, particularly if they can be run sequentially, to guide you in the right direction?

Is the plan too complicated to run? Should the order of trials be changed so that fewer demands are made on the operators? The sequence of runs could be arranged so that a 'difficult or expensive to change' factor is altered only a few times during the investigation. However, the adoption of any of these strategies will remove the benefits of randomisation and a blocking structure may be a good idea.

What procedure will be adopted if a trial does not run or gives rogue values? Will the trial be repeated? If so, how many times?

Who is going to do the work and when? Is it possible to divide the experiment, materials or analytical work into blocks within each of which there will be less variation than over the experiment as a whole?

Are there likely to be potential trends over time?

Are there factors or covariates that will change during the experiment over which we have no control but which could be recorded?

Have you randomised the run order? Are there constraints on the randomisation?

When running the experiment it is vital to monitor the process carefully to ensure that everything is being done according to plan. Upfront planning is crucial. It is easy to underestimate the logistical and planning aspects of running a designed experiment in a complex manufacturing or research and development environment. Don't stop part way through a design unless there are major problems. *Don't be tempted to stop early because you have reached your ideal value for a particular run.*

Analysis of the results. Use a package that makes full use of graphical techniques. Back these up with simple statistics for objective conclusions.

Are you clear what analyses you will perform? Will your analysis take into account variation between blocks? Will your analysis adjust for covariates? Are you clear what model checks you want to perform?

Conclusions from the statistical analysis need to be translated into practical conclusions.

Further experimentation (verification run or new design). Are there still questions that need to be addressed? Do you need to perform a fold-over? Don't forget to verify the results from the experiment; perform a confirmatory experiment at your predicted optimum.

6.9 Effect of not following the design exactly

There are occasions when the results from experimental runs have been lost through carelessness or accident. This could result in an unequal number of replications per trial, or even in the loss of the whole set of results for one or many experimental runs. If a lost value is not the result of the effects of the factors involved, then it is a true 'missing value'. However, if a particular combination of levels led to the destruction of a product unit, and as a result there are no response values available, these cannot be considered as missing values. Important information has been acquired regarding something the experimenter had not known in advance.

In the case of 'proper' missing values, obviously the design will be incomplete if the experimental trial cannot be re-run and, as a result, the analysis could become complicated. The problem with missing data is that the confounding pattern is messy. The degree of the problem depends on the proportion of missing values. If you have relatively few missing values in relation to the overall size of the experiment, the effect should not be too great (so long as there is no pattern to the missing values). However, if you have a large proportion of missing values, then the validity of the experiment may be questionable and it may not be possible to get any useful information from the runs.

6.10 Other potential problems

Here are some common causes of problems:

- Failure to randomise.
- False replication, i.e. taking duplicate readings on an experimental run rather than repeating the whole run, leads to an underestimate of the background variability.
- Overlooking nuisance factors leads to an inflated estimate of background variability.
- Incorrect choice of design. Too few runs means you cannot detect differences against the level of background variability. Too many runs means a waste of resources.
- Overlooking factors. If a critical factor is not included in the list of potential factors for investigation, then it may influence the results of the runs that are made but you will have no direct information about it. The same is also true of traditional, non-experimental design approaches. Fortunately, experimental design provides a set of diagnostic tools, such as lack-of-fit tests and residual analysis, that enable the presence of a lurking factor to be detected.
- Sizing the design space: if the factor ranges are too wide, run sizing or ranging experiments first, then narrow the ranges to avoid catastrophic failures.
- Exploring too small a range of the factors. It may be that the optimum conditions are outside your initial design space—experimental design is a sequential process, so do not spend all your resources on the first experiment.
- Ignoring the possibility of interactions.
- Assuming that all the levels will run and give reasonable responses.

- Designing the experiment to optimise the process when there is no knowledge about which factors are important.
- Failing to sequence trials appropriately: a time trend in the response looks like a significant effect.
- Not consulting relevant people before the experiment.
- No reliable, robust and meaningful way to measure responses. Whether using a DOE approach or the traditional approach, this will cause problems. Good experimentation demands a good measurement technique. DOE will often alert you to the fact that your measurement technique is flawed. It may be some time before you realise this with a traditional approach as you could easily attribute the response noise to an experimental factor. Time spent in establishing a meaningful response measurement before you start is rarely wasted. And, of course, DOE is an ideal way of proving that the response method is robust!

6.11 Screening designs in context

Throughout this chapter we have alluded to the benefits of taking a sequential approach to process or product development. Screening designs, such as full or fractional factorials, are an important component of this approach but they are only the starting point of what can be quite a long process. A schematic of a typical sequential development process

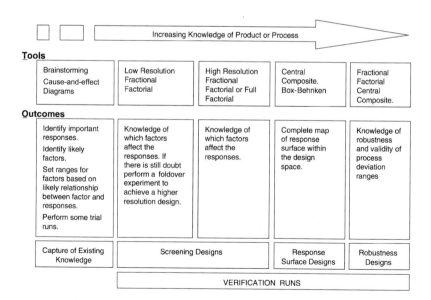

Figure 6.19 Typical sequential development process using DOE.

using DOE is shown in Figure 6.19. It shows how information and knowledge about the product or process increase as the structured sequence is followed. But what is important from a practical point of view is that the quick and cheap actions occur at the beginning of the process and those that can be complex and expensive are deferred to the end of the process.

The scheme might suggest that all investigations have to follow all stages and follow each to completion. This is not the case. Many investigations stop after the first box simply because the act of questioning what is happening throws up critical information that is already known but not appreciated and that is the answer to the investigation.

A real-life example of the structured sequential approach is given here to conclude this chapter. Another example can be found in the chapter by D.R. Pilipauskas in Gadamasetti (1998).

6.11.1 Example: Summary of DOE performed during the development of a beta-lactam

This example is discussed in more detail in the chapter by M.R. Owen in Gadamasetti (1998).

The reaction scheme is shown in Figure 6.20 and the cause and effect diagram in Figure 6.21. The aim of this study is to evaluate the use of automation equipment in conjunction with experimental design to increase understanding of this reaction at process scale. A series of studies were carried out and a summary of each is given below. The studies in this case go through the full sequential process, but it may not be necessary in every situation.

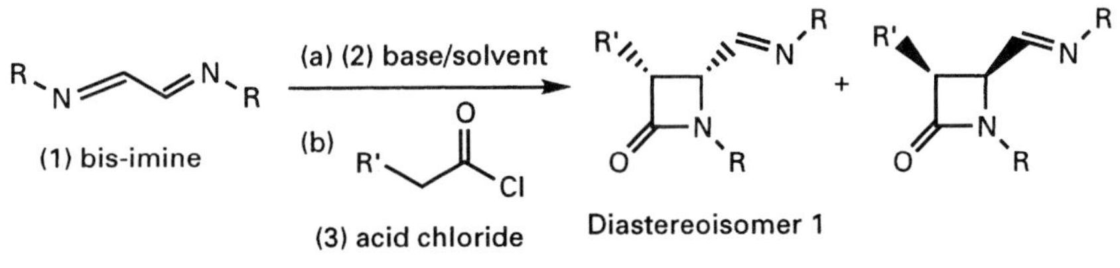

Figure 6.20 Chemical structure for case study.

Study 1: Order of addition (6 reactions)
   Goal. Keeping all other factors constant, look at six permutations of order of addition of bis-imine, base, and acid chloride (as shown in the reaction sequence).

   Outcome. To obtain maximum yield of diastereoisomer 1, the best sequence was found to be (1) imine, (2) base, (3) acid chloride.

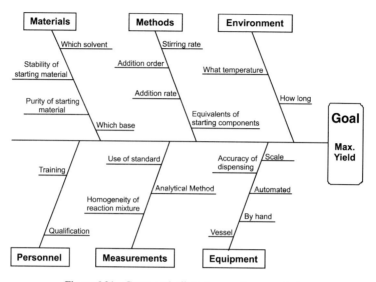

Figure 6.21 Cause-and-effect diagram for case study.

Study 2: Screening of solvents (16 reactions at 2 time-points)
Goal. A range of eight solvents (dichloromethane, tetrahydrofuran, isopropylether, ethyl acetate, chloroform, toluene, acetonitrile, N-methylpyrrolidone) were screened, under four sets of conditions varying time and temperature.

Outcome. This screen indicated that dichloromethane, chloroform and acetonitrile gave the highest yields. For the other solvents, yields were temperature dependent. In general, time was relatively unimportant.

Study 3: Screening of bases (12 reactions at 2 time-points)
Goal. A range of six bases (triethylamine, trioctylamine, triisopro-pylamine, 4-methylmorpholine, pyridine, N-methylpiperidine) were screened, under four sets of conditions varying time and temperature.

Outcome. This screen indicated that triethylamine, trioctylamine and triisopropylamine gave the highest yields.

Study 4: Screening of solvent/base array (18 reactions at 2 time-points)
Goal. An array of three solvents (dichloromethane, chloroform

and acetonitrile) and three bases (triethylamine, trioctylamine and triisopropylamine) were screened, under four sets of conditions varying time and temperature.

Outcome. This screen indicated that the combination of dichloro-methane and triethylamine gave the best result.

Study 5: Screening of factors (20 reactions)
Goal. Keeping the combination of dichloromethane and triethylamine constant, the following factors were screened for relative importance on their effect of influencing yield in solution:

- Equivalents of triethylamine
- Equivalents of acid chloride
- Rate of addition of acid chloride
- Time
- Temperature

Outcome. This screen indicated that the following factors were the most critical in affecting yield:

- Equivalents of triethylamine
- Rate of addition of acid chloride
- Temperature

Studies 6 and 7: Optimisation of the three critical factors (2 × 20 reactions)
Goals
Study 6: To apply a central composite design on the three critical factors (identified by study 5) to maximise the yield.
Study 7: To repeat study 6 using the same factors, but narrowing the range for each factor.

Outcome. These studies revealed the optimal settings for the factors, and indicated the degree of robustness and the effective ceiling for the reaction yield using these conditions.

Overall outcome for studies 1–7
These studies showed the versatility and synergy of using experimental design and the automation equipment to solve typical process investigation targets. Having gained understanding at this scale, thought should be given to the possible scale-up issues and any necessary experimentation on a larger scale would need to be performed.

Acknowledgements

I am indebted to Martin Owen (Glaxo Wellcome) and Dennis Lendrem (TTC Training and Consulting, New Moor House, Alnwick) for their contribution to this work in making it more relevant for chemists, and to the reviewers from Glaxo Wellcome, Richard Lyons, David Robinson and Peter Godbert.

Bibliography and References

Box, G.E.P., Hunter, W.G. and Hunter, J.S. (1978) *Statistics for Experimenters*, 1st edn, Wiley, New York.

Carlson, R. (1992) *Design and Optimization in Organic Synthesis*, 1st edn, Elsevier Science, Amsterdam.

Frigon, N.L. and Mathews, D. (1997) *Practical Guide to Experimental Design*, 1st edn, Wiley, New York.

Gadamasetti, K. (ed.) (1998) *Process Chemistry in the Pharmaceutical Industry*, Marcel Dekker, New York.

John, P.W.M. (1971) *Statistical Design and Analysis of Experiments*, MacMillan, New York.

Logothetis, N. and Wynn, H.P. (1989) *Quality Through Design*, 1st edn, Oxford University Press, Oxford.

Montgomery, D.C. (1991) *Design and Analysis of Experiments*, 3rd edn, Wiley, New York.

7 Designs for response surface modelling—Quantifying the relation between factors and response

I. Langhans

7.1 Introduction

The whole concept of experiment design is one of taking a structured approach to understanding and controlling what is going on in your experiments or processes. Chapter 6 describes how factors that influence the responses in your experiments may be identified. In this chapter we shall look at how a quantitative description of the relationship between the response and the factors can be obtained.

Throughout the chapter, reference is made to a particular case study. The detail of this study is given in the appendix at the end of this chapter.

7.2 Where to go

Information about	Go to Section(s)
What is a response surface?	7.3.1
What is an interaction and how do I recognise it?	7.4.2
Is a concentration range 0–5% ok for a factor?	7.5.1
How many experiments are needed?	7.5, 7.9.8
Why are there several centre points?	7.5.2
How do I choose between the different designs?	7.5.4
How do I get the response surface?	7.7.2
How well can a future response be predicted?	7.8.1
Are there problems if an experiment fails?	7.8.7
I can't do all the experiments on a single batch or in a single day.	7.8.8
Do I have any other options?	7.9
I want a design for a solvent mix	7.9.4

7.3 The basics

7.3.1 What is response surface modelling?

Response is the output of the system under investigation to various factor settings that we apply. Typical and obvious responses are properties like the yield or selectivity of a reaction, the strength or melt flow rate of a polymer, the precision or accuracy of a chemical analysis, and so on. Less obvious but plausible candidates are such things as the cost of a product, its market share or even the value of the chemical company's shares on the stock exchange. Needless to say, establishing the effect of, for example, reaction temperature on the yield is bound to be more successful than determining its effect on the company's share value.

The *surface* in response surface modelling comes in because we usually want to investigate the relation between a response and several factors, leading to a multidimensional problem and requiring a multidimensional representation. Starting with a one-dimensional problem, e.g. the effect of temperature on conversion, the response surface is one-dimensional (and sometimes called a line) but we need two dimensions to represent it: one for the factor and one for the response (Figure 7.1). Keep in mind that only rarely will a problem be truly one-dimensional; slicing your problem up into univariate subproblems may simplify things for a moment, but is not efficient and in most cases it can be very misleading.

Describing the relation between a response and two factors usually requires a three-dimensional representation, e.g. the relation between Conversion and the factors Temperature and IEP in the case study (Figure 7.2). A popular and more useful alternative to the 3-D graph is a contour plot (Figure 7.3). Note that the univariate slice in Figure 7.1

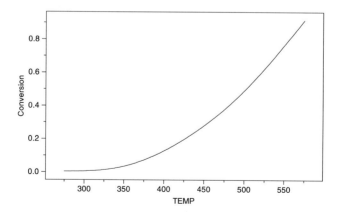

Figure 7.1 Conversion versus reaction temperature (TEMP) at high IEP.

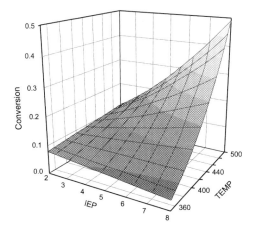

Figure 7.2 3-D plot of conversion versus TEMP and IEP.

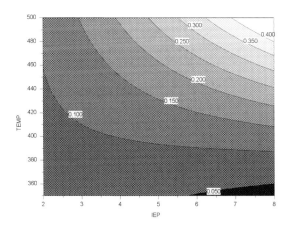

Figure 7.3 Contour plot of conversion versus TEMP and IEP.

(corresponding to an IEP value of 8 in Figure 7.3) gives a misleading or at least a very 'local' view of the relation between conversion and temperature.

Although it is possible to go one step further and visualise the relation between a response and three factors, this is rarely done. The resulting graphs (actually 3-D contour plots) are difficult to construct, interpret and use in communications with others (some examples can be found in Box and Draper, 1987).

Modelling, or in particular the term *model*, is not to be associated with the ideal but unrepresentative mannequins that we see showing off on the catwalk, but rather with a pile of clay or plasticine that we can knead and mould until it fits our needs. In more chemical terms, we are not going to use ideal, theoretically derived 'mannequin' functions (e.g. some equation derived from kinetic theory) to describe the relation between response and factors. Rather, we shall use some flexible, easy to mould empirical function that does what it is supposed to do: to represent the relation between response and factors as it is present in the experimental data. Another term to discriminate between these two approaches is *hard* versus *soft* or *data-driven* modelling.

Put together, *response surface modelling*, or RSM, stands for obtaining a quantitative, *sufficiently detailed* description of the relation between response and the different factors, such that we can use it for determining optima. In more general terms, RSM is used to predict the response for any combination of factor settings within the ranges we have chosen to study (that is, we shall not extrapolate).

7.3.2 *What has this got to do with experimental design?*

Experimental design is all about doing the optimal experiments to accomplish some objective or reach some goal. If we choose as optimality criteria the quantification of the effects of the factors with *maximum precision* and in an *unambiguous* way (i.e. have uncorrelated estimates of the different effects), then we will come to different designs depending on the goal (Table 7.1).

Table 7.1 Overview of study objectives and corresponding optimal designs

Objective of study	Optimal designs
Screening	Fractional factorial or Plackett–Burman
Determining main effects + all interactions	The full factorial
Response surface modelling	RSM designs: multilevel designs

Before we meet these RSM designs, let us first get acquainted with the models we are going to use.

7.4 Soft modelling using polynomials

7.4.1 *Of true models and their approximations*

As stated in Section 7.3.1, we are not going to use theoretically based functions to describe the relation between response and factors. Instead we shall apply models that are *obviously wrong*, so that you and others will only use them in the way they are meant to be used: *local modelling* of the response–factor relation in the area that was explored by the experimental design. These wrong models are usually the polynomial models that originate from a *Taylor approximation*. Any continuous function can be approximated by means of a polynomial and this approximation will be better if we increase the order of the polynomial and/or reduce the interval over which we approximate this function. You might remember the formula for the one-variable situation:

$$y = f(x_1, x_2) \approx \beta_0 + \beta_1 x + \beta_2 x^2 + \cdots + \beta_p x^p + R(p+1) \qquad (7.1)$$

Although reality is truly nonlinear and not some polynomial, very often the second-order approximation from the Taylor series is *nonlinear enough* to describe the relation between factors and response sufficiently well, especially given the narrow ranges over which factors are allowed to vary in most industrial experiments.

The second-order Taylor expansion for a two-variable system is

$$y = f(x_1, x_2) \approx \beta_0 + \beta_1 x_1 + \beta_2 x_2 + \beta_{11} x_1^2 + \beta_{22} x_2^2 + \beta_{12} x_1 x_2 \qquad (7.2)$$

Notice that next to the familiar first-order and second-order components, a new term appears in the polynomial: the *cross-term* $x_1 x_2$. Actually, this little term is all we need to model an *interaction effect* (the 'twist' in the response surface in Figure 7.2).

The next step is to generalise this second-order polynomial to a function of an arbitrary number of factors:

$$y \approx \beta_0 + \sum_i \beta_i x_i + \sum_{i \neq j} \sum_j \beta_{ij} x_i x_j + \sum_i \beta_{ii} x_i^2 \qquad (7.3)$$

For the sake of clarity: we will not use these polynomials for approximating some *known* function $f(X)$ but for modelling the relation between the response and the factors as it is present in the *experimental data*.

Some typical response surfaces that can be modelled by means of these second-order polynomials are shown in Figure 7.4. It should be noted that the β coefficients in equation (7.3) can act as a 'zoom' factor such that any portion of these surfaces can be expanded and used for the experimental region.

Theorems are nice and mathematics is great, but simulations are better! Let us suppose we are looking at a first-order reaction where a species A changes into a species X as shown in equation (7.4).

$$A \xrightarrow{T} X \qquad (7.4)$$

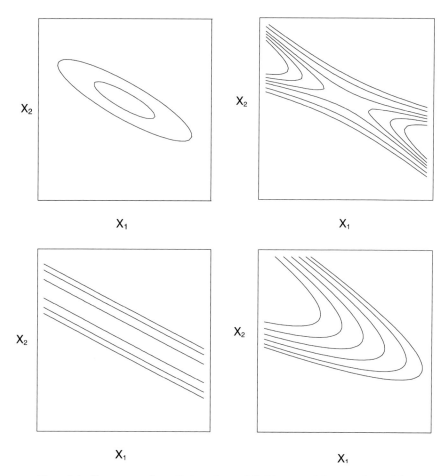

Figure 7.4 Response surfaces that can be modelled by second-degree polynomials.

Clearly, chemical kinetic theory has no problem describing such a system. But suppose we do not know anything about these theories. All we have is a knob labelled 'Start concentration of A' and a button called 'Temperature'. Now we want to model the relation between the response X (amount of X after 10 minutes of reaction) and the factors A (the initial amount) and temperature, T.

Since we do not know anything about experimental design yet, we simply take a grid of factor values and perform an experiment for each combination of A and T, resulting in 25 response values: $A = 1, 2, 3, 4, 5$ combined with $T = 300, 325, 350, 375, 400$. We then fit a second-order polynomial to the responses, X, at these points and find the response surface shown in Figure 7.5.

RESPONSE SURFACE MODELLING

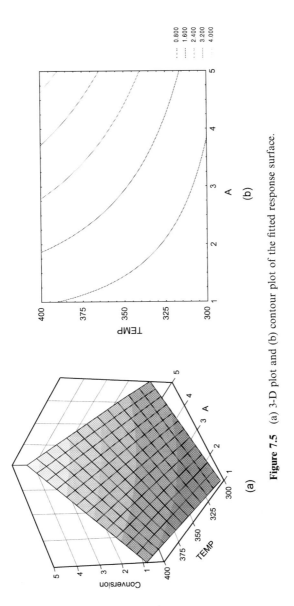

Figure 7.5 (a) 3-D plot and (b) contour plot of the fitted response surface.

The advantage of this example is that we know the 'true' relation between X, A and T (assuming that indeed nothing else is happening besides this first-order reaction—no impurities, no reverse reactions, etc.). The 'true' response surface derived from chemical kinetic theory is shown in Figure 7.6.

The polynomial approximation is pretty much like the true response surface, as long as we do not extrapolate to large initial concentrations of A and high reaction temperatures, because then the polynomial tells us that we can make more X than we put in A. This is shown graphically in Figure 7.7. This extrapolation failure is not a real problem since you should *never extrapolate* (at least not without performing confirmatory experiments). It is also dangerous to extrapolate using 'true' models without confirmation, for the simple reason that 'true' models are often less true than you want them to be.

Note: There is no simple cross term in the true kinetic function but there is an interaction effect present: the effect of T on X in an absolute sense depends on the value of A. This is also called non-additivity of effects. We very often need second-order interaction terms even to model first-order kinetics.

This example may look a little silly since 'everybody' knows about first-order kinetics and we might as well use the information derived from the theoretical model. But the example shows that even with no knowledge of chemical kinetics we could have found an appropriate answer (in this example at the cost of a few too many experiments). For most real problems we do not know the real underlying models but we can use the polynomial modelling in a similar fashion.

7.4.2 Relation between the objective of a study and the complexity of the polynomial

Screening. Remember that screening is a first-order question: does factor x_i contribute to variation in the response y or not? Or put in terms of a polynomial approximation: is the term $\beta_1 x_i$ important in the first-order polynomial? This can be put in a more statistical phrasing for equally scaled variables: is the coefficient β_i significantly different from 0?

$$y = \beta_0 + \beta_1 x_1 + \cdots + \beta_i x_i + \cdots + \varepsilon \qquad (7.5)$$

A screening study, often using fractional factorial designs, corresponds to fitting a first-order model, as in equation (7.5), to the response surface. Figure 7.8, for example, visualises the first-order description of the relation between Conversion and the factors Gas and CR in the case study.

RESPONSE SURFACE MODELLING

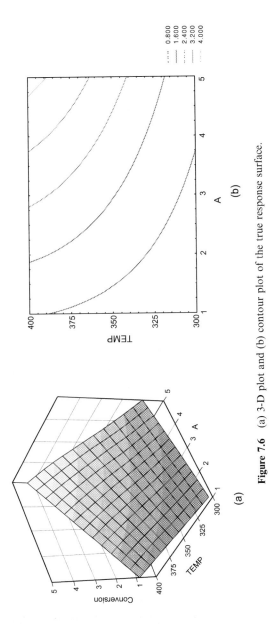

Figure 7.6 (a) 3-D plot and (b) contour plot of the true response surface.

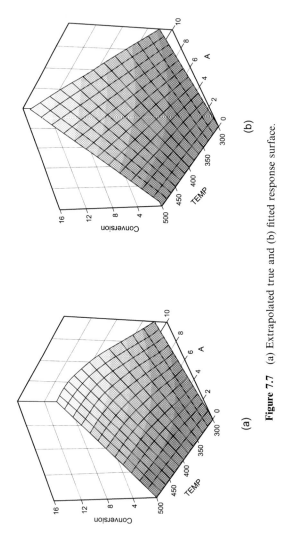

Figure 7.7 (a) Extrapolated true and (b) fitted response surface.

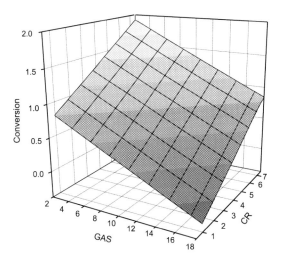

Figure 7.8 Fitted linear relation between Conversion and the factors GAS and CR.

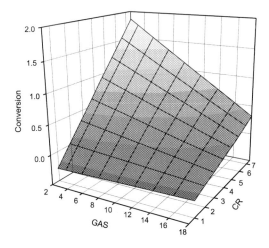

Figure 7.9 Second-order (interactions only) surface for Conversion versus GAS and CR.

Studying interactions. Obviously, if we want to quantify interaction effects, we need to expand the polynomial with cross terms (usually second-order will do, occasionally you might need third-order interactions) as is shown in equation (7.6). An example is shown in Figure 7.9.

$$y = \beta_0 + \sum_i \beta_i x_i + \sum_{i \neq j} \sum_j \beta_{ij} x_i x_j + \varepsilon \tag{7.6}$$

Optimisation and RSM study. Here, we want a detailed description of the *y–x* relation and go for a full quadratic approximation (equation 7.7) which means that all second-order interactions need to be quantified, not only the square terms. Here too, once in a (long) while you might need some higher-order terms.

$$y = \beta_0 + \sum_i \beta_i x_i + \sum_{i \neq j} \sum_j \beta_{ij} x_i x_j + \sum_i \beta_i x_i^2 + \varepsilon \qquad (7.7)$$

The corresponding response surface is shown in Figure 7.10.

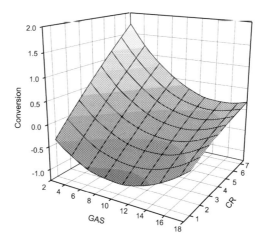

Figure 7.10 Full quadratic (interactions + squares) response surface for Conversion versus GAS and CR.

7.5 Designs for response surface modelling

7.5.1 General considerations

Since the number of coefficients in the RSM models is higher, we will need to perform more experiments than, for instance, in a screening study. Table 7.2 shows the number of coefficients in quadratic models with two to seven factors. Being able to determine all second-order interactions

Table 7.2 Number of coefficients in quadratic models for two- to seven-factor systems

Number of factors	2	3	4	5	6	7
Coefficients in quadratic model	6	10	15	21	28	36

requires a design that is resolution V. (*Note*: The concept of resolution strictly only applies to two-level designs. Later we will encounter RSM designs where the factorial part has a resolution lower than V.)

Including square terms in the model means that two-level designs will no longer be appropriate so we have to look for multilevel designs ('multi' being at least 3). Fortunately, this does not mean we have to go for the 5^2 design that we used in the example in Section 7.4.1, nor does it mean that we have to apply three-level full factorials (costing 3^k experiments).

In Chapter 6 we saw that the differences between two-level designs are very important (full factorial versus resolution IV or Plackett–Burman designs), and the risks involved with choosing a resolution III design can be enormous. In the RSM phase of an investigation it does not really matter all that much. If you stay away from the cheap designs that will be discussed in Section 7.9, nothing really bad can happen to you.

Two last warnings before we really get started:

- Avoid a value of 0 for the concentration of a reactive component. Usually nothing happens at zero concentration, resulting in an extreme nonlinearity. Also, if you want to perform experiments near the 0 value, you might want to consider transforming your factors to, for example, a log scale.
- If you are not sure about the factor ranges, avoid jumping straight on to an RSM design even if you have only a few variables—do a screening design or some exploratory experiments first.

There are two popular designs for response surface modelling: the central composite and the Box–Behnken designs. They form two classes of designs that have numbers of runs much closer to the number of coefficients to be determined in the model (at least when compared to the 3^k full factorial designs).

7.5.2 *Central composite design*

These designs are composed of two parts (see Figure 7.11 also):

- a cube part, which is full factorial, or fractional factorial of resolution V, allowing for determination of main and interaction effects;
- a star design (actually a one-variable-at-a-time design) for quantifying main and quadratic effects.

(See Section 7.9 for cheaper alternatives and their consequences.)

Central composite designs are a flexible tool in that they can be modified in a number of ways. For instance, the length of the star axes, α, can be modified relative to the size of the cube, d, having an effect on several properties (see Section 7.8). The most obvious consequence of changing the axial distance is that the design changes from a three-level to a five-level design.

Thus, the CCF (central composite face-centred) design has three-levels and fills a square (two factors—see Figure 7.11) or a cube (three factors—see Figure 7.12), while the CCC (circumscribed central composite) design is more circular or spherical and has five levels. These structures also allow flexibility for using blocks of runs in the design (see Section 7.8.8).

Another major advantage of these designs is that the structure fits nicely into the *sequential experimentation* that is involved with experi-

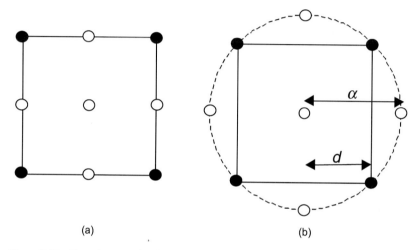

Figure 7.11 Central composite designs: (a) face-centred (CCF), (b) circumscribed (CCC).

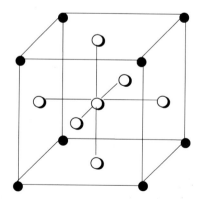

Figure 7.12 Three-factor central composite design.

mental design:

(1) Start screening with some sparse screening design of resolution IV.
(2) Upgrade by adding a fold-over design resulting in a cube part of resolution V or higher.
(3) Add stars to the cube to obtain a central composite design.

Usually steps 2 and 3 are combined since the extra number of runs associated with step 3 is usually not large. The disadvantage of central composite designs is that they can sometimes be quite expensive. The number of experiments needed for central composite designs with numbers of factors from 2 to 7 is given in Table 7.3.

Ninety-two runs for seven factors is really too much. Although it happens only rarely that seven factors survive a screening study (and are identified as important), if we look at just one response, when there are 10 or 15 responses it will happen more often that no variable can be omitted.

Table 7.3 Total number of experiments (N_{tot}) and number of centre points (N_{centre}) for central composite designs for two- to seven-factor systems

	No. of factors					
	2	3	4	5[a]	6[a]	7[b]
N_{tot}	13	20	31	32	53	92
N_{centre}	5	6	7	6	9	14

[a] The factorial points are based on a resolution V fractional factorial.
[b] The factorial points are based on a resolution VII fractional factorial.

Another puzzling part may be the large number of centre points. There is only one point in the centre of the design, so five centre points means five repetitions of this particular combination of factor levels. However, it is important to note that five experiments is not the same as five measurements. The between-repeats variability should be representative of the overall noise level in your data, so you have to *repeat the whole run and not simply make repeated measurements of the response within one run*. The number of centre points listed generates a so-called uniform precision design (Section 7.8), which should mean that the precision with which you predict the response is the same over the whole design region. Uniform precision, however, is, never completely reached and if you carry out four replicate centre points instead of six centre points for the 3-, 4- or 5- factor case, the prediction precision will be uniform enough (see also replicated axial designs, Section 7.8.7).

7.5.3 Box–Behnken designs

Box–Behnken designs can also be considered as composite designs but composed of rotated lower-dimensional designs (Figure 7.13). Box–Behnken designs are globular three-level designs that offer much less flexibility than composite designs, but they are cheaper in most cases (see Table 7.4). Box–Behnken designs do not fit in with the sequential design approach since the fractional factorials used for screening do not form a subset of the Box–Behnken designs (the extreme corners are missing).

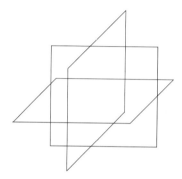

Figure 7.13 Representation of a three-factor Box–Behnken design.

Table 7.4 Total number of experiments (N_{tot}) and number of centre points (N_{centre}) for Box–Behnken designs for two- to seven-factor systems

	No. of factors					
	2	3	4	5	6	7
N_{tot}	–	15	27	46	54	62
N_{centre}	–	3	3	6	6	6

7.5.4 Choosing a design

Although a lot can be said about the differences between these RSM designs, a few pragmatic criteria will help in selecting the most appropriate design.

How well does the design fill the experimental region?
If the design region is *spherical* (investigation around some point), and then the CCC and Box–Behnken are appropriate designs.

If the design region is *cubic* (region limited by lower and upper levels on factors), then the CCF designs are more appropriate.

Often the factor limits are only best guesses and in this case a CCC design can be useful to 'peek' across the 'hard' limits.

The type of nonlinearity.
CCC designs are five-level designs and have some room to check a nonquadratic nonlinearity in one of the variables.

CCF and BB are three-level designs (BB is somewhat better for checking the presence of the third order interaction A^2B).

The available budget?
The upgrade of a screening design will usually be cheaper for a central composite design than for a Box–Behnken design (provided the variable ranges remain the same).

Prediction precision.
CCC designs are superior unless they are shrunk into the cube (a so-called inscribed central composite design), but the differences are not that large (Section 7.8).

Identification of terms.
The least important property of an RSM design is the precision with which you can determine the effect of the individual terms. Although it is always nice to be able to offer some interpretation to the results, the main goal is an adequate description of the response surface, which comes down to a high precision of prediction. This can be accomplished without having precise estimates of the individual effects (cf. the problem of colinearity in Chapter 11).

7.5.5 Case study

For the catalyst problem described in the appendix, a central composite design was obviously the design of choice, since for five factors the Box–Behnken is about 50% more expensive and the central composite designs offer more flexibility.

The ranges, as they are given in the appendix, span the whole range for the first factor and a reasonable range for the remaining factors, leaving some room to check beyond this initially defined area. Also, instead of six centre points, four were included in the design.

Table 7.5 lists the central composite design in standard order: first the 16 cube points (according to a resolution V fractional factorial), then the 10 star points followed by the four centre points.

7.6 Doing the experiments

Just as for screening designs (Chapter 6), it is dangerous to carry out an RSM design in standard order, so the order in which we carry out experiments is randomised or at least less structured. Centre points are not randomised but are distributed more or less evenly over the design. When carrying out the experiments it is important that you perform them with the same attention that you usually use when experimenting. This seems obvious, but experience teaches that, because results are interpreted only after the *whole* design has been completed, experimenters sometimes lose concentration as the number of runs increases, resulting in deviations from the chosen factor settings, or sometimes plain mistakes.

The centre points should be carried out as if they were experiments like any other: not extra carefully nor extra sloppily. Otherwise, they are not a representative estimate of the noise variance.

Also write everything down, measure everything that can be measured and control every variable that is not part of the design and that might have potential influence on the results. The experimental results may not

Table 7.5 Central composite design for the case study

Number	IEP	CR	GAS	TEMP	TIME
1	2	0.5	2	350	50
2	8	0.5	2	350	10
3	2	7.5	2	350	10
4	8	7.5	2	350	50
5	2	0.5	18	350	10
6	8	0.5	18	350	50
7	2	7.5	18	350	50
8	8	7.5	18	350	10
9	2	0.5	2	500	10
10	8	0.5	2	500	50
11	2	7.5	2	500	50
12	8	7.5	2	500	10
13	2	0.5	18	500	50
14	8	0.5	18	500	10
15	2	7.5	18	500	10
16	8	7.5	18	500	50
17	2	4	10	425	30
18	8	4	10	425	30
19	5	0.1	10	425	30
20	5	8	10	425	30
21	5	4	1	425	30
22	5	4	19	425	30
23	5	4	10	300	30
24	5	4	10	550	30
25	5	4	10	425	5
26	5	4	10	425	55
27	5	4	10	425	30
28	5	4	10	425	30
29	5	4	10	425	30
30	5	4	10	425	30

be representative of the usual noise level, but the purpose is to quantify the relation between response and factors as well as possible, not to estimate the noise variance.

A final comment: every experiment should be comparable with every other experiment. For example, if you normally have to clean a reactor every 3 or 4 experiments, clean it after each experiment, otherwise the experiment just before cleaning may be systematically different from the experiment just after cleaning.

7.7 Analysing the data

7.7.1 General considerations

The results are analysed by means of regression analysis and you will find a more detailed treatment in Chapter 11 on regression. What follows is a brief (p)review and some special considerations.

The most important difference between regression analysis for designed as compared to nondesigned data is that in the designed case the analysis becomes almost trivial. *Almost*, not completely! Otherwise you should just be able to plug the data in your PC and click the 'Analyse' button, and this is unfortunately *not* the case.

7.7.2 A step-by-step look at the analysis of designed data

Fitting the model
Fitting the model means finding the best fitting values for the β-coefficients in the model. 'Best fitting' requires some criterion and in most cases the least squares criterion will do the job. Finding the best fitting b-coefficients (note that β denotes the true value of the coefficient and b the least-squares estimate) is not a statistical problem but an algebraic one, namely solving a set of linear equations, the solution being given by (see Chapter 11)

$$\mathbf{b} = (\mathbf{X}'\mathbf{X})^{-1}\mathbf{X}'\mathbf{y} \qquad (7.8)$$

This solution can always be found except when the columns in the design matrix $\mathbf{X}$ are not linearly independent (note that $\mathbf{X}$ consists of all terms currently in the model, not only the first-order terms). Linear dependence occurs if at least one column can be written as a linear combination of the other columns, which in the designed case can be caused by one of the following:

- The model contains more terms than there are distinct data points (distinct: think of 100 centre points and nothing else: you will only be able to fit a constant not even a linear model).
- The model contains identical terms (e.g. two confounded interactions or two square terms when the design is a factorial + centre points).
- The model contains a term of higher order than can be determined by the design (e.g. fitting a term x_i^3 when the design is three-level).

In all other cases, you will always find some nonzero value for the different b-coefficients. That seems good, but it is not really: it would be nice if we would find a value of 0 for b_i when the true value β_i is 0. However, in practice you will never encounter a value of 0 (except in some very rare cases where you have rounded response values) for two reasons:

- The true model is not a polynomial but some nonlinear function and no β will be exactly zero.
- We never measure the true response of the system but always response + noise ($y = \eta + \varepsilon$), so even if the true response η was constant over the whole design region, we would still find nonzero values for the b_i due to variation in the noise part.

Thus, individual b-coefficients will never be zero and the model as a whole will always explain only part of the variation in the response—hopefully a large part. That is why we need to quantify the performance and validate the model.

Checking the quantitative performance of the model
The first things you ask yourself are

- How good is the current model?
- How well does it fit the data?

A simple and sufficient measure is R^2, the percentage of variance in the response, explained by the model. However, R^2 is a nonlinear measure of the quality of the model. An improvement from $R^2 = 0.7$ to $R^2 = 0.73$ is negligible, while an improvement from $R^2 = 0.96$ to $R^2 = 0.99$ is important.

An alternative is to look at $1 - R^2$ or, better, to look at the RMSE (root mean square error):

$$\text{RMSE} = \sqrt{\frac{\sum_{i=1}^{n}(y_i - \hat{y})^2}{n - p}} \qquad (7.9)$$

which corresponds to the standard deviation of the scatter around the fitted surface. The RMSE is in the units of the response and is directly interpretable providing you know the range of the variation in the response. Ideally you have some idea of the noise level, so you can check whether the RMSE is at the expected level (check for overfit or lack of fit).

Other alternatives are R^2_{adjusted} and R^2_{PRESS}. Of these, R^2_{PRESS} is the most sensitive and it is supposed to be an estimate of how well you can predict the response. In general, R^2_{PRESS} is also a bit too optimistic and for designed experiments simply looking at R^2 or RMSE is sufficient, provided that you only keep terms in your model that are clearly significant.

Statistical validation of the model—is it real or is it noise?
R^2 or RMSE or any other performance measure tells you how good the model is at fitting the data but does not answer the questions

- Is the current model really needed?
- Does the current model explain a statistically significant part of the variation in the data or might it all be just noise?

The model *F*-test will check this.

The model *F*-test is a hypothesis test, usually carried out in a classical way (comparing the *p*-values with a pre-chosen significance level). As for any hypothesis test it is essential to know what we are testing. In other words, what is H_0 and what is H_1?

Although the exact formulation of these hypotheses depends somewhat on how we get an estimate of the noise variance, the hypothesis tests can roughly be put as follows:

H_0: the current model can just as well be replaced by the model $y = b_0$, i.e. no effect of the factors (at least in the current degree of model complexity).
H_1: the current model is better than the model $y = b_0$; it is unlikely that the variation in the data is only noise.

Small *p*-values (less than the chosen confidence level, α) indicate that it is too unlikely that the model is determined by chance variation.

Statistical validation of the model—is the model good enough?
Whether or not the model *F*-test indicates a statistically significant model, we always want to know whether there is statistical evidence that it can be improved. Provided that there are replicates, this can be done by a so-called *lack-of-fit F-test*.

The associated hypotheses are:

H_0: the model explains everything except noise.
H_1: the model can be improved.

Small *p*-values indicate that the model can be improved.

Note that the lack-of-fit *F*-test is not very powerful if only a few replicates are available.

Statistical validation of the model—which terms offer a statistically significant contribution?
Although identifying the individual contributions to the model in an RSM study is less important than in a screening study, it is always useful to know the relation between the response and the factors. This allows us to choose the most efficient graphical representation (e.g. if a factor shows only a linear relation to *y* we can leave it out of a 3D or contour representation and visualise the interacting factors in these graphs). Also, removing non-significant terms makes the model more compact and more transparent.

The statistical significance is tested using a *t*-test on the coefficients (H_0: $\beta_i = 0$), so small *p*-values ($< \alpha$) indicate statistically significant effects.

A word of caution is needed (although not very important in an RSM study). Sometimes an α of 0.1 or 0.05 is suggested by software. This means that the probability of a Type I error (keeping a term in when its real contribution is 0) is 10% or 5%, respectively. But since we perform as many tests as there are terms in the equation of the model, the probability that we make at least one Type I error is in fact close to 1. A Bonferroni correction (test at α/p) may be too strict, but using $\alpha = 0.05$ often leads to unrealistically complex models. Keeping an eye on the size of the effect helps in selecting a reasonable model.

Assumptions
As for any hypothesis test, a number of conditions need to be fulfilled for the test to be valid. They can be summarised in the little statistical phrase $\varepsilon_i \sim N(0, \sigma^2)$, which should be read as: the noise component in each experiment comes from a Normal distribution with a constant mean of 0 and some constant variance σ^2. This means the average should not depend on anything (e.g. it is not allowed to 'drift' during the experiments) and also that there should not be any relation between the size of the noise component and, for instance, the settings of a factor or the value of the response. If any of these things happens (and quite often they do), it means that the calculated *p*-values are not correct, in which case we have to apply some corrective action (see regression Chapter 11) or be more careful in the interpretation.

Verifying the assumptions will be the main task of the graphical validation step.

Graphical validation of the model
Whatever the result of the previous tests and statistics, always carry out a full graphical analysis because a few numbers never tell the whole story

and we need to check the assumptions that underlie the tests used. This graphical analysis will mainly consist of all sorts of residual graphs:

- Normal probability plot (do not waste time looking at histograms—probability plots are much more powerful)
- Residuals versus the factors
- Residuals versus $\hat{y}$ or fitted versus measured response values
- Residuals versus the experiment order or time

For all of these plots the same rule applies: *no structure* should be present.

Note: Look at leave-one-out residuals (also known as deleted Studentised or externally Studentised residuals), but also look at raw residuals, since they are in the units of the response.

Reporting results
Some suggestions to aid the clear reporting of results:

- Do not report every test or statistic: after a while these may seem obvious and very informative to you but most of your colleagues will not share that opinion and will stop reading your reports.
- Use the fitted vs measured plot to illustrate the quality of your model. Be careful with R^2: a value of 0.9 or 90% sounds quite good but is unsatisfactory for most chemical applications (use RMSE instead).
- An even better way to illustrate the quality of your model is to show prediction plots (effect plots + confidence intervals, e.g. Figure 7.20).
- Use contour plots for quantitative interpretation and 3-D graphs for providing qualitative insight. Be aware that if you have linked interactions $A*B$ and $A*C$, the A–B contour or 3-D plot will alter if you change C.
- If you include the polynomial model (including the best fitting values of the coefficients) in your report, make sure that you have rescaled everything to the original factor units (instead of a $[-1, 1]$ scaling). Put strong emphasis on the fact that the resulting equation is only to be used for predicting response values for factor combinations that do not involve extrapolation! Also warn the reader that they should not try to interpret the coefficients of this polynomial directly but rather use the graphs you made for that purpose.

Using the model—finding optima
If you find an optimum or some other interesting set-point, calculate the uncertainty on that prediction. Note that most software gives you

confidence intervals that should contain the true response value with given probability. This does not mean that 95% of your individual future experiments will fall within this interval. You need to construct a *prediction interval* for this.

Carry out some *confirmation experiments* before you start building that new plant and do not be surprised if these confirmatory runs do not correspond exactly to what you predicted (they might even fall outside the prediction intervals). This can happen for various reasons:

- The true relation is not well described by a quadratic polynomial.
- The design was carried out in a nicely controlled environment with one batch of each material and the confirmatory experiments were performed using other batches.
- You have done some overfitting (unavoidable since you have a limited number of experiments).

7.7.3 Case study

As sometimes happens, there is a slight difference between the design that was suggested (Table 7.5) and the design that was actually carried out (Table 7.6). Instead of four centre points (experiments 27–30) only two (experiments 27 and 28) were carried out. The last two experiments were performed using other inorganic oxides having the same IEP, namely TiO_2 and ZrO_2, but at settings of the other factors that did not correspond to any of the other design points.

To complicate things a little bit more, the two centre points have exactly the same response value (owing to rounding) so lack of fit testing is not possible. The modified design along with the response values is shown in Table 7.6. We will not discuss the analysis in detail (which would double the length of this chapter) but show some typical tables and graphs. Also note that the analysis was performed without the two odd oxide runs: they will be used as checkpoints at the end. A full quadratic model explains 97% of the variance in the response ($R^2 = 0.97$) but this includes some noise modelling (due to several nonsignificant terms). A model containing only the significant terms still explains 95%, which corresponds to an RMSE = 0.36 (standard deviation of the response equals 1.24).

There are, however, some problems. Figure 7.14 shows that the responses at the star points for the variable CR cover a wide range, suggesting a higher-order relation. In Figure 7.15 we see a problem that we could have predicted just by looking at the response: the response has a lower limit of 0 and most points are relatively close to this limit except for two points at the other end of the scale. In other words, we have a rather skew distribution of responses. This is no problem if we are

Table 7.6 Case study central composite design with measured conversion values

Number	IEP	CR	GAS	TEMP	TIME	Conversion
1	2	0.5	2	350	50	0.07
2	8	0.5	2	350	10	0.63
3	2	7.5	2	350	10	4.30
4	8	7.5	2	350	50	0.88
5	2	0.5	18	350	10	0.44
6	8	0.5	18	350	50	0.09
7	2	7.5	18	350	50	0.33
8	8	7.5	18	350	10	1.26
9	2	0.5	2	500	10	1.71
10	8	0.5	2	500	50	0.24
11	2	7.5	2	500	50	1.18
12	8	7.5	2	500	10	5.18
13	2	0.5	18	500	50	0.09
14	8	0.5	18	500	10	1.40
15	2	7.5	18	500	10	1.09
16	8	7.5	18	500	50	1.48
17	2	4.0	10	425	30	0.04
18	8	4.0	10	425	30	0.20
19	5	0.1	10	425	30	0.05
20	5	8.0	10	425	30	0.11
21	5	4.0	1	425	30	1.77
22	5	4.0	19	425	30	0.09
23	5	4.0	10	300	30	0.12
24	5	4.0	10	550	30	0.73
25	5	4.0	10	425	5	1.75
26	5	4.0	10	425	55	0.06
27	5	4.0	10	425	30	0.09
28	5	4.0	10	425	30	0.09
29	5	4.0	10	350	30	0.24
30	5	0.55	10	350	30	0.07

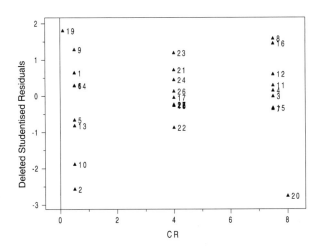

Figure 7.14 Deleted studentised residuals versus the factor CR, for the quadratic model.

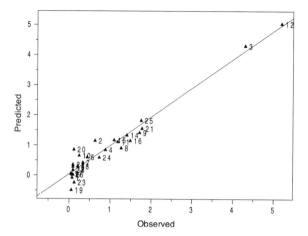

Figure 7.15 Fitted versus observed response values for the quadratic model.

interested only in finding the highest conversions, but it disturbs our modelling quite a bit if we are interested in describing the response surface both for high and low values of the response (which is the case).

A skewed distribution of responses is a typical situation in which a response transformation might be useful. In this particular case a compressing transformation such as square root, log or a reciprocal transformation will make the differences between the small response values larger.

A technique to test whether the improvement obtained from a transformation is statistically significant is a so-called Box–Cox analysis. In this case a square root transform of the response resulted in a statistically significant improvement (95% confidence level). Repeating the analysis but now with the square root of the entered response values leads to the model in Table 7.7 which explains 92% of the (transformed!) response, corresponding to an RMSE of 0.2 (standard deviation of the transformed response equals 0.56).

The fitted–measured plot (Figure 7.16) looks a little better now and, although there are still things that can be improved, the current model gives a reasonable description of the relation between response and factors (Figures 7.17a–d): reasonable, but not perfect, as can be seen from the rather wide confidence intervals in Figure 7.17d.

How about those two checkpoints (the two oxides TiO_2 and ZrO_2) that were left out of the analysis? Table 7.8 shows that their response values fall outside the corresponding confidence intervals (interval for the true value of the response at these settings) but are inside the prediction intervals (for individual values). Note that these intervals are a little awkward since a negative value for a square root has been set equal to zero. Also the

Table 7.7 Coefficients for the quadratic model with transformed response values and scaled factors

Conversion$^{1/2}$	Coefficient	Standard error	p-Value
Constant	0.375	0.069	4.593×10^{-5}
IEP	0.070	0.047	0.157
CR	0.278	0.046	1.466×10^{-5}
GAS	−0.215	0.046	2.354×10^{-4}
TEMP	0.140	0.043	4.624×10^{-3}
TIME	−0.347	0.046	7.176×10^{-7}
GAS∗GAS	0.344	0.085	8.449×10^{-4}
TIME∗TIME	0.259	0.080	4.878×10^{-3}
IEP∗TEMP	0.116	0.050	0.033
CR∗GAS	−0.125	0.050	0.022
GAS∗TIME	0.128	0.050	0.020

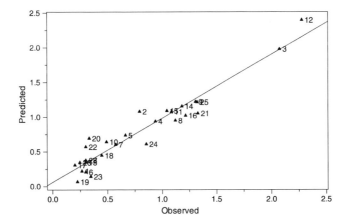

Figure 7.16 Fitted versus observed (transformed) response values for the quadratic model.

Table 7.8 Measured and predicted values for the two checkpoints with 95% confidence and prediction intervals

	Conversion		Intervals	
Run	Measured	Predicted	95% confidence	95% predicted
29	0.24	0.06	0.004–0.166	0.00–0.48
30	0.07	0.001	0.000–0.025	0.00–0.18

Bonferroni correction would have been appropriate (construct intervals at the 97.5% confidence level), but this does not affect the conclusion.

Maybe now is a good time to read the rest of the book or (if you are eager) to build your first RSM design.

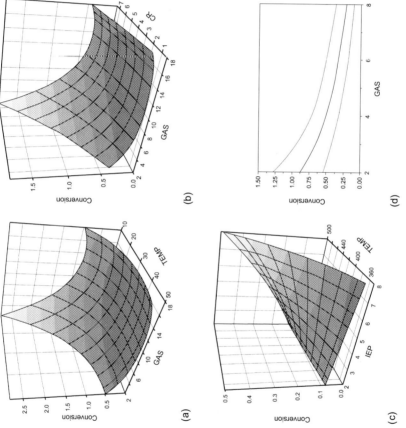

Figure 7.17 Visualisation of the fitted relations between the response and the factors.

7.8 A closer look at the properties of RSM designs

7.8.1 Prediction error

One of the most important prerequisites for a RSM design is that it should give a sufficiently precise characterisation of the response surface. We are not talking about how well the polynomial approximates the true response surface but about how precise our reconstruction would be supposing that the true model was a second-order polynomial. This comes down to the question 'How well can we predict the response for an arbitrary combination of factor settings (within the design region of course)?' In statistical terms, we want to know the size of the prediction variance assuming zero bias.

Some relevant design characteristics are then:

- What is the maximum prediction variance?
- What is the average prediction variance?
- How variable is the prediction variance (is it the same everywhere in the design region or does it have high and low values)?
- How does it depend on the location in the design region?

Note that all the following characteristics are *standard errors* and are *relative* to the standard deviation of the noise in the measured response.

7.8.2 Maximum prediction error

Table 7.9 shows the maximum prediction error within the cube, for different designs and different numbers of factors. The conclusion we can draw from Table 7.9 is that the differences are never really large; so if the choice is between these three designs, the maximum prediction error is not a very important criterion.

Table 7.9 Maximum prediction error for CCC, CCF and BB designs for two- to seven-factor systems

	No. of factors					
	2	3	4	5	6	7
CCC	0.79	0.82	0.76	0.94	0.79	0.74
CCF	0.89	0.89	0.81	0.98	0.82	0.69
BB	—	0.87	0.76	0.71	0.75	0.79

The number of runs follows Tables 7.3 and 7.4 and is different for CC and BB: α is chosen such that a rotatable design is obtained (Section 7.8.5).

7.8.3 Average prediction error

This criterion is less important in a sense, since we are more interested in the worst case value (the maximum prediction error). Again, the differences are small.

7.8.4 Uniform precision

Ideally, the prediction error would be the same all over the design region. In practice, this is not attainable but we can try to obtain an almost uniform prediction error or uniform precision. This is the main reason for the sometimes large number of centre points in Table 7.3. The effect of these extra centre points is illustrated in Figures 7.18 and 7.19, where the prediction error is plotted versus the two factors for a two-factor CCC design with respectively 5 centre points and 1 centre point.

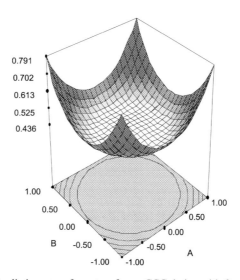

Figure 7.18 Prediction error for a two-factor CCC design with five centre points.

The five-centre-point CCC is fairly flat in the middle, whereas the one-centre-point CCC has the largest prediction error in the middle of the design. Table 7.10 shows the prediction error in the middle of a five-factor CCC as function of the number of centre points. This behaviour is a typical picture for most designs: the difference between the ideal number of centre points and four centre points is often quite small. Four centre points also give a reasonable estimate of the pure error variance.

RESPONSE SURFACE MODELLING

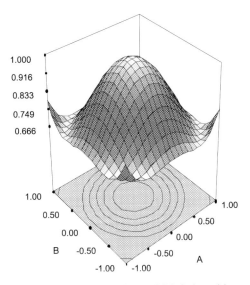

Figure 7.19 Prediction error for a two-factor CCC design with one centre point.

Table 7.10 Prediction error in the centre for a five-factor CCC design with one, two, four and six centre points

Centre points	1	2	4	6
Prediction error	0.88	0.66	0.48	0.40

7.8.5 Rotatability

For rotatable designs the prediction error depends only on the distance from the centre point, not on other coordinates. The idea is that you do not know in which direction an optimum might occur, but you assume it not to be too far off-centre (another idea behind rotatability has to do with the estimation of the different components in the model, Section 7.8.6). Given a proper choice for the star distance, CCC designs are rotatable and CCF designs are not. Also Box–Behnken designs are rotatable or nearly rotatable.

The uniform precision and rotatability criteria are usually over-emphasised for two reasons:

- If you predict a value, you should confirm it. If you add this point to the design, the prediction error would change anyway (becoming lower at that point). Because you now have an asymmetric design, rotatability is lost.

- Uniform precision and rotatability are only attained for prediction with the full model. Usually only a fraction of the complete model is retained (only the significant terms) and, if you predict using this model, all the prediction properties of the design change, so you lose uniform precision and rotatability anyway. Remember that the full quadratic model is not the most complex model you can fit: you could choose to add terms like A^2B or A^3 if the data would indicate this, so the full quadratic model is not chosen *a priori* either.

Since you do not know beforehand which model will result, the resulting prediction error estimate might be an underestimation of the true prediction error, but using the full model may result in an overestimation.

7.8.6 Estimation of the individual effects

Since the main goal is to describe the response surface 'as a whole', the estimates of individual effects are often less important (except in the case where there are only a few factors and the RSM design is the first design to be carried out). However, as a researcher you will always ask yourself which variables contribute in which way to the response. For this reason, we might want to know how well you can estimate these individual terms; in other words, what is the standard error of these estimates and how correlated are they with other terms?

Table 7.11 lists the standard errors of main effects, second-order interactions and square terms (again if the full quadratic model is fitted) for the central composite and Box–Behnken designs. As can be seen, rotatable designs (CCC and the four-factor Box–Behnken) show comparable standard errors for interaction and square terms. This is, in fact, an illustration of the fundamental idea behind rotatability, namely that you do not know beforehand how the principal axes of the response surface are going to be oriented. Are they parallel to the factor axes (no interaction), parallel to the diagonals of the factor axes (no square terms) or in-between?

Table 7.11 Standard errors of effects for respectively CCC/CCF/Box–Behnken designs for two-, four- and six-factor systems

	No. of factors		
	2	4	6
Main effects	0.35/0.41/–	0.20/0.24/0.29	0.15/0.17/0.20
Interactions	0.50/0.50/–	0.25/0.25/0.50	0.18/0.18/0.35
Squares	0.38/0.60/–	0.19/0.62/0.39	0.13/0.65/0.31

In each situation a rotatable design will be equally precise in the characterisation of the surface.

For central composite and Box–Behnken designs, only the estimates of the square terms are correlated with each other and with the estimate of the constant. These correlations are not important for describing the response surface, but they are important if you want to draw conclusions about individual contributions.

Table 7.12 shows the correlation between squares for four-factor CCC, CCF and BB designs.

Table 7.12 Correlation between coefficient estimates for squares for four-factor CCC, CCF and BB designs

	CCC	CCF	BB
Correlation	0.11	−0.30	0.19

7.8.7 *Robustness towards missing or 'wild' responses—replicated axial designs*

In the whole statistical analysis, it is assumed that the noise part in the response follows some distribution, usually a Normal distribution. Larger deviations are then interpreted as indications that the true relation is not well modelled by the polynomial used. However, in practice it happens once in a while (if not more often) that some mistakes are made during the experimentation or in the measurement that pass unnoticed until the statistical analysis.

Box–Behnken and central composite designs are fairly robust to these kind of rogue values except in the case of a CCC design, where an extreme value in one of the star points might indicate a higher-order relation between a factor and the response or might be an outlier. For these designs it might not be a bad idea to repeat the star points beforehand, resulting in a so-called *replicated axial design*. This increases the cost of the design somewhat (by $2k$ experiments), but this can be partly compensated by reducing the number of centre points to 2 (and still have lower maximum prediction error than the original CCC design).

The same goes for missing values. If one of the factorial points is missing, this has less effect than if one of the star points is missing. Again, a replicated axial design will be more robust towards these situations.

7.8.8 *Blocking*

Just as in screening designs, we may be forced to add a blocking factor to our model. A blocking factor is used to correct for a random effect (sometimes the term blocking factor is used for a fixed categorical variable, in which case it should be called categorical variable, Section 7.9).

The best example is a batch-to-batch effect. If you expect the differences between batches to be large, and if you cannot perform all experiments with the same batch, it is best to make sure that this batch-to-batch effect does not affect the estimate of a factor effect, in other words, ensure an *orthogonal blocking*. Note that this does not take away the effect on the prediction error, since future runs will be performed using some batch with unknown properties. Most software has facilities for blocking designs but very few packages offer a correct statistical analysis in that these blocking effects are treated as fixed effects and the effect on the prediction error is not taken into account.

Owing to high signal-to-noise ratios available in chemistry experiments, blocking is often ignored without too much harm.

From a design point of view, central composite designs are more interesting than Box–Behnken designs because of the flexibility in the number of blocks that can be used. For central composite designs, orthogonal blocking can be achieved by assigning the star points to one block and fractions of the factorial part to other blocks. Centre points are divided over the blocks. Table 7.13 shows the options for CCC designs for 2 to 7 factors. The number before the slash denotes the number of blocks, the number after the slash gives the total number of experiments excluding replicates of the centre point. Note that orthogonal blocking and rotatability are not always attainable at the same time (in which case priority should be given to orthogonal blocking).

Table 7.13 Orthogonal blocking schemes for CCC designs for two- to seven-factor systems

	No. of factors					
	2	3	4	5	6	7
Blocks/runs	2/10	2/16	2/26	2/28	2/46	2/80
		3/17	3/27	5/47	3/47	5/83

For Box–Behnken designs, only a few orthogonal blocking schemes exist: four factors in three blocks (27 runs) and five factors in two blocks (46 runs).

7.9 A glimpse of what else is out there

To conclude, there follows a list of problems you might encounter and topics you might run in to when using software. The main idea is to give a very brief opinion and direct you to additional reading.

7.9.1 Optimal designs

Although these designs have been around for some decades now, it is still possible to distinguish two extreme groups of statisticians: one group who tackle every problem using optimal designs and the other group, typically of an older generation, who get goose bumps every time they hear the word optimal.

The principle of optimal designs is that you choose a model you want to fit and then construct a design that is optimal for quantifying this model.

Optimal actually means '*optimal according to some criterion*'. Popular optimality criteria are D and G optimality but many more criteria exist. Roughly put, D-optimal designs minimise the maximum standard error of the individual model coefficients and G-optimal designs minimise the maximum prediction error (D-optimal designs are often also G-optimal or close to it).

In fact, the designs we have come across in this book (so-called classical designs) are all optimal or near optimal. Resolution III designs are (near) D-optimal for linear models, resolution V designs are (near) D-optimal for second-order interaction models, and central composite designs are near D-optimal for a quadratic model. The main difference is that all classical designs are *symmetric*, which in turn leads to *orthogonality* between at least a number of terms (e.g. main effects and second-order interactions), and that their properties (e.g. standard errors of prediction and coefficient estimates) are known beforehand.

Optimal designs are also called algorithmic designs because they have to be generated for each individual problem and also have to be evaluated since you do not know beforehand what the properties (standard errors and correlations between the effect estimates) of the designs are going to be.

Why, then, would we need these so-called optimal designs? Because sometimes extra constraints are put on the problem that we want to investigate. For example, one constraint may be the budget, another may be the time or amount of material available to perform experiments. You might have to optimise a seven-factor system but only have budget for 50 experiments (which is sufficient since the quadratic polynomial consists of 36 coefficients), so neither the central composite neither the Box–Behnken design is possible. Other types of constraints are restrictions on the accessible factor space, which covers the situations where you have categorical variables or constrained regions (cf. Section 7.9.3).

As general advice, I would suggest not using optimal designs unless you really have to (certainly not if you just want a cheaper alternative to a classical design) and after you have had some experience with and read some more about optimal designs.

7.9.2 Small composite designs

In small composite designs (e.g. Draper–Lin designs) the cube part of a central composite design is a resolution IV design or sometimes a Plackett–Burman design and therefore they can be much cheaper than other designs. But there is no free lunch. If we take a look at the standard errors of prediction for a seven-factor Draper–Lin design we see that they can be more than 10 times larger than the prediction errors for the corresponding 'normal' central composite design (which consists of a resolution VI fractional factorial design).

Small composite designs are therefore only sensible if you have extremely high signal-to-noise ratios, and even then you might want to go for optimal designs.

7.9.3 Constrained regions

Sometimes constraints are put on the design region, which has as a main consequence that cubic or spherical designs no longer provide an optimal filling of the accessible space. An example is shown in Figure 7.20. In this case optimal designs or space-filling designs provide a solution. Note that sometimes an 'ellipsoidal' CCC design (different values for the star distance α for different factors) may not reach every corner but can provide acceptable coverage of the design region.

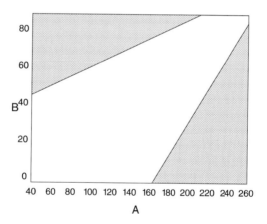

Figure 7.20 Visualisation of a constrained experimental region for two factors.

7.9.4 Mixture problems

We have a mixture problem when the sum of levels for at least two factors has to be constant. Typical examples are polymer blends, cocktails and all sorts of patties where you add different amounts of the constituents together. If you double each of the amounts you have twice the total amount (chemists know well that this holds for weights, not for volumes). This is not of interest in most cases: you want to know how changes in the *relative* amounts affect the properties of the mixture.

In a sense, the majority of all chemical problems are mixture problems. Whenever we talk about concentrations in aqueous solutions, we assume a fixed total amount or volume. We usually do not think of water as an important factor, it is just the 'filler' that you add to obtain the same fixed total, in which case we can treat it as a standard design problem.

Why are mixtures different from an experiment design point of view? Because the sum of the components has to be constant, thus creating a set of linearly dependent or collinear variables. As can be seen in Figure 7.21, we no longer have three independent dimensions but only two (the well-known ternary mixture triangle).

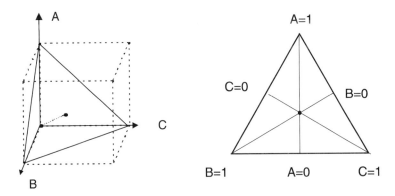

Figure 7.21 Dimension reduction for three-component mixture due to constraint $A + B + C = 1$.

The cubic and related designs we have seen so far do not fit very well in these constrained spaces, so we need special (classical) mixture designs. But in most problems there are extra constraints on the components: they have to be within some (often quite narrow) ranges, in which case only optimal or space-filling designs offer a solution. The shaded region in Figure 7.22 is a typical constrained region for a mixture design.

This seems complicated, but actually designs are the easy part (assuming you have mastered the art of optimal design): it is the analysis and more specifically the *interpretation of individual effects* that is the real problem. Again, if you just want to build a response surface model that you can use for finding optima or mixtures with desirable properties, this is not too difficult.

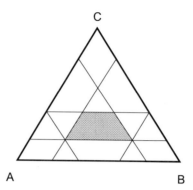

Figure 7.22 Example of a constrained three-factor mixture problem.

7.9.5 Categorical variables

Categorical or nominal or qualitative variables are factors that can only take discrete levels—type of solvent, reactor type, etc. An m-level categorical factor corresponds actually to $m - 1$ independent factors (the $m - 1$ independent differences between these levels). These factors are highly constrained and, in fact, this is similar to a mixture problem, except that only the vertices are accessible.

Usually optimal designs are the only alternative to repeating a classic design for each level of the categorical variable; an exception is for categorical factors with only two levels.

If the categorical variable is of chemical nature (type of solvent, type of substrate, etc), interactions with other factors are almost certain to be very important. Since it is not possible to construct true resolution IV designs for multilevel categorical variables, screening becomes a very tricky thing. Principal properties designs are an elegant alternative for these situations (and not only for screening).

7.9.6 Principal properties or multivariate designs

Principal properties designs offer an interesting alternative for problems with chemical categorical variables. The approach is as follows:

- Find a set of relevant physicochemical descriptors, typically in the order of 10 variables.
- Perform a principal components analysis (see Chapter 12). Since the descriptors are often strongly correlated, this usually results in a sufficient set of only a few independent variables.
- Build a design treating the principal components as regular design factors.

Since an *m*-level categorical variable corresponds to $m-1$ underlying variables, this can in some cases result in an enormous reduction in the number of experiments or in a strong increase in the number of types that can be tested. Suppose you want to test the effect of different solvents. It is not hard to come up with 100 different solvents but in practice you will probably select a few 'typical' solvents. With the principal properties approach you can actually consider all 100 solvents and select those that 'span the solvent space' in a (near) optimal way. Although the usual choices made by the chemist often give a reasonable coverage of the different solvents, the principal properties approach may lead to alternative and less obvious choices with possibly more interesting results.

7.9.7 *Space-filling designs*

One of the main criticisms of optimal designs is that they are optimal for a pre-specified model, e.g. a quadratic model. Space filling designs, also known as distance based or model-free designs, simply fill up the available design space as well as possible. The problem is that 'well' implies a criterion and here too there are several criteria. Think about using space filling designs if you expect strong nonlinear behaviour and you want to go beyond a quadratic approximation. One problem at present is that space filling designs are not well (or not at all) implemented in most software.

7.9.8 *Power considerations—what are the smallest effects you can estimate?*

Although this is more relevant in a screening investigation, you might wonder in an RSM study how strong an interaction or curvature must be before you can detect it as being statistically significant. Put as a more active question: how many experiments or which design do you have to use to detect an effect of a prespecified size? Some software has this as a built-in option, but there is a rough-and-ready way to get an approximate answer to this question if you do not have that type of software.

Proceed as follows:

1. Select a potentially suitable design.
2. Calculate for each of the design points a response that is a function of all terms in the model and add Normally distributed noise to the response of size comparable to the noise level you expect in your experiments.

3. Estimate the model coefficients and standard errors from the calculated responses using the experiment design analysis software.
4. Look at the standard errors of the coefficient estimates. You will be able to detect effects that are roughly 4 times the size of these standard errors.
5. Select a different or larger design (if necessary replicating the whole design) if the standard errors are too large.

Notes: (i) Step 2 will give a slightly optimistic estimate of the standard error. (ii) In some software you can skip step 2 since the standard errors are given directly assuming a noise variance of size 1. Do not forget to multiply these standard errors with the standard deviation of the experimental noise you expect.

7.9.9 Robustness modelling

In most cases you should not only wonder how to obtain the highest conversion or the strongest polymer but also look for conditions in which you can produce in the most stable way; that is, look for a *robust* process (a Taguchi-like investigation). For further reference on this very useful but too often neglected subject, consult Myers and Montgomery (1995) (and references therein), who treat both the traditional Taguchi approach as well as the RSM approach.

7.9.10 Multiresponse optimisation

Most problems involve not just one but several and sometimes as many as 10 or 20 responses. In these cases it is always a good idea to check whether all these responses are really different. Often you find very strong correlations between them, implying that there are only a few fundamental properties, so again a principal components analysis might not be such a bad idea (Chapter 12).

Let us assume that you have different responses: how do you find the overall optimum or settings that satisfy the constraints on these responses? If you have only two factors, an overlay of the contour plots might help, but even in this simple situation this might fail if you have many responses. The most powerful technique is Derringer's desirability approach (or some variant), in which a desirability parameter that quantifies the distance from target is attached to each response. The individual desirability parameters are then combined into an overall desirability in a multiplicative way and this overall desirability is maximised (cf. Box and Draper, 1987).

7.10 Further reading

There are numerous books on experiment design but only a small number are relevant to response surface modelling in chemical work. Two all-round favourites are Myers and Montgomery (1995) and the book by Box and Draper (1987) (the first book is a little easier to read and covers most of the topics covered here). A nice introduction into the fundamental ideas behind experiment design, from a modelling point of view, can be found in Part 1 of the book by Atkinson and Donev (1992) (the remainder offers a thorough discussion of optimal design).

If you want to know more about mixture designs, you have a problem; it is treated either superficially or very elaborately (Cornell, 1990). Myers and Montgomery give a fair amount of detail.

More on principal properties can be found in Carlson (1991).

7.11 Appendix: a case study in the use of response surface modelling

The case we will be looking at is special in the sense that it is industrial research but it describes an investigation that was carried out at a university, resulting in fewer of the confidentiality problems that prevent many interesting cases from becoming public. As will become clear, it is not a typical textbook example where everything runs smoothly and according to the book—it is a *real-life* case.

The subject of this study is the catalytic nonoxidative dehydrogenation of isobutane by supported chromium oxide catalysts. A mixture of isobutane and nitrogen is passed over an inorganic oxide doped with Cr^{3+} in a flow system. Isobutane is selectively transformed to dehydrogenated products on the surface of the catalyst. The amount of isobutane in the gas flow is monitored by gas chromatography (GC). For more details see Weckhuyzen *et al.* (1998).

The goal is

- to find optimal process conditions for the efficient conversion of isobutane;
- to understand and quantify the relation between the activity of the catalyst and the processing variables.

Based on previous experience, the researchers chose to investigate the effects of the following variables:

IEP: The isoelectric point (IEP). This is a characteristic of inorganic oxides (SiO_2–Al_2O_3) that depends on their composition.

Alternative functions could be used to describe the properties, such as %SiO_2 or %Al_2O_3 or the ratio or difference of the two species, but using a fundamental characteristic like the isoelectric point offers some possibilities for checking oxides that are not based on SiO_2 or Al_2O_3.

CR: the amount of Cr^{3+}.
GAS: the composition of the gas mixture (% isobutane in N_2).
TEMP: the reaction temperature.
TIME: the cumulative time the catalyst has been in contact with isobutane. This is not the contact or reaction time between a 'plug' of gas and the catalyst.

The ranges over which the variables had to be investigated (chosen by the researchers and again based on previous experience) were,

Variable	IEP	CR	GAS	TEMP	TIME
Range	2–8	0.5–7.5	2–18	350–500	10–50

The choices about the experiment designs, the results of the experiments and the interpretations drawn from the results are described in the main body of the chapter.

Bibliography

Atkinson, A.C. and Donev, A.N. (1992) *Optimum Experimental Designs*, Clarendon Press, Oxford.

Box, G.E.P. and Draper, N.R. (1987) *Empirical Model Building and Response Surfaces*, Wiley, New York.

Carlson, R. (1991) *Design and Optimisation in Organic Synthesis*, Elsevier Science, Amsterdam.

Cornell, J.A. (1990) *Experiments with Mixtures*, Wiley, New York.

Myers, R.H. and Montgomery, D.C. (1995) *Response Surface Methodology*, Wiley, New York.

Weckhuysen B.M., Abdelhamid Bensalem and Schoonheydt, R.A. (1998) *Journal of the Chemical Society, Faraday Transactions* **94** 2011.

8 Analysis of variance. Understanding and modelling variability
M. Porter

8.1 Introduction

Almost all decisions are uncertain, due to the variability inherent in the available data and to the impossibility of obtaining all relevant data. In addition to this noise or *common cause* variability, within the data there is *special cause* variability due to the controlled variables or factors about which decisions are to be made or whose effects are to be estimated from the available data. The first aim of this chapter is show how the types and causes of *variability* can be identified, modelled, estimated, eliminated or used as part of the decision-making process.

General ideas about modelling variability, least squares estimation and the analysis of variance are introduced. These concepts or ideas are fundamental to understanding many applied statistical methods. A full understanding of the concepts relies on mathematics beyond the scope of this book (the reader wishing to tackle the mathematics is referred to Stuart and Ord, 1991, 1994). The second aim of this chapter is to explain these concepts with only a necessary minimum of mathematics. This will give you an understanding of the principles on which these statistical methods are based, so that you can confidently apply them. This will be reinforced by modelling variability in three practical situations that are likely to be met by most scientists:

- Fitting a line to describe the effect of a quantitative predictor variable on a response (simple linear regression).
- Estimating the effects of changes in a single factor on a response (one-way ANOVA).
- Modelling the effects on a response of simultaneous changes in two factors (two-way ANOVA)

These are described in more detail elsewhere in the book and provide the basis for much of statistical modelling and the design of experiments. Diagnostic techniques for testing the appropriateness of models are also considered. A framework for modelling variability using *general linear models* is outlined, and extensions and alternatives to the methods are presented. You may find that in order to understand the general concepts in this chapter that you need to refer both forwards from the concepts to the applications and backwards from applications to the concepts a number of times.

Although sophisticated techniques are available for modelling variability, in general whenever appropriate and practical, you will obtain best results by

- using simple easily understood models,
- planning investigations using simple balanced experimental designs.

In practice, when a complex model is required, the advice and assistance of an expert should be sought.

8.2 Where to go

The list shows where information on the analysis of variance can be found.

Information required	Go to Section(s)
Alternatives to least squares and the analysis of variance	8.4.3
Introduction to models, estimates and the analysis of variance	8.4.2
Blocking	8.7.1, Chapters 6 and 7
Common and special causes of variability	8.3.2, Chapter 1
Crossed and nested classifications	8.7
Errors in variables models	8.5.4
Properties of estimates	8.3.3, Chapters 4, 7, 11, 12
Fixed and random effects in analysis of variance	8.6.1 and following
General linear models	8.8
Interactions	8.7.2, Chapters 6 and 7
Least Squares	8.4.2
One-way or single factor analysis of variance	8.6
Variability in simple linear regression	8.5, Chapter 11
Using residuals to assess models	8.5.3, Chapters 2, 11, 12, 13
A model for total variability	8.4.1
Variables and factors	8.3.1

8.3 Preliminaries

8.3.1 Variables and factors

Variables are quantitative measurements or observations, for example the purity of a sample of a material (a continuous variable) or the number of particles counted on a square centimetre of a filter plate (a discrete variable). The latter can be only an integer, or whole number. Some information connected with a result is nonnumerical. Examples are the technician carrying out an assay, the site and the batch assayed. The last of these usually looks like a variable, but the values taken simply give positions in the production sequence with probably variable time intervals between consecutive batches. For convenience or security or to avoid bias, an arbitrary numeric code may be assigned to the technician carrying out the assay or to the site. To distinguish between these arbitrary numeric quantities and measured variables, the former are called *factors*. Particular statistical methods are only appropriate for particular types of variables or factors; for example, it is meaningless to calculate an average batch number.

8.3.2 The types and causes of variability

Statistics was defined in Chapter 1 as the science of using data to improve the odds of correct decisions. It was implicit in this definition that decisions have to be made under conditions of uncertainty due to variability in processes that produce the data and to the impossibility of obtaining all relevant data.

All measurements are subject to variation or error (see Section 1.4 for a more detailed discussion and Chapter 4 for examples). That is, if two apparently identical measurements are made at different times or places, then the two results will not be identical. For example, two measurements of the purity of a bulk chemical made using different samples or even the same sample at different times or in different places are unlikely to give identical results due to small variations between the material assayed or to the set-up of the analytical equipment.

This variability is one reason why statistical methods are needed in order to obtain reliable estimates of parameters, to increase the chance of making correct decisions or to provide reliable models that can be applied to the real world. Statistical methods are required if you wish to model and estimate the variability in the system due to the factors and variables that can be controlled and those that cannot be controlled. The former include both those variables and factors about which inferences are to be made and also nuisance parameters that would influence or bias the results if not controlled.

Some of the potential causes of variability are:

- Heterogeneity in materials, both between batches and within batches. The former may be a nuisance parameter that needs to be included in the design of the investigation and in the model. The latter may contribute to the apparently random variability in results.
- Inherent variability in measuring devices or between measurement devices due to the process or processes by which samples are passed through the device and the measurements are made. For example, variable amounts of material may be left behind in a tube, ampoule, container or vessel, or be carried over to the next sample. There may also be differences between operators in technique, eyesight or training. Both of these may contribute to the apparently random variability in results. They can have an impact particularly in cross-site or multisite validation of methods.
- Differences in operating conditions, e.g. in actual temperature profile or in ambient conditions.

In Section 1.4, descriptions of the different types or causes of variability were given:

- *Special* causes of variation are effects, factors, variables or predictors known to affect the measurement.
- Special causes can be *systematic* or *random* and there are statistical methods appropriate for both.
- *Common* causes of variation are random, uncontrolled or uncontrollable effects, or errors; sometimes called *noise*.
- *Systematic errors* are present when a sequence of measured values consistently deviate from the true, or expected, value; also called *bias*.

8.3.3 Properties of estimates

Calculated statistics are used to estimate unknown quantities or *parameters*. However, as was shown in Section 1.5.3 and in Chapter 4, a *statistic* is both a value calculated from data and a random variable describing the set of values that the statistic could have taken for all possible samples (the population distribution). As for any random variable, this population distribution can be described by a mathematical function. Furthermore, equations for population statistics such as the average or standard deviation can be derived. For some statistics or estimates, the distribution of possible values will have properties that make these statistics good estimates. These properties are usually intuitively appealing, though deriving the equations is beyond the scope of this book. The most commonly sought properties are:

- *Unbiasedness*—the population average of the possible values is equal to the quantity or parameter being estimated (see Chapter 4 and Stuart and Ord, 1991, p. 609).
- *Minimum variance*—among possible estimates, it has smallest variability as measured by the population variance. This is often restricted to minimum variance among unbiased linear estimates (see Stuart and Ord, 1991, p. 614).
- *Maximum likelihood*—estimates obtained by maximising with respect to the unknown quantities or parameters the likelihood of obtaining the observed sample (see Stuart and Ord, 1991, p. 649).

8.4 Variability in data

If it is accepted that a set of measurements is subject to variability and that potential causes of this variability are identified, then the scientist requires:

1. A method for measuring the total variability in the data set.
2. Methods for modelling the effects of the different causes, i.e. for subdividing the total variability into portions which can be ascribed to the causes.
3. Methods for assessing the significance of different causes.
4. Methods for assessing the appropriateness of the models, often called diagnostic tools.

These are considered in turn in the following sections, presenting the most common methods and their alternatives.

8.4.1 A model for total variability in a data set

Consider a data set consisting of n measurements or observations $y_1, y_2, \ldots, y_n$ drawn at random from some population. Clearly, any measurement of variability must be independent of arbitrary changes in location of the data set; for example, variability in a set of temperature measurements should be independent of whether they are taken in Kelvin or °C. The measure of variability should be a function of the deviations $(y_i - a)$, where a is an appropriate quantity, statistic or parameter that locates the data set.

One suitable function for measuring the total variability is

$$\text{Sum of squared deviations} = \sum (y_i - a)^2$$

which increases as data becomes more variable and has been found to have many good properties, both intuitive and mathematical. It is also easy to manipulate mathematically. Thus, sums of squares dominate methods for modelling variability. Only recently have other less mathematically tractable methods become possible with the growth of computer power.

However, how should a be chosen? Gauss is usually credited with developing least squares estimation (see for example Stuart and Ord, 1991, p. 712). Least squares estimates are unbiased and have minimum variance amongst unbiased linear estimates. Under certain conditions, least squares also gives maximum likelihood estimates.

The least squares estimate of the location parameter a is given by minimising $\sum (y_i - a)^2$ with respect to a, and is

$$\bar{y} = \frac{\sum y_i}{n}$$

This is the arithmetic mean of the observations, y_i, which is the unbiased, minimum variance estimate of the unknown population mean. It is also the maximum likelihood estimate of the mean of Normally distributed (and other) populations for random samples. It is the most commonly used measure of location of the distribution underlying the data. Note also that the deviations sum to zero:

$$\sum (y_i - \bar{y}) = 0$$

Thus, the total variability in the data set can be measured by

$$\text{Sum of squared deviations} = \sum (y_i - \bar{y})^2$$

which is called the *total sum of squares* and is equal to $(n-1)s^2$ where s is the sample standard deviation. The sample variance s^2 is an unbiased estimate of the population variance. Since once $\bar{y}$ is known, only $(n-1)$ of the deviations $(y_i - \bar{y})$ are independent, with the last one fixed because all of the deviations must sum to zero. Thus, there are said to be $(n-1)$ *degrees of freedom* for variability, which can be subdivided between possible causes. Also, for random samples from a Normally distributed population, the distribution of the total sum of squares is

$$\sigma^2 \chi^2_{n-1}$$

where σ^2 is the population variance and χ^2_{n-1} is the chi-squared distribution with $(n-1)$ degrees of freedom (see Chapter 4). Johnson and Kotz (1970, chapter 17) give a detailed description of the chi-squared distribution, its properties and applications. Under appropriate assumptions, the sums of squares considered in this chapter and in the chapters on regression and on the analysis of experiments have chi-squared distributions with given degrees of freedom. For simple models and balanced experimental designs, the degrees of freedom of the sums of squares have intuitively appealing explanations like that given above. This is not necessarily true for complex models or unbalanced designs. The extensive mathematical statistics theory supporting these comments is beyond the scope of this book. It is, however, given in many books on linear models, see Stuart and Ord (1991), Searle (1971) and Searle *et al.* (1992).

8.4.2 Models, estimates and the analysis of variance

The effect of special $(X_1, X_2, \ldots, X_k)$ and common causes of variation $(\varepsilon_1, \varepsilon_2, \ldots, \varepsilon_m)$ on a response Y can in general be written as

$$Y = F(X_1, X_2, \ldots, X_k) + f(\varepsilon_1, \varepsilon_2, \ldots, \varepsilon_m)$$

where F and f represent some particular mathematical functions.

In the applications considered in this chapter, the common causes of variation are represented by a single *error* or *residual* ε that has a Normal distribution with zero mean (no bias) and constant variability, with separate observations being independent. The effects of the special causes are assumed to be adequately describable by some linear function or *linear model* in the unknown parameters. For a data set $(y_i, x_{1i}, \ldots, x_{mi})$ the linear model can be written

$$y_i = \sum \alpha_j x_{ji} + \varepsilon_i$$

where ε_i are the unknown residuals and the α_j the unknown parameters.

The estimates a_j of α_j are obtained by least squares, i.e. by choosing those estimates that minimise the sum of squared observed residuals $\sum e_i^2$ where

$$e_i = y_i - \hat{y}_i = y_i - \sum a_j x_{ji}$$

are the observed errors or residuals and, thus, estimate the 'true' residuals ε_i.

This definition of a best-fit model is intuitively appealing and also gives minimum variance unbiased linear estimates and under certain conditions gives maximum likelihood estimates. The family of models based on this structure for common cause variability has been applied widely. Illustrations are given in Sections 8.5 (simple linear regression) and 8.6 (single-factor or one-way analysis of variance). However, this model will not apply to all practical situations and the appropriateness of fitted models should always be assessed. The most commonly applied methods for assessing fitted models are based on testing the assumptions using plots of and statistics based on the estimated residuals e_i. These methods are described in more detail below in Section 8.5.3 and in Chapters 11 and 12.

The fitted or estimated residuals are a combination of two causes:

1. The true residual or error from common cause variability
2. The differences between the form of the true model and the form of the fitted model

Consider an example. The true relationship between a response y (such as the purity of a compound) and a predictor (such as a peak area) is an unknown, probably complex function. If a linear calibration model is fitted, the residuals will be an amalgam of the true common cause variability and the differences between the true and fitted models. Often in practice over a restricted range of purities a linear calibration may be adequate for measuring the purity of a sample, though over a wider range a more complex function would be required to represent the 'true' relationship sufficiently accurately.

The fitted values,

$$\hat{y}_i = \sum a_j x_{ji}$$

estimate (or predict) the combined effects of special causes about which the observed values vary due to common causes or noise.

The total variability,

$$\sum (y_i - \bar{y})^2$$

or *total sum of squares* (denoted SST) can be split into components that represent the contributions of the special and common causes of variability. This is called the analysis of variance. Put simply,

Total sum of squares = sum of squares due to special causes +
sum of squares due to common causes

where

1. For a single error or residual ε, the sum of squares due to common causes is $\sum e_i^2$, the *residual* or *error sum of squares* (SSE), which was minimised to obtain the least squares estimates. Sometimes SSE can be subdivided into sums of squares describing the components of variability due to different common causes.
2. For simple models with a single special cause of variation, there is usually a simple equation for the *sum of squares due to special causes* (denoted SSR), which describes the variation due to that special cause. Illustrations are given in Sections 8.5 (simple linear regression) and 8.6 (single-factor or one-way analysis of variance).
3. If there are no special causes of variation, the residual sum of squares equals the total sum of squares. Thus, total sum of squares SST is the residual sum of squares for the minimal model where all variability about the mean is due common causes. It follows that the sum of squares due to special causes is the reduction in the residual sum of squares due to fitting a more complex model, because

$$SSR = SST - SSE$$

4. For more complex models with multiple special causes of variation, the sum of squares due to special causes can be split into sums of squares that represent the contributions to overall variability from each of the special causes.
5. The special case in (3) above extends to a general procedure for subdividing the sum of squares for special causes into sums of squares for each special cause included in the linear model. This procedure is given in Section 8.8. A complex model is a hierarchy in which simpler models are embedded. Thus, the sum of squares for a special cause can be seen as the reduction in the residual sum of squares obtained by adding that cause to the model. That is,

$$SSC = SSR1 - SSR2$$

where SSC is the sum of squares for the special cause, SSR1 is the residual sum of squares for the model excluding the special cause and SSR2 is the residual sum of squares for the model including the special cause. Examples of subdividing the sum of squares due to special causes can be found in Chapter 11 on multiple regression and Chapters 6 and 7 on the analysis of designed experiments.

If specific properties such as statistical distributions can be assumed for the different components, methods are available for assessing the importance of the different components representing the special causes against the background noise of common cause variability. Extensive mathematical statistics theory has been developed to support these methods. This theory is beyond the scope of this book, but is given in many books including Stuart and Ord (1991), Searle (1971) and Searle *et al.* (1992).

8.4.3 Alternatives to least squares and the analysis of variance

Least squares and the analysis of variance are widely used for the many reasons given above, but are not appropriate for all problems and practical situations. The weaknesses of least squares and the analysis of variance, as with any modelling method, relate to the appropriateness to the application of the structure of the chosen model of special cause variation and the assumptions about common cause variation. Thus, it is important that the appropriateness of models and assumptions should be rigorously justified as at least a good approximation to the unknown 'truth'. This is a theme that recurs throughout this book, in many different guises, and is crucial to the development of useful and reliable models. A comprehensive survey of alternatives and extensions is given by Miller (1997).

In most standard models the structure for special cause variation is linear in unknown parameters, which are then easily estimated by least squares. Even before 1970 when this methodology was the only one available for wide application, authors were aware of the limitations (see Acton, 1959; Sprent, 1969). Since 1970, alternative methods have been introduced (see Aitken *et al.*, 1989; Searle *et al.*, 1992; Miller, 1997). Two examples illustrating differences between standard models and alternatives are given later in this chapter (other examples may be found in Chapters 11 and 12): first, fitting a straight line when the predictor or *x*-variable is also subject to error (Section 8.5.4); second, the modelling of the effects of a factor on a response when the levels of the factor are themselves the outcomes from a random selection not fixed values (see Section 8.6.1).

The most frequently use model for common cause variation assumes that it can be aggregated into a single random variation fluctuating about the true values given by the model for special cause variation. This random variation or 'noise' is assumed to have the following properties:

- A Normal distribution.
- Common variance σ^2; i.e. variability about the true response is independent of the size of that response, of the special causes or of any other cause.
- Independent occurrences; i.e. the 'true' errors ε_i between the observed data y_i and their 'true' values are independent of each other.

The first of these is often justifiable by reference to the Central Limit Theorem, which shows that under appropriate assumptions sums of many effects have approximately Normal distributions (see Stuart and Ord, 1994, p. 319, and Chapter 4). The others may also be open to question in particular applications, for example in modelling chromatography data where random variation in measurements may be proportional to the size of the responses or where there may be carry-over from one sample to the next. You should be aware of the limitations of the methods that you use and of the alternative methods that are available (see Aitken *et al.*, 1989, on generalised linear models; Searle *et al.*, 1992, on variance components; and Miller, 1997, on a wide range of topics). It is generally advisable to seek expert assistance when using complex models for the first time.

One extension of least squares will be considered here to show the necessary link between practical application and development of mathematical statistics. Least squares estimates have been shown to be highly influenced by *outliers*, i.e. observations that appear not to fit the general pattern. These observations should more sensibly and less emotively be called *discordant*. Such observations can be easy to identify, as in the following data,

$$95.4, \quad 95.5, \quad 95.2, \quad 95.1, \quad 59.4$$

where the first two characters in the last number have been reversed. However, unambiguous identification is not so easy for

$$95.4, \quad 95.5, \quad 95.2, \quad 95.1, \quad 94.5$$

in which the last two characters in the last number have been reversed. When discordant values are identified, the classical approach is to check and correct them, or, if this is not possible, to delete them. In the eighteenth and nineteenth centuries, scientists were much more willing to delete or ignore observations that did not fit their theories than is currently acceptable. Deletion of data without good and well-documented reasons is now unacceptable to regulators. Transcription and calculation errors are relatively easy to find and, thus, changes are easy to justify. Errors created by automatic analysis equipment cause more problems. Discordant values may be mistakes, but could also be a natural part of the process by which the data was obtained. Often the only solution is to refer to historical data on method variability and to repeat the analysis at higher replication to confirm or refute the data originally obtained.

Whatever their causes, least squares is highly influenced by outliers in both the response and predictor variables, because they have high *influence* or *leverage* on $\sum e_i^2$, the residual or error sum of squares, which is minimised to obtain the least squares estimates. The estimates obtained will be biased. When using least squares the only practical approach is to identify 'true' outliers and delete them. This is not easy to do in practice (see Chapter 2 for a fuller discussion). An alternative is to use a different method, one that is *robust* to outliers. Minimising the sum of absolute values of the residuals $\sum |e_i|$ produces estimates that are more robust to outliers than least squares which minimise the sum of squared residuals $\sum e_i^2$. The former is one *robust regression* method. However, unlike least squares, mathematical solutions minimising $\sum |e_i|$ are often unattainable, though some mathematical properties are known. For example for simple linear regression, the straight line that minimises the sum of absolute values of the residuals $\sum |e_i|$ must pass through two of the points. Thus, for moderately sized data sets, a computer can be programmed to find this line by calculating $\sum |e_i|$ for all lines joining two points in the scatter of data. In general, methods for fitting robust regression models rely on the availability of computer power. Such methods are being developed, but have not yet found their way into commonly available software.

One group of methods robust to outliers available in many software packages is that of *nonparametric methods*. Miller (1997) says that 'the sign test effectively obliterates the effect of outliers'. Nonparametric methods are based on different, and usually fairly minimal, assumptions about the distribution of the errors in the linear model. They are less powerful than the equivalent methods for Normally distributed errors, but can be optimal for non-Normally distributed errors. Robust and nonparametric alternatives to least squares methods are discussed in Chapter 5, where an extensive set of references is given. Miller (1997) is again a useful reference.

The variability in a data set may be damped, increased or masked by other effects. For example, differences in heating or cooling may cause neighbouring cells or wells on a plate to give more similar results than separated cells. Such nuisance effects can be reduced or eliminated by *blocking* cells into relatively homogeneous groups (see Section 8.7.1). Also, results can be affected by the order in which samples are taken from the cells, due to there being longer incubation periods for cells later in the sequence. This is an example of *autocorrelation*, with each observation correlated in some way with those immediately before it in the sequence. Most analysis runs will to some extent be affected by such *shift* and *drift effects*, which should be estimated so that the effects of interest can be estimated without bias. If possible, the causes of such effects should then be identified and eliminated. A useful reference on these methods is again Miller (1997).

One further extension is to models where the response **Y** is *multivariate*; for example, when the analyst considers not just the purity of a material but the multivariate profile of purity and impurities. Multivariate techniques are covered in Chapter 11 and, particularly, Chapter 12.

8.5 Modelling variability in simple linear regression

In this section, simple linear regression is used as an example of modelling variability using a linear model. Simple linear regression provides methods for fitting a straight-line relationship between a response and a single predictor. Simple linear regression is also an example of a linear model, because the prediction model for the response is linear in the unknown parameters β_0 and β_1. Both response and predictor are continuous variables, such as the purity of a compound and the area of a trace peak. Analytical chemists are often concerned with the practical problems of fitting models or calibration curves to such data. Often simple linear model may be used because good enough predictions are obtained within a range of interest, even though the 'true' model is more complex. Thus, predictions should be made only within the range of observed values and should not be extrapolated. The methods and applications of regression and correlation analysis are covered in detail in Chapter 11. The simple linear regression model can be written as

$$y = \beta_0 + \beta_1 x + \varepsilon$$

with y as the response, x as the predictor (special cause) and ε as a single residual or error, which is used to aggregate the common cause variability. The residuals are assumed to be independent with common variance σ^2.

The relationship between the response y and the predictor x is assumed to be linear, though with unknown slope β_1 and unknown intercept on the y-axis β_0. The common cause variability ε masks the true relationship and measurements give data (y_i, x_i) scattered in some way about the unknown 'true' model as in Figure 8.1. A line could be drawn through the data by eye, though different people would obtain different lines, sometimes radically different lines.

The purposes of these statistical methods are

- to estimate β_0 and β_1 in a consistent manner;
- to obtain estimates with known good properties;
- to be able to obtain confidence intervals for and test the significance of the estimates obtained;
- to examine the appropriateness of the model when fitted.

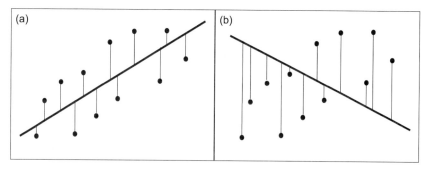

Figure 8.1 Linear regression: (a) good fit, (b) poor fit.

8.5.1 Residuals and estimation

Using least squares, a simple linear model is fitted to data (y_i, x_i) by choosing estimates b_0 of β_0 and b_1 of β_1 by minimising the sum of squared errors or residuals e_i:

$$\text{RSS} = \sum e_i^2 = \sum (y_i - b_0 - b_1 x_i)^2$$

This is done by differentiating RSS separately with respect to b_0 and to b_1, and solving the resulting *normal equations*, to obtain the estimates.

8.5.2 Analysis of variance

The least squares regression line passes through the centroid, $(\bar{x}, \bar{y})$ of the bivariate data. As is shown in Figure 8.2, the deviation of an observed y_i from the mean of the y_i can be written as the sum of the residual for that observation (the deviation of the observed y_i from the value predicted by the model) and the deviation of the predicted value from the mean, i.e.

$$(y_i - \bar{y}) = (y_i - \hat{y}_i) + (\hat{y}_i - \bar{y})$$

Then, after some mathematics,

$$\sum (y_i - \bar{y})^2 = \sum (y_i - \hat{y}_i)^2 + \sum (\hat{y}_i - \bar{y})^2$$

i.e. the sum of squared deviations, SST, of the observed y_i from the mean (the *total sum of squares*) is the sum of

- the sum of squares of residuals or errors, SSE (the *residual sum of squares*), and
- the sum of squared deviations of the predicted values from the mean, SSR, 'the *sum of squares due to regression*'.

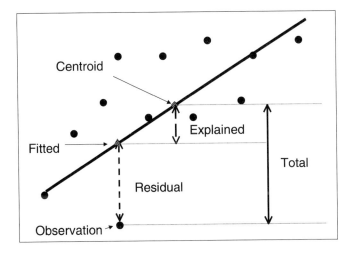

Figure 8.2 Causes of variability in linear regression models.

Just as SSE is the residual or error variability, or noise, SSR is the variability explained by the model, or signal. SSR can be shown to be a function of the slope of the line, a single statistic based on the data. In passing, this result is mathematically a generalisation of Pythagoras' theorem.

In Section 8.4.1 the total sum of squares, SST, was shown to have $(n-1)$ degrees of freedom. The sum of squares due to regression SSR is dependent on the slope and so has a single degree of freedom. Thus, by subtraction, the residual sum of squares SSE has $(n-2)$ degrees of freedom. The analysis of variance for linear regression is usually presented in the analysis of variance table in Table 8.1.

Table 8.1 Analysis of variance for linear regression

Source	Degrees of freedom	Sums of squares	Mean squares	F-Ratio
Regression	1	SSR	SSR	SSR/s^2
Residual or error	$n-2$	SSE	$s^2 = SSE/(n-2)$	
Total	$n-1$	SST		

In Chapter 11, these quantities are used

- To estimate the error variance σ^2 (The mean residual sum of squares $SSE/(n-2)$ is an unbiased estimate of σ^2).
- To test hypotheses about the slope and intercept of the fitted line.
- To construct confidence intervals for parameter estimates and for the fitted line.

8.5.3 Assessing the appropriateness of the model

This section summarises some of the simple ways that the appropriateness of linear regression models and their assumptions can be assessed. Similar methods can be used for all linear models. Many of these methods use simple plots of the residuals. In general, plots of the residuals will show up patterns more strongly than plots of the original data (see Cleveland, 1993, p. 103). Special care needs to be taken when interpreting plots for rounded data. Applications of these, and methods used to assess the fit of particular models, are given in other chapters.

Plots of the residuals against the fitted values $\hat{y}_i$. This should look like the plot (a) in Figure 8.3. A single outlier will appear as in (b). If the true relationship is curved, this plot will look like (c). However, even with a plot of residuals like this, the curvature may not be enough to stop a simple linear regression from being adequate for predictive purposes (see also Chapter 11). If the variability increases with the response, the plot will be as in (d). There are other patterns that may indicate that the model is inappropriate. Clustering of points may indicate the effects of some biasing factor. Outliers in the horizontal direction may indicate points with high influence or leverage. Plotting against the fitted values is better than plotting against the response, because high and low responses will

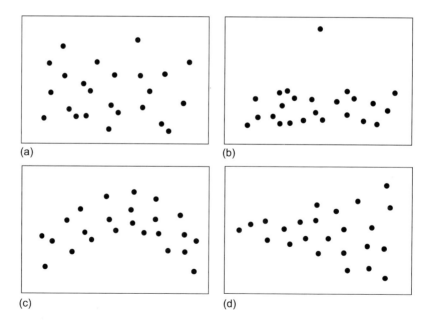

Figure 8.3 Patterns in plots of residuals.

correlate with larger residuals whereas the fitted values and residuals should be uncorrelated.

Plots of the residuals against the predictor. The analysis of patterns is similar to that given above. In addition, if there are multiple responses for given predictor values, the standard deviations within these groups should be similar. This also applies if the predictor is a factor. Such groups can be used also to obtain an estimate of the true residual or error (see discussion in Section 8.4.2 above) and hence the true error sum of squares. This can then be split out of the residual sum of squares to give a sum of squares due to the *lack of fit* between the fitted model and the true model that can be assessed against the true error (see Chapters 7 and 11).

Normal probability plots. A Normal probability plot can be used to assess the assumption that the residuals have a Normal distribution. This plot can be used also for each level of a factor or groups for common predictor values. Other methods for assessing the Normality of a data set can also be used (see Chapter 2).

Time sequence plots. Plots of response, predictor and residuals in time sequence should show no patterns, trends, step changes, outliers or autocorrelation. The presence of any of these could bias estimates and significance tests. Statistical process control (SPC) methods can be used for a more detailed assessment (see Wetherill and Brown, 1991, and Bissell, 1994, for descriptions of SPC methods).

8.5.4 Variability in the predictor

In the simple linear regression model given above, the response is assumed to be subject to random variation and it is assumed that the predictor is known exactly. However, consider a chemical analysis in which the response is proportional to the weight of material used. A technician attempts to take x grams of material for analysis. He will not get exactly x grams, but records it as x grams. A more careful technician will check the weight and use the precise weight $(x \pm d)$ in the calculation. However, this will also not be the precise weight used; a small amount of material may be lost and the measuring device will be subject to error. Fortunately, if the variability in the predictor is small relative to the variability in the response, using simple linear regression does not lead to gross errors. However, for any statistical method, users should be sure that the method is valid for their practical application by careful consideration of the likely causes and magnitude of variability. For a more extensive discussion of methods for fitting such *errors-in-variables models*, see Miller (1997).

8.6 One-way or fully randomised ANOVA

In this section, the one-way analysis of variance is used as a second example of modelling variability. The one-way analysis of variance (ANOVA) considers the estimation and significance tests used in a simple experimental design for exploring the effects of a single factor on a response. The different levels of the factor are often called *treatments*, because of early application of experimental design in agricultural field trials for which the levels corresponded to different treatments applied to plots of land. In this section, the levels will be called groups. Experimental design and the related analyses of variance are covered in detail in Chapters 6 and 7.

The model for the one-way ANOVA can be written as

$$Y = \mu + \tau_i + \varepsilon$$

with Y as the response, μ as the overall mean, τ_i as the deviation of the ith of the g groups from the overall mean, and ε as the residual or error that is used to aggregate the common cause variability. This is also a linear model, because the prediction model for the response is linear in the unknown parameters μ and τ_i. There are alternative ways of writing the parameters in this linear model (see Section 8.6.2 below). The different forms can be shown to be mathematically equivalent but they involve estimating different parameters and thus are interpreted differently.

When using analysis of variance and linear modelling packages, ensure that you understand the model used by the software. The residuals are assumed to be independent with common variance σ^2. The common cause variability ε masks the true relationships between the response and the treatments. Measurement gives data y_{ij} for the jth *replicate* of the ith group scattered in some way about the unknown 'true' value. The groups need not be equally replicated (see discussion on allocation of experimental units to groups in Section 8.6.4). The ith group has r_i replicates, with $\sum r_i = n$. The design can be represented as in Figure 8.4.

8.6.1 Fixed and random effects

In defining the model above, nothing was said about the properties of the groups as a special cause of variability. Thus, the properties of the parameters τ_i are not yet defined. If the experimenter is, for example,

- comparing the three different types of LC column used in an instrument,
- comparing four new standard lots for an HPLC method against that in current use for assaying samples, or
- comparing the purities of three lots of a raw material,

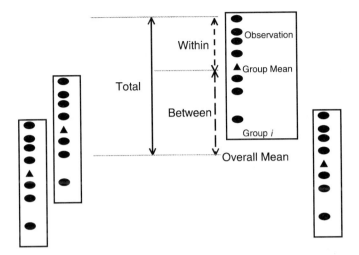

Figure 8.4 Single-factor analysis of variance.

she can take the groups as fixed entities and, thus, the τ_i as unknown constants. This is called the *fixed effects model*. The estimates for this model are considered in the next section. However, the groups could, under different circumstances, be considered to be a random sample from a population. This is especially true in the last of the above, where the experimenter may be more interested in the potential variability in the raw material than in the particular differences between the specific lots. This is called the *random effects model*. The estimates for this model are considered in Section 8.6.3. The appropriate model is defined by the purposes of the experiment. More complex models may include both fixed and random effects (mixed effects models).

8.6.2 *Estimation of fixed effects*

For the fixed effects model using least squares, a simple linear model is fitted to data y_{ij} by choosing estimates m of μ and t_i of τ_i that minimise over i and j the sum of squared errors or residuals e_{ij}. This residual sum of squares is

$$\text{RSS} = \sum e_{ij}^2 = \sum (y_{ij} - m - t_i)^2$$

This is done by differentiating RSS separately with respect to m and to t_i, and then trying to solve the resulting *normal equations*. The normal equations for this one-way analysis of variance model, are not solvable without imposing an extra condition on the t_i. Given that the τ_i were described as the deviations of the groups from the overall mean, the same condition is imposed on the t_i to obtain the estimates by setting $\sum r_i t_i$ as zero. This gives intuitively appealing estimates:

$$m = \bar{y}$$
$$t_i = \bar{y}_{\bullet\bullet} - \bar{y}_{i\bullet}$$

where $\bar{y}_{\bullet\bullet}$ is the mean of all observations, and $\bar{y}_{i\bullet}$ is the mean of observations in group i.

No extra conditions are required for an alternative form of the model for the one-way ANOVA,

$$Y = T_i + \varepsilon$$

with T_i $(= \mu + \tau_i)$ the effect of the ith group, because there are only g parameters to estimate. The estimate of T_i is

$$T_i = \bar{y}_{i\bullet}$$

However, this model is not as easy to extend to include multiple factors and covariates as the original form.

A second alternative form takes μ as the mean for the first level or control and τ_i ($i > 1$) as the deviation of the ith level from this control ($\tau_1 = 0$). For this second alternative there are also only g parameters to estimate and no extra conditions are required. This form for the model is used by many linear modelling packages.

For the different ways of writing the model, certain quantities are constant, namely:

- The predicted value for an observation y_{ij} is the mean for the ith group $\bar{y}_{i\bullet}$.
- The estimate of the differences between two groups a and b is $(\bar{y}_{a\bullet} - \bar{y}_{b\bullet})$. Thus, these are often called *comparative experiments*.

8.6.3 Analysis of variance

As shown in Figure 8.4, the deviation of an observation y_{ij} from the overall mean can be written as the sum of the residual for that observation (the deviation of the observed y_{ij} from the value predicted by the model) and the deviation of ith group from the overall mean, i.e.

$$(y_{ij} - \bar{y}_{\bullet\bullet}) = (y_{i\bullet} - \bar{y}_{i\bullet}) + (\bar{y}_{i\bullet} - \bar{y}_{\bullet\bullet})$$

Then, after some mathematics,

$$\sum(y_{ij} - \bar{y}_{\bullet\bullet})^2 = \sum(y_{i\bullet} - \bar{y}_{i\bullet})^2 + \sum(\bar{y}_{i\bullet} - \bar{y}_{\bullet\bullet})^2$$

where all the summations are over j and i. Thus,

$$\text{SST} = \text{SSE} + \text{SSR}$$

where SST is the *total sum of squares*, SSE is the error, residual or *within-group* sum of squares and SSR is the *between group of squares*. The one-way or single-factor analysis of variance is usually presented in the analysis of variance table in Table 8.2.

Table 8.2 One-way or single-factor analysis of variance

Source	Degrees of freedom	Sums of squares	Mean squares	F-Ratio
Groups	$g-1$	SSR	MSR = SSR/$(g-1)$	MSR/s^2
Residual or error	$n-g$	SSE	$s^2 = $ SSE/$(n-g)$	
Total	$n-1$	SST		

As for simple linear regression, SSR is the variability explained by the model and is a function of the g group means of which $(g-1)$ are independent (the average group mean is the overall mean). Thus, SSR has $(g-1)$ degrees of freedom. By subtraction, the residual sum of squares SSE has $(n-g)$ degrees of freedom. These quantities are used

- To estimate the error variance σ^2 (the *mean residual sum of squares* SSE/$(n-g)$ is an unbiased estimate of σ^2).
- To test hypotheses about the τ_i.
- To construct confidence intervals for parameter estimates and for differences.

8.6.4 Estimation of random effects

Consider the following situation, which has been simplified to aid understanding of the principles. A bulk chemical is produced in lots, each of which is sampled and assayed r times (constant for all lots). These replicate assays make up a group. The sampling process and the assay method are subject to common cause variation (within-group, replicate or residual variation). These causes might be identifiable and possibly eliminated, but the laboratory staff are too busy to do so. The process is also subject to variability (between-group, lot-to-lot or process variation). The differences between observations can be described by the one-way analysis of variance model

$$Y = \mu + \tau_i + \varepsilon$$

with Y as the response, μ as the overall mean, τ_i as the deviation of the ith of the g groups from the overall mean and ε as the residual or error that is used to aggregate the common cause variability. However, though the group effects can be estimated, of greater interest is obtaining an overall measure of lot-to-lot variability. For example, what are the relative contributions of process and sampling/assay variation? Thus, the lots are assumed to vary randomly about μ with variance σ_τ^2. This model is called a *random effects* or *component of variance* model. How is σ_τ^2 to be estimated? Searle et al. (1992, p. 58) show that the expected value of the *mean group sum of squares* is

$$E[SSR/(n-t)] = r\sigma_\tau^2 + \sigma^2$$

Thus, lot-to-lot variability can be estimated by the difference between the mean treatment sum of squares and the mean residual sum of squares, divided by the constant replication r. Note that the differences between lots and the between-group sum of squares are influenced by both the process and the residual variation.

The data in Table 8.3 comes from replicate samples/assays for five lots of a chemical. From these data, the analysis of variance in Table 8.4 is obtained. The estimates of the components of variance are

$$\hat{\sigma}^2 = 0.14143$$

$$\hat{\sigma}_\tau^2 = (0.61163 - 0.14143)/3 = 0.15673$$

Table 8.3 Purity data

Lot	Replicates			Average	Standard deviation
1529	97.25	97.31	97.94	97.50	0.382
1530	97.76	97.44	98.22	97.84	0.445
1531	98.52	98.14	97.77	98.23	0.375
1532	97.02	97.38	96.78	97.23	0.302
1533	97.56	96.84	97.14	96.99	0362

Table 8.4 Analysis of variance for purity data

Source	Degrees of freedom	Sums of squares	Mean squares
Between groups (lot-to-lot)	4	2.4465	0.61163
Within groups (residual)	10	1.4143	0.14143
Total	14	3.8608	

If SSR/$(g-1)$ is less than SSE/$(n-g)$, the estimate of σ_τ^2 is negative. This may indicate that

- the model is wrong and there are extra causes of variation biasing the analysis, or
- that σ_τ^2 is zero (or small enough to be effectively zero). In this case its estimate is set as zero in any subsequent calculations.

Searle *et al.* (1992) also give estimates for unbalanced data.

8.6.5 Allocation of experimental material

For a single-factor experiment, the experimenter usually has

- to estimate the differences between groups to a desired precision (equivalently, to establish that the differences are unlikely to be larger than some amount), or
- to use the available experimental material, money or other resources in the best way to obtain estimates of or to test hypotheses about the differences between groups.

The former requires the calculation of the required replication and then completion of the design (see Chapters 6 and 7). The solution for the latter is dependent upon the purposes of the design. If all comparisons are equally important, the groups should be as nearly as possible equal. If comparisons with a control are the most important (for example of new lots of a standard with that currently in use), the control should have a proportionally higher replication. See the discussion in chapter 8 of Cox (1958) on the choice of the number of observations. The optimal design of experiments is discussed in Chapter 7 and in Atkinson and Donev (1992).

8.7 Two-way, two-factor or randomised blocks ANOVA

The two-factor analysis of variance is used to present three concepts used in modelling and the analysis of variability. These are *blocking*, *crossed classifications* including *interactions* between factors, and *nested classifications* with one factor nested within another. The analyses for these experimental designs are given in Chapters 6 and 7 and also in books on design of experiments such as Gardiner and Gettinby (1998). More complex designs can be built using these three components.

8.7.1 Blocking

In the single-factor analysis of variance model, it is assumed that the experimental units are 'identical'. More precisely, the assumption is that the experimental units are sufficiently homogeneous that any differences can be aggregated into the common cause variation and after randomisation will not bias the analysis sufficiently to affect the conclusions. In many practical situations, this assumption will not be valid; for example, when assays are undertaken in different laboratories, in runs at different times, using different solvent lots or by different technicians. Thus, variation between experimental units can be a significant contribution to overall variability. The larger part of the variability between experimental units can be modelled by grouping them into *blocks* of relatively homogeneous units. This will reduce the residual or common cause variation (noise) and thus, improve the precision of estimates of differences between treatments. The model is

$$Y = \mu + \beta_j + \tau_i + \varepsilon$$

with Y as the response, μ as the overall mean, β_j as the deviation of the jth of the b blocks from the overall mean, τ_i as the deviation of the ith of the a treatments from the overall mean and ε as the residual or error that is used to aggregate the common cause variability. If exactly a units are available for each block, a *randomised block design* can be used.

The degrees of freedom in the analysis of variance table for this randomised (complete) block design are given in Table 8.5 and an example of its use is given in Section 8.9.

Table 8.5 Analysis of variance table for a randomised block design

Source	Degrees of freedom
Treatments	$a-1$
Blocks	$b-1$
Residual	$(a-1)(b-1)$
Total	$ab-1$

8.7.2 Crossed classifications

The randomised block analysis of variance model could also be applied if rather than blocks there were a second experimental 'treatment', grouping or classification *crossed* with the first (see Figure 8.5). The randomised blocks model assumes that the effects of the two classifications are additive and independent. This will not be true in many practical experiments involving two factors or treatments. Consider an experiment on the robustness of an assay to changes in the amounts of two solvents. There is no reason why the effects of changes in one solvent will always be independent of the level of the other. On the contrary, it is likely that the effects of the solvents will *interact*, i.e. the effects of changing the amount of one solvent will be dependent on the amount of the other present.

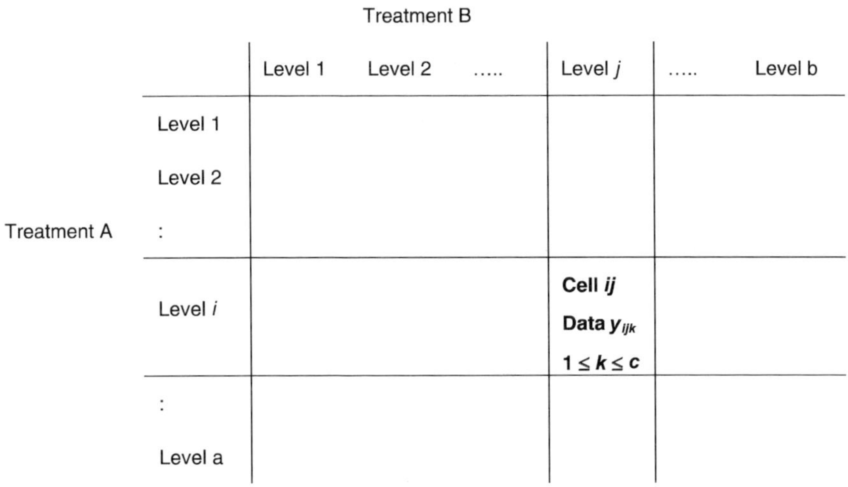

Figure 8.5 Data structure for a full factorial experiment.

The data structure is as in Figure 8.5 and the model is

$$Y = \mu + \alpha_i + \beta_j + \alpha\beta_{ij} + \varepsilon$$

with Y as the response, μ as the overall mean, α_i as the deviation of the ith of the a levels of the first treatment from the overall mean, β_j as the deviation of the jth of the b levels of the second treatment from the overall mean, $\alpha\beta_{ij}$ (the *interaction* term) as the deviation of the ijth cell from $(\mu + \alpha_i + \beta_j)$ and ε as the residual or error that is used to aggregate the common cause variability. If sufficient units are available so that each cell is replicated a constant number of times $c > 1$, a *full factorial analysis of variance design* can be used.

The degrees of freedom for the analysis of variance for this full factorial design are given in Table 8.6 and an example of its use is described in Section 8.9.

Table 8.6 Analysis of variance for a full factorial design

Source	Degrees of freedom
Treatment A	$a - 1$
Treatment B	$b - 1$
Interaction	$(a-1)(b-1)$
Residual	$ab(c-1)$
Total	$abc - 1$

8.7.3 Nested classifications

In some applications, one factor is *nested* within another. The hierarchy of factors will usually be evident. For example, if in an enzyme-based assay as in Figure 8.6 the replicates for a sample are plated out into cells, replicates are nested within samples. In this case, the residual sum of squares is the sum of squares within replicates (i.e. between cells). The degrees of freedom in the analysis of variance table are given in Table 8.7.

	Replicate 1	Cell 1	Cell 2	Cell 3	Cell 4
Sample 1	Replicate 2	Cell 1	Cell 2	Cell 3	Cell 4
	Replicate 3	Cell 1	Cell 2	Cell 3	Cell 4
	Replicate 4	Cell 1	Cell 2	Cell 3	Cell 4
	Replicate 1	Cell 1	Cell 2	Cell 3	Cell 4
Sample 3	Replicate 2	Cell 1	Cell 2	Cell 3	Cell 4
	Replicate 3	Cell 1	Cell 2	Cell 3	Cell 4
	Replicate 4	Cell 1	Cell 2	Cell 3	Cell 4
	Replicate 1	Cell 1	Cell 2	Cell 3	Cell 4
Sample 3	Replicate 2	Cell 1	Cell 2	Cell 3	Cell 4
	Replicate 3	Cell 1	Cell 2	Cell 3	Cell 4
	Replicate 4	Cell 1	Cell 2	Cell 3	Cell 4

Figure 8.6 A nested structure.

Table 8.7 Analysis of variance for nested factors

Source	Degrees of freedom	
	Example	Formula
Between samples	2	$a-1$
Between replicates (within samples)	9	$a(b-1)$
Between cells (residual)	36	$ab(c-1)$
Total	47	$abc-1$

The model is

$$Y = \mu + \alpha_i + \beta_{j|i} + \varepsilon$$

with Y as the response, μ as the overall mean, α_i as the deviation of the ith of the a samples from the overall mean, $\beta_{j|i}$ as the deviation of the jth of the b replicates for the ith sample from the mean for the ith sample and ε

as the residual or error that is used to aggregate the common cause variability. There are c cells per replicate.

8.8 A general approach

The analysis of variance is a numerical procedure for dividing the variability in a data set into different parts that can be ascribed to

(1) *Special cause variability*, i.e. systematic differences caused by differences in

- The experimental units, which are often grouped into relatively homogeneous *blocks*.
- The experimental conditions or treatments applied to the units; these may be expressed as *variables* or *factors*, and may include *products* and *powers* of variables, *interactions* between factors and *nested* factors.

These special causes may be modelled by fixed or random effects.

(2) Uncontrolled, uncontrollable or unrecognised random *common cause variability*, for example due to method variability, sampling variation or heterogeneity in raw materials.

In this section, general approaches are outlined to linear models and the analysis of variance. The regression and experimental design models covered above and in Chapters 6, 7 and 11 are special cases of these general linear models. An aim of this section is to illuminate the general concept supporting these practical applications. A second aim is to give you a general understanding of modelling with these general and powerful tools that can facilitate the understanding of many complex situations and the analysis of almost any experiment. However, though much is possible, it is usually not wise or cost effective to use such methods except when forced to do so. It is more *efficient, cost effective* and likely that *reliable conclusions* will be obtained if the experimenter uses as simple a model as possible (so long as it is valid) and a simple well-designed experiment. As always, time should be spent on defining the purposes of the work and planning what is to be done. If complex general linear models have to be used, expert assistance should be called upon (see also comments on software packages in Section 8.8.4).

In Section 8.8.1, standard matrix notation is used to describe a general linear model and the related analysis of variance. However, the notation is only used as convenient shorthand and no mathematical proofs or derivations are given. For an $r \times s$ matrix (or a vector) $\mathbf{A}$, $\mathbf{A}'$ is used to denote the $s \times r$ transpose matrix. In Section 8.8.2, some of the possible model components are described. Finally in Section 8.8.3, generalised linear models are described briefly.

8.8.1 A general model and analysis

The fitted form of a general linear model with a single residual or error can be written as

$$\mathbf{y} = \mathbf{D}_1\mathbf{t}_1 + \mathbf{D}_2\mathbf{t}_2 + \cdots + \mathbf{D}_{k-1}\mathbf{t}_{k-1} + \mathbf{D}_k\mathbf{t}_k + \mathbf{e}_k$$

In this equation,

- $\mathbf{y}$ is an $n \times 1$ vector with elements $(y_i \quad \bar{y})$ where $\bar{y}$ is the average of the observed responses y_i.
- $\mathbf{t}_i$ is a $r_i \times 1$ vector of estimated effects for the ith model component.
- $\mathbf{D}_i$ is a $n \times r_i$ design matrix for the ith model component. If the component is a factor or an interaction of factors, $\mathbf{D}_i$ is a matrix with all elements in each row equal to zero except for the element indicating which factor level applies to that unit. This last element is 1. If the component is a predictor variable or covariate, the elements are $(x_i - \bar{x})$ where $\bar{x}$ is the average of the observed predictor values, x_i.
- $\mathbf{e}_k$ is the $n \times 1$ vector of estimated residuals or errors for this model with k components.

The estimates of the effects are obtained by least squares by minimising the residual sum of squares,

$$\mathbf{e}'_k \mathbf{e}_k = \mathbf{y}' \mathbf{H}_k \mathbf{y}$$

where $\mathbf{H}_k$ is a $n \times n$ matrix, the *residuals matrix*. The estimated residuals are

$$\mathbf{e}_k = \mathbf{H}_k \mathbf{y}$$

with $\mathbf{H}'_k \mathbf{H}_k = \mathbf{H}_k$.

If the kth component is eliminated from the model (i.e. compounded into the residual), the model becomes

$$\mathbf{y} = \mathbf{D}_1\mathbf{t}_1 + \mathbf{D}_2\mathbf{t}_2 + \cdots + \mathbf{D}_{k-1}\mathbf{t}_{k-1} + \mathbf{e}_{k-1}$$

The estimates of the remaining effects ignoring the kth component are obtained by least squares by minimising the residual sum of squares,

$$\mathbf{e}'_{k-1} \mathbf{e}_{k-1} = \mathbf{y}' \mathbf{H}_{k-1} \mathbf{y}$$

where $\mathbf{H}_{k-1}$ is the $n \times n$ residuals matrix for this model. The estimated residuals are

$$\mathbf{e}_{k-1} = \mathbf{H}_{k-1}\mathbf{y}$$

Thus, the sum of squares for the kth component adjusted for the other components in the model is

$$\mathbf{y}'\mathbf{H}_{k-1}\mathbf{y} - \mathbf{y}'\mathbf{H}_k\mathbf{y} = \mathbf{y}'(\mathbf{H}_{k-1} - \mathbf{H}_k)\mathbf{y}$$

which is always nonnegative.

If this process is repeated, sequentially removing the $(k-1)$th, $(k-2)$th, ..., 2nd, 1st components, the *sequential fit* or *Type I sums of squares* are obtained. These are adjusted for those components fitted earlier, but not for those fitted later in the model. The Type I sums of squares add to the total sum of squares. They are useful when there is a natural ordering of terms, in model building and in understanding the influence on the analysis of imbalance in the design. When estimating the effects of each of the components independently of the others, the *Type II sums of squares* are often used. These are obtained by fitting each of the components as the last in the sequence. Type II sums of squares do not add to the total sum of squares. They are mainly of use when there is no natural ordering of components, in particular when there are no interaction terms. In practice, the appropriate analysis sequences are often predicated by the practical application. However, for complex models it can be difficult to identify the appropriate analysis, and specialist help will probably be needed. There are other types of sums of squares available (see Gardiner and Gettinby, 1998). The problems of analysis can be minimised by good design, in particular the use of orthogonal designs.

Quadratic forms like $\mathbf{y}'\mathbf{H}_k\mathbf{y}$ and $\mathbf{y}'(\mathbf{H}_{k-1} - \mathbf{H}_k)\mathbf{y}$ under appropriate assumptions about the distribution of $\mathbf{e}_k$ are independent and have chi-squared distributions. The degrees of freedom for these distributions may or may not be as intuitively expected, depending on the level of confounding or aliasing between the components. See Chapters 6 and 7 and Gardiner and Gettinby (1998) for explanations of confounding and aliasing in factorial experiments. Thus, ratios of the mean sums of squares (sums of squares divided by degrees of freedom) can be used to test hypotheses about the components.

8.8.2 Classification of model components

The common types of model components have been introduced in this chapter. These are:

- Variables or *covariates*.
- Blocks.
- Factors or treatments that can be *crossed* with interactions between the individual factors, or *nested* within a higher-level factor.

Wilkinson and Rogers (1973) introduced a symbolic logic for structuring complex models using these components (see also McCullagh and Nelder, 1989). Similar symbolic structures are used for fitting and analysing general linear models in most statistical software packages.

8.8.3 Generalised linear models

The model presented in Section 8.8.1 is usually called a general linear model. The components of this model are:

- A *linear predictor*; in Section 8.8.1 this is the sum $(\mathbf{D}_1\mathbf{t}_1 + \mathbf{D}_2\mathbf{t}_2 + \cdots + \mathbf{D}_k\mathbf{t}_k)$.
- A *link function* that relates the linear predictor to the mean response; in Section 8.8.1 this is the identity function, i.e. the mean response is given by the linear predictor.
- An *error distribution* that defines the variability about the mean; in Section 8.8.1 this is the residual or errors $\mathbf{e}_k$ that are usually assumed to have independent Normal distributions with zero mean and common variance.

These three components are used in *generalised linear models* (GLMs), but with different forms available for each. Aitken *et al.* (1989) and McCullagh and Nelder (1989) give introductions to generalised linear models and the software for fitting and analysing GLMs.

8.8.4 Software for general linear models

Many statistical software packages include general linear model fitting routines, and some offer generalised linear model fitting routines. However, there are many methods for fitting models and many forms of the analysis of variance. Each method is appropriate for some situations, but not for others. Thus, care should be taken to understand exactly what is calculated, and to ensure that the method is appropriate for your need. If in doubt, get help.

8.9 Examples of two-factor analyses of variance

More detail on the analyses for the experimental designs given in this section can be found in books on design of experiments such as Gardiner and Gettinby (1998).

8.9.1 Randomised blocks design

In an inter-laboratory comparison, you wish to establish whether there are any consistent biases in obtaining purity by HPLC between three sites. Four samples of material of different purity are each well blended and then split into three parts. One part of each sample (chosen at random) is sent to each of the three sites for assay. The sites randomise the sequence in which the parts are analysed. In practice, a larger number of samples would be required in order to obtain an acceptably powerful test. The results obtained are shown in Table 8.8.

Table 8.8 Data from a randomised blocks design

Site	Sample				Site average
	1	2	3	4	
A	96.62	97.64	97.26	95.96	96.87
B	97.15	98.03	98.01	96.47	97.42
C	97.01	98.31	98.03	97.24	97.65

In this experiment the differences between sites are the factor of main interest (the treatments). However, to measure the significance of these differences the differences between the samples (blocks) must also be removed from the residual variability.

The sums of squares for the analysis of variance are

$$\text{Total SST} = \sum (y_{ij} - \hat{y}_{\bullet\bullet})^2$$
$$\text{Blocks SSB} = \sum (\hat{y}_{\bullet j} - \hat{y}_{\bullet\bullet})^2$$
$$\text{Treatments SSR} = \sum (\hat{y}_{i\bullet} - \hat{y}_{\bullet\bullet})^2$$

Residual SSE obtained by subtraction

The easiest way of obtaining these sums of squares and completing the analysis of variance is to use a statistics package. If you have to do the calculations without a computer or by programming the analysis on a spreadsheet, there are computationally more efficient equations that you should use. The analysis of variance is as in Table 8.9.

The p-value gives the probability of obtaining the observed differences in averages between sites assuming that there are in truth no differences. In this case, the observed differences will occur due to random variation at a frequency of less than 1 in 100 (when there are, in truth, no differences and the assumptions of the model of variation are correct). Thus, it is likely that the observed differences result from genuine

Table 8.9 Analysis of variance for a randomised blocks design

Source	Degrees of freedom	Sums of squares	Mean squares	F-Ratio	p-Value
Blocks	3	4.170	1.390		
Treatments	2	1.274	0.637	11.98	0.008
Residual	6	0.319	0.053		
Total	11	5.763			

differences between the sites. Inspection of the averages shows that the major differences are between site A and the other two. These differences can be shown to be significant, and that between sites B and C as non-significant.

> *Warning.* When comparing many averages it can appear sensible to use t-tests for pairwise comparisons. This should not be done, however, unless the overall comparison in the analysis of variance is significant. This is because the true risk of at least one Type I error (i.e. saying that there are significant differences when there are none) is $1-(1-\alpha)^m$ where α is the significance level for the individual tests and m the number of comparisons. For even moderate numbers of treatments, this risk can become large. Experimenters may think that just testing the significance of the difference between the largest and smallest averages avoids this problem. It does not, because in a multiple comparison there will always be a largest and a smallest. Thus, the risk of a Type I error for this comparison is as given above $1-(1-\alpha)^m$. Methods for multiple comparisons can be found in books on the design of experiments such as Gardiner and Gettinby (1998).

The differences between the averages for the blocks (samples) are as expected also highly significant. If these differences had been ignored and a one-way analysis of variance used, the comparison of the sites would have been not statistically significant (p-value 0.325). However, in this case the differences are of practical significance and further work would probably have been undertaken (at unnecessary cost). This may not always be true. Also, carefully blending and then testing all samples on all sites removed significant biases. Thus, it is important to use a good experimental method and design, *and* the correct analysis for the design.

8.9.2 *Replicated two-factor design*

In assessing the homogeneity of the output from a bulk blender/drier, the following sampling experiment was used. For a blend, the four kegs (numbered in off-load order) were sampled from the top, middle and bottom, with four samples taken at each level. This is a four-replicate complete two-factor experiment. The water content of each sample was measured. The results obtained for this 4 × 3 factorial design are given in Table 8.10. The analysis of variance is as in Table 8.11.

Table 8.10 Data from a two-factor experiment

Level	Keg				Average
	1	2	3	4	
Top	10.10	10.87	10.47	11.12	10.59
	10.65	11.08	10.74	10.73	
	9.57	10.39	10.86	9.97	
	10.23	10.38	11.49	10.72	
Middle	10.81	11.05	10.98	10.42	10.79
	11.05	11.12	10.82	10.63	
	10.58	10.92	11.10	10.41	
	10.09	10.55	11.17	10.95	
Bottom	10.37	11.89	11.02	10.58	11.17
	10.94	11.25	11.41	11.52	
	11.02	11.67	11.28	11.49	
	10.21	11.85	10.77	11.51	
Average	10.47	11.09	11.01	10.84	10.85

Table 8.11 Analysis of variance for a randomised blocks design

Source	Degrees of freedom	Sums of squares	Mean squares	F-Ratio	p-Value
Level	2	2.852	1.426	10.67	< 0.001
Keg	3	2.717	0.906	6.77	0.001
Interaction	6	1.186	0.198	1.48	0.213
Residual	36	4.813	0.134		
Total	47	11.567			

There is no significant interaction between keg and level. There are, however, significant differences both between kegs and between levels. This indicates that the material is not homogeneous and may also indicate settling within kegs. The patterns in the averages show drier material coming off at the start and end of the offload, and also being at the top of kegs. Significant interactions would also have indicated a lack of homogeneity but would probably have been more difficult to interpret.

8.9.3 Nested factors

In introducing and validating a new enzyme-based assay, the quality assurance manager wishes to estimate the relative magnitudes of variability due to sampling, due to the process of preparing samples and due to random noise from the incubation and measurement process. He also hopes to identify ways of reducing the variability in the assay process in order to obtain more precise results and consequently reduce costs by reducing replication.

Three independent samples are taken from the same bulk lot, for each sample four replicate preparations are made, each of which is split into four cells for incubation and measurement. This is a two-level nested design with replicates nested within sample and cells within replicate. The samples, replicates and cells are all random effects, with the last being the residual or error term. The results obtained are shown in Table 8.12. The analysis of variance is as in Table 8.13.

Table 8.12 Data from a two-factor experiment

Sample 1	Replicate 1	105	141	69	113
	Replicate 2	157	171	126	125
	Replicate 3	95	114	122	163
	Replicate 4	134	109	57	107
Sample 2	Replicate 1	123	139	108	75
	Replicate 2	167	173	159	134
	Replicate 3	128	117	136	140
	Replicate 4	85	99	85	121
Sample 3	Replicate 1	42	79	85	31
	Replicate 2	171	129	157	169
	Replicate 3	79	105	96	62
	Replicate 4	71	133	131	133

Table 8.13 Analysis of variance for a two-level nested design

Source	Degrees of freedom	Sums of squares	Mean squares	F-Ratio	p-Value
Sample	2	3 368	1 684	0.45	0.653
Replicates	9	33 880	3 764	6.37	<0.001
Cell (residual)	36	21 280	591		
Total	47	58 528			

In this analysis of variance, each level of the nested design is compared against that immediately below it. Thus, the F-ratio for samples is obtained by dividing the mean square for samples by that for replicates. The resulting test uses the F-distribution with (2,9) degrees of freedom. From the mean squares, the component variances can be estimated. The detail on how to carry out this estimation is given in Gardiner and Gettinby (1998) and in Searle *et al.* (1992). These components of variance are as given in Table 8.14.

Table 8.14 Components of variance for a two-level nested design

Source	Component variance	Percentage contribution
Sample	0	0%
Replicates	793	57%
Cell (residual)	591	43%

The estimate for the component for Sample was negative and, thus, is set to zero. Thus, in this experiment the major contribution to variability is the replicate preparation, closely followed by cell incubation and measurement, with little variability from sampling. Thus, dependent on availability of ideas on how to reduce variability, effort should focus on the sample preparation or cell incubation and measurement. Repeating the experiment with other lots and under different conditions will allow the robustness of these conclusions to be tested.

References

Acton, F.S. (1959) *Analysis of Straight-Line Data*, Dover, New York.
Aitken, M., Anderson, D., Francis, B. and Hinde, J. (1989) *Statistical Modelling in GLIM*, Clarendon Press, Oxford.
Atkinson, A.C. and Donev, A.N. (1992) *Optimal Experimental Designs*, Clarendon Press, Oxford.
Bissell, D. (1994) *Statistical Methods for SPC and TQM*, Chapman and Hall, London.
Cleveland, W.S. (1993) *Visualising Data*, Hobart Press, New Jersey.
Cox, D.R. (1958) *Planning of Experiments*, 2nd edn, Wiley, New York.
Gardiner, W.P. and Gettinby, G. (1998) *Experimental Design Techniques in Statistical Practice: A Practical Software Based Approach*. Horwood, Chichester.
Johnson, N.L. and Kotz, S. (1970) *Distributions in Statistics: Continuous Univariate Distributions 1*, Wiley, New York.
McCullagh, P. and Nelder, J.A. (1997) *Generalised Linear Models*, 2nd edn, Chapman and Hall, London.
Miller, R.G. (1997) *Beyond ANOVA: Basics of Applied Statistics*, Chapman and Hall, London.
Searle, S.R. (1971) *Linear Models*, Wiley, New York.
Searle, S.R., Casella, G.R. and McCulloch, C.E. (1992) *Variance Components*, Wiley, New York.
Sprent, P. (1969) *Models in Regression*, Methuen, London.
Stuart, A. and Ord, J.K. (1991) *Kendall's Advanced Theory of Statistics*, Volume 2, *Classical Inference and Relationship*, Edward Arnold, London.
Stuart, A. and Ord, J.K. (1994) *Kendall's Advanced Theory of Statistics*, Volume 1, *Distribution Theory*, Edward Arnold, London.
Wetherill, G.B. and Brown, D. (1991) *Statistical Process Control: Theory and Practice*, Chapman and Hall, London.
Wilkinson, G.N. and Rogers, C.E. (1973) Symbolic description of factorial models for analysis of variance. *Applied Statistics*, **22** 392-399.

9 Optimisation and control
T. Kourti

9.1 Introduction

Optimisation and control are two very general terms. Subjects such as automatic process control, statistical process control, control charts, process monitoring and fault diagnosis, process optimisation, optimisation of an analytical method or a procedure, real-time process optimisation may come to mind, depending on the technical background of the reader. There is a common misconception that process optimisation and control are purely the province of large-scale chemical plants. They are not. Most laboratory-based research in physical chemistry, synthetic organic and inorganic chemistry and analytical chemistry depends on having optimised and controlled instruments and reactions for consistent and reproducible experiments. This chapter focuses on statistical process control (SPC) and optimisation using experimental data and/or empirical models.

This chapter first provides an overview of both traditional and new multivariate SPC methods for monitoring a process or procedure, to ensure that it is in a state of statistical process control and to diagnose sources of deviation. The methods can be applied to laboratory-scale experiments, analytical procedures and pilot plant operations as well as to full-scale operations where hundreds of variables may be involved. They can be applied to data collected and analysed off-line and to on-line data for real-time monitoring and diagnosis of process performance in both continuous and batch processes. The optimisation procedures most commonly used in analytical laboratories and process industries when fundamental models do not exist are then discussed. The chapter concludes with a brief discussion on two stochastic search methods based on simulated natural processes that are being used to solve high-dimensional optimisation problems.

9.2 Terminology

The terms and concepts used in process optimisation and control are introduced in this chapter as they first appear in each section.

Subsequently, throughout the chapter we have tried to be consistent and use only one form where different phrases are used.

9.3 Where to go

Information needed	Go to Section(s)
Monitor repetitive operations for consistency.	9.5
Check that product is sold within specifications; Routine product quality tests.	9.5
Check calibration of instruments with reference samples of known properties.	9.5
Assess quality of product from process measurements only.	9.5
Charts to monitor one variable only.	9.5.1
Learn basics of control charts, how to collect samples and how often, how to set limits.	9.5.1
Charts to detect small shifts in mean values for one variable.	9.5.3, 9.5.4
Why not use many univariate charts to monitor many variables?	9.5.5
Charts to monitor many variables simultaneously.	9.5.6
Other traditional multivariate charts to detect small shifts.	9.5.7, 9.5.8
I have quality measurements once a day; how do I know that the process is in control the rest of the time?	9.5.9–9.5.13
I have an intermediate step in a process with no quality measurements on this step; can I assess, from process measurements only, that during this intermediate step the product was not spoiled?	9.5.9–9.5.13
Product is produced by three reactors in parallel and blended before sold. Can I assess, from process measurements only, if any one of the reactors produced bad product so that I do not blend it?	9.5.9–9.5.13
What happens with batch reactors? Can I monitor such processes?	9.5.13, 9.5.14
What happens with multi-stage processes?	9.5.13, 9.5.14
There has been a problem in the production. I have two years of data stored as daily averages on 400 variables and 16 units of the plant when production was very good. Can I find which unit has the problem now?	9.5.16
Process and Product Improvement to Maximise Yield, Minimise Cost, Reduce Pollutant Emissions.	9.6

Information needed	Go to Section(s)
Finding the best combination of values for manipulated variables to achieve a specific objective with an analytical procedure (i.e. identifying the combination of the mobile phase composition, temperature, etc. that maximises resolution between adjacent solutes under time and cost constraints).	9.6
Conformational analysis of macromolecules to fit specific receptor shapes	9.6
I suspect three variables are important and will affect the performance of my analytical procedure. I will manipulate only one for the time; why not?	9.6.2
I want to maximise yield but at the same time keep two other properties within the specifications that the customer demands	9.6.3
I want to find better operating conditions than now but with not too many experiments.	9.6.4, 9.6.5
I have a process producing material within specifications; environmental demands require lower energy consumption. I have to keep production going. How can I find new better operating conditions?	9.6.5
Are designed experiments necessary for optimisation?	9.6.4, 9.6.6
Can I find better operating conditions by analysing the stored data of the last two years in the plant?	9.6.6

9.4 Why process control and optimisation?

9.4.1 Process control

To run a process in either pilot-scale or full-scale operation, or to perform an analytical procedure in the laboratory, we have to follow a recipe. This entails using the proper type of equipment and materials (e.g. type of packing and column length in chromatography, type of reactor in a process) and making sure that the values of certain controlled variables are maintained close to their set points. Set points are determined either by experimental design or in some other way, for example by process simulation. A couple of examples illustrate the basics.

- In a polymerisation process we have to use a specific type of reactor (continuous or semibatch), feed precise amounts of ingredients (monomer, emulsifier, catalyst, etc.), maintain predetermined feed

rates and run the reaction maintaining the temperature close to a predetermined set point.
- When performing a chromatographic analysis we have to prepare the mobile phase with a given composition, use certain packing material in the column and perform the separation using a certain flowrate and keeping the temperature close to a set point.

Although we manually control some inputs in both examples (e.g. weigh proper amounts of ingredients, pick a column with right length), automatic process control is essential to maintain the values of controlled variables at their prescribed set points. For example, once it has been decided that the reaction or chromatography should run at, say, 50°C, automatic process control will be used to keep the temperature in the reactor or column at $50 \pm 0.5°C$, despite temperature fluctuations of the fresh fed ingredients or carrier gas coming from the storage tanks and changes in the heat release (or absorption) of the reaction.

Even when controllers are used to maintain the controlled variables at their set points, the product quality variables (purity, particle size, yield, chromatographic peak shape) and other process variables (reaction temperature and pressure, agitation power) will still show fluctuations around their target values due to the noise of the system. This is called common cause variation (see also Chapter 1), that is, variation that affects the process all the time and is essentially unavoidable within the current setup of the process. If the process experiences only common cause variation, we say that the process remains in a 'state of statistical control'. Statistical process control charts are used to monitor key product or process variables over time to verify they are close to their target values and to detect the occurrence of any event having a 'special' or 'assignable' cause.

By finding assignable causes, long-term improvements in the process and in product quality can be achieved by eliminating the causes, or improving the process, or by changing the operating procedures.

It is important to note that the concepts and methods of SPC are completely different from those of automatic process control, but the two approaches are totally complementary.

Automatic process control should be applied wherever possible to reduce variability in important process and product variables. Automatic process controllers compensate for the predictable component of the disturbances in important variables by adjusting other process variables and thereby transferring the variability into these less important manipulated variables. In this way, consistent product quality is achieved and safe operating conditions are maintained.

SPC monitoring methods are applied on top of the process and its automatic control system to detect process behaviour that indicates the occurrence of a special event. By diagnosing causes for the event and removing them (rather than simply continuing to compensate for them), the process is improved.

A full discussion on automatic process control is beyond the scope of this chapter. For further study, readers are referred to Marlin (1995) and Stephanopoulos (1984).

9.4.2 Optimisation

Although we use process control to maintain the process conditions dictated by a recipe, this recipe may not be the optimal one. SPC may indicate deviations in the product that are due to assignable causes and therefore that operating procedures must change. It is also possible that a process is producing a product within specifications but not at its optimal point regarding cost or current environmental demands (Wold et al., 1989). There may be room for further improvement of the process by increasing the yield, lowering energy consumption or minimising emissions of pollutants to the environment. We can search for better, and even the optimal, operating conditions.

There are several steps we could take to search for the optimum. If fundamental mathematical models exist to describe the process, we can search in the parameter space of the models and solve sets of equations for our optimality criteria. In this way, we find the optimal process variable set points. The task of solving the model equations for the optimum is not trivial and there are numerous optimisation algorithms depending on the kind of model, the form of the constraints and the type of objective function (Edgar and Himmelblau, 1988; Fletcher, 1987). If fundamental models do not exist, then we have to rely on experimentation and possibly on empirical models to describe the process. Finding an optimum is limited by the available data and the one found may not be the best, or global, optimum.

Every time the process operating conditions change because optimisation dictates so, the calibration for the SPC monitoring procedures needs to be recalculated for these new conditions. SPC will then continue to work on top of the automatic process control to detect any unusual events that may affect the product.

9.5 Statistical process control (SPC)

Statistical process control concepts and methods have become very important in the manufacturing and process industries. Statistical process control can also be useful in a laboratory environment where consistency of sets of experiments is required over a period of time.

The aim of SPC is to monitor the performance of repetitive operations over time to verify that they are remaining in a state of statistical control. Such a state is said to exist if certain variables remain close to their desired values and the only source of variation is common cause variation, that is variation that affects the process all the time and is essentially unavoidable within the current operation. Various control charts are used to monitor key variables in the process or the product to detect the occurrence of any event having a special or assignable cause. Long-term improvements in the process and in the product quality can be achieved by finding assignable causes and eliminating the causes.

Until recently, the usual practice in industry was to examine only a small number of variables, independently of one another, using univariate control charts—Shewhart, cumulative sum (Cusum), exponentially weighted moving average (EWMA). When several variables (properties) describe the quality of a product, it is not correct to monitor them independently and to do so will generate serious decision errors. Multivariate control charts should be used. They can be based on the traditional approach of using the measured data directly or the data can first be transformed into latent variables. The latent variable approach allows for a large number of process variables as well as product variables to be included in the process-monitoring scheme. In this section, we look at all three approaches: univariate, traditional multivariate and latent variable multivariate.

9.5.1 Univariate Shewhart charts

A typical control chart is shown in Figure 9.1. Data from a single variable Q are plotted in time order. The chart contains a centre line (CL) that represents the average value of Q expected when the process is in-control. Two other lines are the upper control limit (UCL) and the lower control limit (LCL). These are chosen such that when the process is in-control nearly all the data points fall between them. If a point falls outside the limits (observation 40 in Figure 9.1), the process is investigated to identify the cause of the deviation. The charts are named after Dr W.A. Shewhart of the Bell Telephone Laboratories, USA, who first proposed them in the 1920s (Shewhart, 1931).

The measured variable can be a product quality variable (the molecular weight or melt index from a polymerisation process) or a process variable (temperature of reactor). The value Q at each point on these charts can be either the mean value of measurements on a variable for n samples

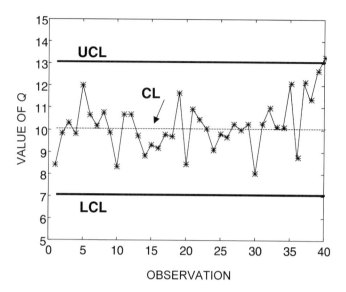

Figure 9.1 A typical control chart.

($\bar{X}$-chart), or the range (R-chart) of the variable or its standard deviation (S-chart) over the n samples.

When $n=1$, the $\bar{X}$-chart is simply a plot of the individual values of the variable. However, range and standard deviation are not defined for $n=1$. To overcome this, the moving range or moving standard deviation can be used (Montgomery and Runger, 1994; Montgomery et al., 1998).

Samples may be collected at periodic intervals (every 15 minutes, every hour or every day) for continuous processes, or after each batch is completed for products manufactured by batch and semibatch processes, or after several completed batches or number of parts produced when many such batches or parts are produced per day. The number of samples collected per interval depends on individual circumstances. For example:

- The molecular weight of a polymer product reported for 4 batches selected at random from those produced in a day ($n=4$).
- The diameter of metal rods reported every 2 hours from 5 randomly chosen rods ($n=5$).
- Melt index measured off-line on a single sample every 8 hours ($n=1$).
- The chromatographic peak area of a reference sample inserted every 15 test samples ($n=1$).

Sample data should be collected using what Shewhart called the rational subgroup concept. Each point on the control chart corresponds to a subgroup and it may be one sample or a collection of n samples. Samples should be collected such that variability within each subgroup excludes assignable causes. This concept, together with the objective of the monitoring, dictates the frequency of the subgroups (points on plot) and the part of the process from which the samples are taken.

Consider the case in which polymer from two reactors running in parallel is blended in a mixer before being sold. If our objective is to follow the quality of the final product, it may be sufficient to construct one chart using samples taken from the mixing vessel. However, if this chart indicates a problem it can be difficult to identify which reactor is responsible. Having one chart per reactor is much more helpful, although it may be more expensive as more samples are analysed.

Similarly, the sampling frequency needs to be considered carefully. Suppose that when a specific event occurs in a unit it causes such an upset in the process if it goes undetected, and that the operators must shut down the unit 12 hours from its onset. The sampling frequency should be fast enough to catch such an occurrence early, perhaps every 1 hour. Certainly a 6-hour interval is not a good idea for such a system.

Finally, when calculating the limits for control charts from past historical data, one should be careful about the data used. Points corresponding to assignable causes should not be included, as this will result in wide control limits and reduced ability to detect small shifts.

9.5.2 A general model of Shewhart charts

If Q is the sample statistic ($\bar{X}$, or R) of the variable we wish to monitor, we can define μ_Q as the population mean of Q, and σ_Q as the population standard deviation of Q. For example, suppose we wish to monitor the relative viscosity of a product produced by a batch process, by plotting on an $\bar{X}$ control chart the mean value from five random batches a day. If σ is the population standard deviation of the relative viscosity measurements, then the estimated standard deviation of Q, which is the average of five measurements, is $\sigma_Q = \sigma/\sqrt{5}$. When σ and μ of the measurement populations are not known, they must be estimated from samples, see Montgomery and Runger (1994).

The centre line and the control limits of the chart are given as

$$\text{UCL} = \mu_Q + \lambda \sigma_Q, \quad \text{CL} = \mu_Q, \quad \text{UCL} = \mu_Q - \lambda \sigma_Q \qquad (9.1)$$

where λ gives the distance of the control limits from the centre line. A common choice is $\lambda = 3$, that is the control limits are located at 3 standard deviations from the target (3-sigma control limits). For Q Normally distributed, we expect $100(1-\alpha)$ percent of the plotted Q values to fall within the limits when the process is in-control. Here, α is the proportion of values that fall in the tails of the Normal distribution outside $\mu \pm \lambda\sigma_Q$. For an $\bar{X}$-control chart, $\alpha = 0.0027$ when $\lambda = 3$, that is we expect 99.73% ($= 100 - 0.27$) of the points to fall within the limits. Note that for an in-control process the points outside the control limits are part of the same distribution as the points inside the limits, they just happen to have a low probability of occurring. It is a convention to say that as the prob-ability is low we should regard points outside the control limits as unusual and worthy of investigation. This is also the probability of rejecting good points as bad (also called Type I error).

It is useful to calculate the average number of points that must be plotted before a point indicates an out-of-control condition simply because a low-probability value has appeared (i.e. nothing unusual has happened in the process). This is called the average run length (ARL). With $\alpha = 0.0027$, the ARL is $(1/0.0027) \approx 370$. With samples taken every 10 minutes, a false alarm will occur approximately every 61.5 hours. The ARL can be used also to decide on sample frequency and sample size in order to detect a shift in mean. As a rule, to detect a small shift early we must either sample more frequently or use larger sample sizes, as illustrated with the following example. Suppose that for an $\bar{X}$-chart the sample size is $n = 1$ and $\sigma_Q = \sigma$. When the process is in control the LCL is at $-3\sigma_Q = -3\sigma$ from the targeted process mean. If the process average shifts downwards by 1.5σ units and remains in this new position with the same standard deviation, the LCL is now only -1.5σ away from the new process mean. This gives $\alpha = 0.0668$ and ARL $= 15$, which means that we expect the shift in the average to be detected by an out-of-limit value after about 15 time points. We can adjust the time interval so that the average time for detecting a shift in the process average by 1.5σ is acceptable, or we can change the sample size. If, at each time point, we take four samples and plot their average, then $n = 4$ and $\sigma_Q = \sigma/\sqrt{n} = \sigma/2$ and LCL $= -1.5\sigma$. If the process average again shifts downwards by 1.5σ units, the LCL is now at the new process mean and we find that $\alpha = 0.5$ and ARL $= 2$. That is, each of the next points has 50% probability of being above or below the new mean, so by the second point we should have at least one below LCL. Thus, increasing the sample size at each time point shortens the average time it takes to detect a shift in the process average.

Very frequently, the 1-sigma and 2-sigma limits, are drawn on the Shewhart chart in addition to the 3-sigma limits. These limits should contain 68.3% and 95.5% of Q values, respectively, if Q has a Normal distribution. The additional limits allow us to do more detailed checking of the Q values to determine unusual conditions in the process by looking for patterns in the sequences of values. Many patterns have been described, the most common being those given in the Western Electric handbook rules (Western Electric, 1956); check for an unusual event, if any of the following occurs:

1. One point plots outside 3-sigma control limits.
2. Two out of three consecutive points plot beyond the same 2-sigma limit.
3. Four out of five consecutive points plot beyond the same 1-sigma limit.
4. Eight consecutive points plot on one side of the centre line.

Other charts are available (*p*-charts for percent defective items, *c*-charts for count data, *m*-charts for medians of the subgroups, etc.). Almost any statistic computed from repeated samples can be displayed as a Shewhart chart. Variable sample size is also possible, but a weighted-average approach is used to calculate the mean and standard deviation for the $\bar{X}$- and S-charts (Montgomery, 1991).

9.5.3 Univariate Cusum charts

The Shewhart chart is relatively insensitive to shifts in the measured value that are smaller than approximately 1.5σ when the sample size is small. In Figure 9.1, for example, the process mean was shifted by 1σ after observation 30. This shift was not detected with the Shewhart chart ($n=1$) until observation 40. The Western Electric rules may solve this problem to some extent; notice that in Figure 9.1, applying rule (2) would have detected the problem at point 36, since two (34, 36) out of three consecutive points are beyond 2σ. However, applying these rules tends to increase the false alarm rate. An alternative to Shewhart charts is the cumulative sum (Cusum) chart, which plots the cumulative sum of the deviations of the Q values from a target value. The chart is very effective for plotting individual measurement data ($n=1$). The assumption in this chart is that observations of the variable are independent with fixed mean and constant variance. (Montgomery, 1991).

Suppose that $\bar{x}_j$ is the average value of the *j*th sample (size $n \geq 1$). If the objective is to run the process such that the mean value of the property is at a target value μ, then the following quantity is plotted for the Cusum chart, for the current sample k:

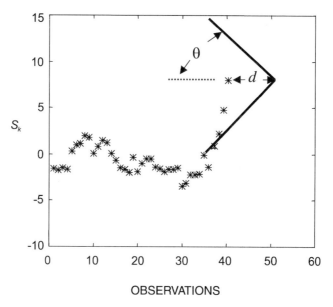

Figure 9.2 A Cusum chart with V-mask.

$$S_k = \sum_{j=1}^{k}(\bar{x}_j - \mu) = (\bar{x}_k - \mu) + S_{k-1} \qquad (9.2)$$

S_k is the cumulative sum up to sample k. The Cusum chart for the same data as in Figure 9.1 is shown in Figure 9.2.

When the process is on target, the value $(\bar{x}_j - \mu)$ is a random error with mean zero and the cumulative sum should fluctuate around zero. If the process mean shifts upwards, then a positive drift will develop for S_k, as happens after observation 30 in Figure 9.2. Sometimes, however, even when the process is at target, S_k can wander remarkably far from zero and give the appearance that there has been a process shift, as for example between observations 5 and 10. This is a symptom of the random walk. A V-mask is used to differentiate between random walk and a shift in process mean. The V-mask is applied to each new point on the Cusum chart with its vertex at distance d from the current point with its axis parallel to the time axis. If all previous points lie within arms of the V-mask the process is in control. In Figure 9.2 the mask is placed on observation 40, to illustrate an out-of-control case.

The construction of the V-mask is essentially the calculation of the angle θ, shown in Figure 9.2, which is given by $\theta = \tan^{-1}(\Delta/2A)$. Here Δ

is the shift we wish to detect and A is the scale factor of the S_k axis. The leading distance d is calculated as

$$d = \left(\frac{2}{\delta^2}\right) \ln\left(\frac{1-\beta}{\alpha}\right) \quad \text{with} \quad \delta = \frac{\Delta}{\sigma_Q}$$

where σ_Q is the standard deviation of Q, α is the probability of false alarm (Type I error) and β is the probability of failing to detect the shift in mean (Type II error). Even without the V-mask, the Cusum is a powerful visual device for informing the user about the status of the process and can be used as a companion to the Shewhart chart. Cusum charts can be constructed for historical data (see Chapter 2) and these are particularly useful for checking the validity of data to be used to estimate control limits for Shewhart charts.

An alternative technique (that sets the limits without a V-mask) is the tabular Cusum, which is attractive for computer implementations and is described in Montgomery (1991). One-sided Cusum charts are also possible (Hunter, 1986). They are useful when only deviations on one side of the target value are important (e.g. purity should never fall below 95% or concentration of an impurity should be kept below 0.02%). Finally, it should be noted that the Western Electric procedures do not apply for the Cusum chart.

9.5.4 Univariate EWMA charts

For each point on a Shewhart chart only the current observation is important; all past observations are ignored. In Cusum charts, all observations, old and new, are weighted equally at each point. In the exponentially weighted moving average (EWMA) charts, past observations are accounted for but they are given a smaller weight as they become older. The exponentially weighted moving average is defined as

$$Z_k = r\bar{x}_k + (1-r)Z_{k-1} \qquad (9.3)$$

where $0 < r \leq 1$. The starting value for EWMA at $k=0$ is $Z_0 = \mu$ (the target value for the property).

Notice that an alternative way for constructing a univariate EWMA is to use the following statistic instead that of equation (9.3), with $Z_0 = 0.0$:

$$Z_k = r(\bar{x}_k - \mu) + (1-r)Z_{k-1} \qquad (9.3a)$$

The values of Z_k (from either equation (9.3) or (9.3a)) are plotted on a chart with a centre line at Z_0 and control limits calculated from

$$\sigma_{Z_k}^2 = \sigma_Q^2 \left(\frac{r}{2-r}\right)[1 - (1-r)^{2k}] \qquad (9.4)$$

The variance of the EWMA chart eventually converges to $\sigma_{EWMA}^2 = \sigma_Q^2[r/(2-r)]$. The limits for the chart are at $Z_0 \pm \lambda \sigma_{Z_k}$, with $\lambda = 3$ for values of r between 0.1 and 0.25, which are typically used in practice. An EWMA chart is plotted in Figure 9.3 for the data set of Figure 9.1, using equation (9.3), $Z_0 = \mu$, $\lambda = 3$ and $r = 0.2$. For this set of data the 1σ shift

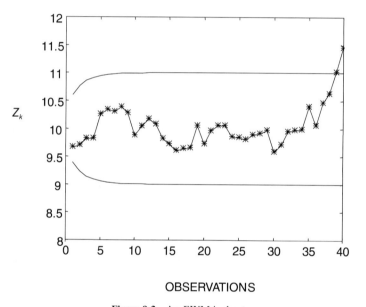

Figure 9.3 An EWMA chart.

was detected at observation 39, only one observation earlier than that detected by the Shewhart chart. Of course the users can fine tune the detection mechanism (values of λ, r) to achieve a desired ARL for small shifts for the particular system they are working with. The choices for the value of r and λ of EWMA charts to detect small shifts are discussed in Montgomery (1991).

As $r \to 1$, $Z_k \to \bar{x}_k$ and the chart of equation (9.3) becomes a Shewhart chart. When r has very small values ($r \to 0$), the most recent observation has a very small weight and previous observations near equal (though very low) weights. In these cases the EWMA takes on the appearance of a Cusum. The EWMA chart for $0 < r < 1$ stands between the Shewhart and the Cusum chart in its use of historical data.

Notice that Z_k contains all the information on past data needed for the next interval $k+1$, and there is no need to keep information on previous observations and their weights. A detailed discussion can be found in Hunter (1986), where it is also shown how the EWMA can be used to provide a forecast of where the process will be in the next instance in time and thus provide a mechanism for dynamic process control.

Harris and Ross (1991) discuss the impact of serially correlated data on the performance of these charts. It is shown that serious errors concerning the 'state of statistical process control' may result if the correlation structure of the observations is not taken into account.

9.5.5 Multivariate charts for statistical quality control

Most industries use univariate charts to monitor product characteristics or key process variables that in some way affect the quality of the final product. The problem with using univariate control charts for separately monitoring key variables on the final product is that most of the time the variables are not independent of one another, and none of them adequately defines product quality by itself. Product quality is defined by the simultaneously correct values of all the measured properties, that is it is a multivariate property.

Figure 9.4 illustrates the problem with using separate control charts for two quality variables (y_1, y_2). In this figure, the two variables are plotted against each other (upper left of the figure). The same observations are also plotted as individual Shewhart charts for y_1 (the horizontal plot) and y_2 (the vertical plot) with their corresponding upper and lower control limits.

Suppose that, when only common cause variation is present, y_1 and y_2 follow a multivariate Normal distribution; the dots in the joint plot represent a set of observations from this distribution. Notice that y_1, and y_2 are correlated. The ellipse represents the 99% joint confidence limit of the distribution (i.e. when the process is in control, 99% of the points will fall inside the ellipse).

The point indicated by the $\otimes$ is clearly outside the joint confidence region, and it is different from the normal in-control population of the product. However, neither of the Shewhart charts gives any indication of a problem for point $\otimes$; it is within limits in both of the charts. The individual Shewhart charts effectively create a joint acceptance region shaped like a square, shown under the ellipse. This will lead to accepting

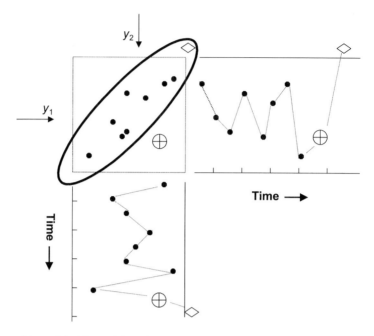

Figure 9.4 The problem with using separate control charts per variable.

wrong products as good (point ⊗), but also showing a good product as bad (point ◇).

In spite of the misleading nature of univariate quality control charts, they continue to be almost the only form of monitoring used by most industries. However, several multivariate extensions of the Shewhart, Cusum and EWMA based on Hotelling's T^2 statistic have been proposed in the literature (see review articles by Wierda, 1994).

9.5.6 Hotelling's T^2 and chi-squared multivariate charts

Hotelling's T^2 and chi-squared charts are the multivariate equivalents of Shewhart charts used when a vector of q variables $\mathbf{y}_k(q \times 1)$ is observed at each time period $k(q \geq 2)$. Details can be found in Montgomery (1991).

Given a $(q \times 1)$ vector of measurements $\mathbf{y}_k$ on q Normally distributed variables with an in-control covariance matrix $\mathbf{\Sigma}$ one can test whether the vector of the means of these variables is at its desired target $\boldsymbol{\mu}$ by computing the statistic

$$\chi_k^2 = (\mathbf{y}_k - \boldsymbol{\mu})^T \mathbf{\Sigma}^{-1} (\mathbf{y}_k - \boldsymbol{\mu}) \tag{9.5}$$

The χ^2 statistic in equation (9.5) represents the directed or weighted distance (Mahalanobis distance) of any point from the target μ. When $q = 2$ as in Figure 9.2, $\mathbf{y}_k$ and $\mathbf{\Sigma}$ are respectively the vector and matrix, with σ_1^2, σ_2^2, the variances of y_1 and y_2, respectively, and σ_{12}^2 their covariance,

$$\mathbf{y}_k = \begin{vmatrix} y_{1,k} \\ y_{2,k} \end{vmatrix}, \quad \mathbf{\Sigma} = \begin{vmatrix} \sigma_1^2 & \sigma_{12}^2 \\ \sigma_{12}^2 & \sigma_2^2 \end{vmatrix}$$

and equation (9.5) represents the weighted distance of each point on the joint plot from the centre of the ellipse.

This statistic is distributed as a central chi-squared distribution with q degrees of freedom if the mean is on target μ. A multivariate chi-squared control chart can be constructed by plotting χ^2 versus time with an upper control limit (UCL) given by $\chi_\alpha^2(q)$ where α is an appropriate level of significance for performing the test (e.g. $\alpha = 0.01$). A χ^2 statistic plot for the points of Figure 9.4 is shown in Figure 9.5. Notice that the

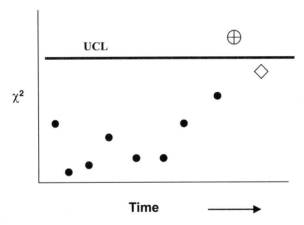

Figure 9.5 A χ^2 statistic plot.

time history is preserved. This multivariate test overcomes the difficulty illustrated in the example of Figure 9.4, where univariate charts were incapable of detecting the special event denoted by $\otimes$. All points lying on the ellipse in Figure 9.4 have the same value of $\chi^2 = \chi_\alpha^2(q)$. Hence, a chi-squared chart will detect any point lying outside of the ellipse as a special event.

When the in-control covariance matrix Σ is not known, it must be estimated from a sample of n past multivariate observations as

$$\mathbf{S} = (n-1)^{-1} \sum_{i=1}^{n} (\mathbf{y}_i - \bar{\mathbf{y}})(\mathbf{y}_i - \bar{\mathbf{y}})^{\mathrm{T}} \qquad (9.6)$$

When a new multivariate observation (**y**) is obtained, Hotelling's T^2 statistic is given by

$$T^2 = (\mathbf{y} - \boldsymbol{\mu})^{\mathrm{T}} \mathbf{S}^{-1} (\mathbf{y} - \boldsymbol{\mu}) \qquad (9.7)$$

This can be plotted against time in a similar way to χ^2. An upper control limit on this chart is given by

$$T^2_{\mathrm{UCL}} = \frac{(n-1)(n+1)q}{n(n-q)} F_\alpha(q, n-q) \qquad (9.8)$$

where $F_\alpha(q, n-q)$ is the upper $100\alpha\%$ critical point of the F-distribution with q and $n-q$ degrees of freedom (Tracy et al., 1992).

The above charts are for a single new multivariate observation vector at each time. If an average of m new multivariate observations is used at each time, or if the estimate of the variance **S** is based on pooling estimates from rational subgroups, then the chi-squared and T^2 charts and their UCLs must be correspondingly redefined (Wierda, 1994). Furthermore, if the charts are utilised to examine past data that are also used in computing **S**, then the distributional properties of T^2 are different from the above (Tracy et al., 1992; Wierda, 1994).

Once an out-of-control signal is detected, the challenge is to determine which variables are responsible for it. Several approaches have been suggested for this. Kourti and MacGregor (1996) review these approaches and suggest a computationally fast and easy algorithm based on principal components to determine the variables that contribute numerically to the out-of-control signal.

9.5.7 Multivariate Cusum charts

There have been several suggestions for multivariate Cusum (MCusum) charts by various researchers. These are reviewed by Wierda (1994) where the author investigated whether a multivariate Cusum chart had a smaller ARL than a Hotelling's T^2 chart in detecting small shifts, and ranked various MCusum charts based on their performance in simulated cases. The best performer was that of Pignatiello and Runger (1990) described below. The charts were evaluated for their ability to detect a shift and the interpretability of the out-of-control signal. We describe briefly some of these charts and refer the reader to Wierda (1994) and Lowry et al. (1992) for more discussions and comparisons of their performance.

Multivariate Cusum charts can be thought of as sequential probability ratio tests. Suppose we have a vector $\mathbf{y}_k$ of measurements on q variables, taken at each interval k, so we have a sequence of $\mathbf{y}_1, \mathbf{y}_2, \mathbf{y}_3, \ldots$ The $\mathbf{y}$ variables are distributed independently with an in control covariance $\mathbf{\Sigma}$, as $N_q(\boldsymbol{\mu}, \mathbf{\Sigma})$, and it is desired to test the Null Hypothesis H_0: the mean is at $\boldsymbol{\mu}$ versus H_1: the mean is at $\boldsymbol{\mu}_1$. Let

$$d^2 = (\boldsymbol{\mu}_1 - \boldsymbol{\mu})^T \mathbf{\Sigma}^{-1} (\boldsymbol{\mu}_1 - \boldsymbol{\mu})$$

define the squared length of the shift in mean we wish to detect. Then the sequential probability ratio test rejects the Null Hypothesis whenever

$$\sum_{i=1}^{k} \{d^{-1}(\boldsymbol{\mu}_1 - \boldsymbol{\mu})^T \mathbf{\Sigma}^{-1} (\mathbf{y}_k - \boldsymbol{\mu}) - d/2\} > -\frac{\log \alpha}{d} \quad (9.9)$$

where α is the level of significance chosen (the probability for Type I error). Healy (1987) used this result to propose plotting the Cusum:

$$C_k = \text{Max}\{0, C_{k-1} + d^{-1}(\boldsymbol{\mu}_1 - \boldsymbol{\mu})^T \mathbf{\Sigma}^{-1}(\mathbf{y}_k - \boldsymbol{\mu}) - d/2\} \quad (9.10)$$

Recall that with this Cusum an out-of-control signal is interpreted as a shift in mean from $\boldsymbol{\mu}$ to $\boldsymbol{\mu}_1$.

Crosier (1988) proposed the COT scheme (Cusum of T). This consists of computing T_k^2 (equation 9.7) at each point in time k, and then forming a Cusum of the scalars T_k as

$$C_k = \text{Max}\{0, C_{k-1} + T_k - \varphi\} \quad (9.11)$$

with initial condition $C_0 \geq 0.0$ and $\varphi > 0.0$, where φ is a shrinkage factor (the updated C_k is shrunk towards zero by φ). The value of φ is chosen based on ARL considerations. This Cusum scheme signals an out-of-control situation when $C_k > h$, where h is the control limit for the Cusum.

Crosier (1988) also proposed replacing the scalar quantities of the univariate Cusum by their vector counterparts and computing the vector Cusum:

$$\begin{aligned} \mathbf{s}_k &= 0, \text{ if } C_k \leq \varphi \\ &= (\mathbf{s}_{k-1} + \mathbf{y}_k - \boldsymbol{\mu})(1 - \varphi/C_k), \text{ if } C_k > \varphi \end{aligned} \quad (9.12)$$

where C_k is the weighted length $\{(\mathbf{s}_{k-1} + \mathbf{y}_k - \boldsymbol{\mu})^T \mathbf{\Sigma}^{-1} (\mathbf{s}_{k-1} + \mathbf{y}_k - \boldsymbol{\mu})\}^{1/2}$, $\mathbf{s}_k$ is the vector of the individual S_k of each variable, and $\mathbf{s}_0$ is a vector of zeros.

The scheme signals an out-of-control situation whenever

$$\max\{0, C_k - \varphi\} > h \qquad (9.13)$$

A reference value $\varphi = d^2/2$ is usually chosen, to detect any shift in the mean vector yielding square length d. This choice minimises the average run length or ARL at deviation d for a given on-target ARL. The on-target ARL is determined by the choice of the control limit h.

Pigniatello and Runger (1990) suggest a chart based on the following vectors of cumulative sums:

$$\mathbf{d}_k = \sum_{j=k-n_k+1}^{k} (\mathbf{y}_j - \boldsymbol{\mu})$$

n_k is the number of observations since the most recent renewal (the last zero value) of the Cusum chart. The chart is constructed for the statistic

$$\mathrm{MC}_k = \max\{0, (\mathbf{d}_k^\mathrm{T} \boldsymbol{\Sigma}^{-1} \mathbf{d}_k)^{1/2} - \xi n_k\}$$

where $\xi > 0$. $n_k = n_{k-1} + 1$ if $0 < \mathrm{MC}_{k-1} < h$, and $n_k = 1.0$ otherwise. An out-of-control signal is given when $\mathrm{MC}_k > h$, where $h > 0$ is the control limit.

9.5.8 Multivariate EWMA

Lowry et al. (1992) extended equation (9.3a) to the multivariate exponentially weighted moving average (MEWMA). Suppose that random vectors $\mathbf{y}_1, \mathbf{y}_2, \ldots$ on q process variables are observed over time with covariance matrix $\boldsymbol{\Sigma}$. At time k, for the observed vector $\mathbf{y}_k$,

$$\mathbf{z}_k = \mathbf{R}(\mathbf{y}_k - \boldsymbol{\mu}) + (\mathbf{I} - \mathbf{R}) \mathbf{z}_{k-1} \qquad (9.14)$$

where $\mathbf{z}_k$ is essentially the vector of univariate Z_k calculated at interval k for each variable; the vector $\boldsymbol{\mu}$ gives the targets of the individual variables and $\mathbf{R} = \mathrm{diagonal}\{r_1, r_2, \ldots, r_q\}$ and $0 < r_j \leq 1; j = 1, \ldots, q$. The MEWMA gives an out-of-control signal when

$$Q_k^2 = \mathbf{z}_k^\mathrm{T} \boldsymbol{\Omega}_k^{-1} \mathbf{z}_k > h \qquad (9.15)$$

where the control limit h is chosen to achieve a specified in-control ARL and $\boldsymbol{\Omega}_k$ is the covariance matrix of $\mathbf{z}_k$. When all the r_j are equal to each other ($r_j = r$, for $j = 1, 2, \ldots, q$), the covariance matrix is given by

$$\boldsymbol{\Omega}_k = \frac{r[1 - (1-r)^{2k}]}{2 - r} \boldsymbol{\Sigma} \qquad (9.15\mathrm{a})$$

Lowry et al. showed that, depending on the type of process and the shifts to be detected, this multivariate chart, although it is simple to construct, performs as well as, and sometimes better than, the Cusum of Crosier (1988)—not the COT and Pignatiello and Runger (1990), based on ARL comparisons.

9.5.9 Multivariate control charts based on latent variables

Control charts based on latent variables have been introduced in the last few years and their use in industry is increasing. A detailed discussion can be found in Kourti and MacGregor (1996). The charts answer the need of process industries for a tool that allows them to utilise the massive amounts of data being collected on hundreds of process variables, as well as the spectral data collected from modern analysers.

Latent variable control charts can be constructed to monitor either a group of response variables Y (e.g. product quality variables), or a group of predictor variables X (process variables). However, a very important advantage of latent variables is that they can be used to monitor predictor variables taking into account their effect on the response variables. A model is built to relate X and Y using available historical, or specially collected, data. Monitoring charts are then constructed for future values of X. This approach means that the process performance can be monitored even at times when the product quality measurements, Y, are not available.

The main approach of statistical quality control (SQC) methods developed throughout the statistical literature has been to monitor only product quality data (Y) and, in some cases, a few key process variables (X). However, often hundreds of process variables are measured much more frequently (and usually more accurately) than the product quality data. Thus monitoring of the process data is expected to supply much more information on the state of the process and to supply this information more frequently. Furthermore, any special events that occur will also have their fingerprints in the process data. So, once a special event is detected, it is easier to diagnose the source of the problem as we are dealing directly with the process variables. On the contrary, control charts on the product variables only indicate that the product properties are no longer consistent with specification but they do not point to the process variables responsible for this.

Control charts on process variables are useful also in multistep operations when quality data are not available between successive steps. For example, if a catalyst is conditioned in a batch process before being used for polymer production, the quality of the catalyst (success of conditioning) is assessed by its performance in the subsequent polymer production. It would be useful to know whether the catalyst will produce good product before using it; monitoring the batch process variables with a latent variable chart would give early detection of poor-quality product. Similarly, the few properties measured on a product sometimes are not sufficient to define product performance for several different customers. For example, if only the viscosity of a polymer is measured, end-use applications that depend on chemical structure (e.g. branching, composition, end-group concentration) are unlikely to receive good material. In these cases the process data may contain much more information about events with special causes that affect the hidden product quality variables.

The traditional multivariate charts (Hotelling's T^2, etc.) that have been used to monitor product quality can be applied to monitor process variables too (Kourti and MacGregor, 1996). However, as shown later, with a large number of correlated process variables these methods are impractical. Furthermore, they offer no way of relating the X and Y to form a model. Equally, the methods cannot handle missing data arising from sensor failure, etc. The most practical approaches to multivariate SPC appear to be those based on multivariate statistical projection methods, or latent variable methods, such as principal component analysis and partial least squares. The methods are ideal for handling the large number of highly correlated and noisy process variable measurements that are being collected by process computers on a routine basis and they can handle missing data.

9.5.10 Principal component analysis (PCA) for multivariate monitoring

When the number of measured variables is large, it is frequently the case that they are highly correlated with one another and their covariance matrix Σ (required for the chi-squared statistic calculation, equation (9.5)) is nearly singular. Principal component analysis (PCA) is a procedure for reducing the dimensionality of the variable space by representing it with a few orthogonal (uncorrelated) variables that capture most of its variability. New variables are calculated as linear combinations of the original variables. The new variables are independent of one another, and are calculated such that the first one explains the highest amount of variation in the system, the second the next highest amount, and so on. A few principal components may explain a very high percentage of variation of the system. Control charts using these first few variables can then be developed.

Details on principal components can be found in Chapters 2, 10 and 12. Here we present a brief summary necessary to explain their use in multivariate control charts.

Suppose we have a sample of mean centred and scaled measurements with n observations on q variables, $\mathbf{Y}$. The first principal component (PC) is defined as the linear combination $\mathbf{t}_1 = \mathbf{Y}\mathbf{p}_1$ that has maximum variance subject to $|\mathbf{p}_1| = 1$. The second PC is the linear combination defined by $\mathbf{t}_2 = \mathbf{Y}\mathbf{p}_2$ that has next greatest variance subject to $|\mathbf{p}_2| = 1$, and subject to the condition that it is uncorrelated with (orthogonal to) the first PC ($\mathbf{t}_1$). Up to q PCs are similarly defined. The sample principal component loading vectors $\mathbf{p}_i$ are the eigenvectors of the covariance matrix of $\mathbf{Y}$ (in practice the covariance matrix is estimated by $(n-1)^{-1}\mathbf{Y}^T\mathbf{Y}$). The corresponding eigenvalues give the variance of the PCs (i.e. var $(\mathbf{t}_i) = \lambda_i$). In effect, PCA decomposes the observation matrix $\mathbf{Y}$ as

$$\mathbf{Y} = \mathbf{T}\mathbf{P}^T = \sum_{i=1}^{q} \mathbf{t}_i \mathbf{p}_i^T \tag{9.16}$$

PCA is scale-dependent in that, if the $\mathbf{Y}$ matrix has variables that differ numerically in orders of magnitude (e.g. temperature in the range 200–300°C and viscosity in the range 0.1–0.3 Poise), the first few PCs are dominated by the numerically large variables. So, the $\mathbf{Y}$ matrix must be scaled in some meaningful way to remove the effect of numerically large values. The commonest approach is to scale all variables to unit variance. This means that PCA is carried out on the correlation matrix rather than the covariance matrix $\mathbf{\Sigma}$.

In practice, one rarely needs to compute all q eigenvectors, since most of the predictable variability in the data is captured in the first few PCs. The NIPALS algorithm (Wold et al., 1987) is ideal for computing the principal components in a sequential manner when the number of variables is large. The number of PCs that provide an adequate description of the data can be assessed using a number of methods (Jackson, 1991), with cross-validation (Wold, 1978) being perhaps the most reliable. By retaining only the first A PCs the $\mathbf{Y}$ matrix is approximated by

$$\hat{\mathbf{Y}} = \sum_{i=1}^{A} \mathbf{t}_i \mathbf{p}_i^T \tag{9.17}$$

Now, having defined the principal components, we can illustrate some of the problems with using T^2 when the variables are highly correlated and $\mathbf{\Sigma}$ is very ill-conditioned. The traditional Hotelling's T^2 on the original variables (equation (9.7)) can be written in terms of the PCs of the variables (Kourti and MacGregor, 1996):

$$T^2 = \sum_{i=1}^{q} \frac{t_i^2}{\lambda_i} = \sum_{i=1}^{q} \frac{t_i^2}{s_{t_i}^2} = \sum_{i=1}^{A} \frac{t_i^2}{s_{t_i}^2} + \sum_{i=A+1}^{q} \frac{t_i^2}{s_{t_i}^2} \tag{9.18}$$

In effect, each t_i^2 is scaled by the reciprocal of its variance (λ_i), bringing them to the same numerical scale. Thus, each PC term plays an equal role in the computation of T^2 irrespective of the amount of variance it explains in the **Y** matrix. The last PCs ($i > A+1$) are divided with very small values of λ, and very small deviations in their value (sometimes due to round-off error) are magnified and can cause large changes in T^2. As a consequence, if these PCs, which explain very little variance in **Y** and generally represent random noise, are used to calculate T^2, they may cause the generation of out-of-control signals. By using only the first A important PCs and calculating T^2 only from these,

$$T_A^2 = \sum_{i=1}^{A} \frac{t_i^2}{\lambda_i} = \sum_{i=1}^{A} \frac{t_i^2}{s_{t_i}^2} \qquad (9.18a)$$

the distorting effect of the minor PCs is removed.

9.5.11 Partial least squares (PLS) for multivariate monitoring

Suppose that two matrices are available, an $(n \times m)$ process variable data matrix, **X**, and an $(n \times q)$ matrix of corresponding product quality data, **Y**. It would be very useful to extract latent variables that explain the high variation in the process data, **X**, which is most predictive of the product quality data, **Y**. Then we can create charts to monitor the process variables but with such control limits that an alarm signals when a change in the process variables will affect the product. Partial least squares (PLS) is a method (or really a class of methods) that accomplishes this by working on the sample covariance matrix $(\mathbf{X}^T\mathbf{Y})(\mathbf{Y}^T\mathbf{X})$.

In the most common version of PLS (Geladi and Kowalski, 1986), the first PLS latent variable $\mathbf{t}_1 = \mathbf{X}\mathbf{w}_1$ is the linear combination of the x-variables that maximises the covariance between it and the **Y** space. The first PLS weight vector $\mathbf{w}_1$ is the first eigenvector of the sample covariance matrix $\mathbf{X}^T\mathbf{Y}\mathbf{Y}^T\mathbf{X}$. Once the scores for the first component have been computed, the columns of **X** are regressed on $\mathbf{t}_1$ to give a regression vector, $\mathbf{p}_1 = \mathbf{X}\mathbf{t}_1/\mathbf{t}_1^T\mathbf{t}_1$, and the **X** matrix is deflated (the $\hat{\mathbf{X}}$ values predicted by the model formed by $\mathbf{p}_1$, $\mathbf{t}_1$ and $\mathbf{w}_1$ are subtracted from the original **X** values) to give residuals $\mathbf{X}_2 = \mathbf{X} - \mathbf{t}_1\mathbf{p}_1^T$. The second latent variable is then computed from the residuals as $\mathbf{t}_2 = \mathbf{X}\mathbf{w}_2$ where $\mathbf{w}_2$ is the first eigenvector of $\mathbf{X}_2^T\mathbf{Y}\mathbf{Y}^T\mathbf{X}_2$ and so on.

As in PCA, the new latent vectors or scores ($\mathbf{t}_1, \mathbf{t}_2, \ldots$) and the weight vectors ($\mathbf{w}_1, \mathbf{w}_2, \ldots$) are orthogonal. Details on PLS are given in Chapter 12. PCA and PLS are frequently referred to as projection methods because the initial information is projected on to a lower-dimensional space.

9.5.12 Control charts based on latent variables

The philosophy applied in developing multivariate SPC procedures based on projection methods is the same as that used for the univariate or multivariate Shewhart charts. An appropriate reference set is chosen that defines the normal operating conditions for a particular process. Future values are compared against this set. A PCA or PLS model is built based on data collected from periods of plant operation when performance was good. Periods containing variations due to special events are omitted at this stage. The choice and quality of this reference set is critical to the successful application of the procedure.

When we are interested in developing a chart on a set of variables $\mathbf{Z}$ ($\mathbf{Z}$ could be either the $\mathbf{X}$ predictor or process space or, the $\mathbf{Y}$ response or quality space), a PCA model can be used. From historical data we develop the model

$$\hat{\mathbf{Z}} = \sum_{i=1}^{A} \mathbf{t}_i \mathbf{p}_i^T \qquad (9.19)$$

We calculate the variance of the scores from the model scores per component $s_{t_i}^2 (= \lambda_i)$ and use them together with $\mathbf{P}_A = [\mathbf{p}_1 \ \mathbf{p}_2 \ \cdots \ \mathbf{p}_A]$ for future monitoring calculations. Future behaviour now can be referenced against this 'in-control' model. Using equation (9.20) we construct a chart for T_A^2, for each new multivariate observation $\mathbf{z}_{new}$ ($q \times 1$).

$$T_A^2 = \sum_{i=1}^{A} \frac{t_{i,\,new}^2}{s_{t_i}^2} \qquad (9.20)$$

by calculating scores and residuals as

$$t_{i,\,new} = \mathbf{p}_i^T \mathbf{z}_{new} \quad \text{and} \quad \mathbf{e}_{new} = \mathbf{z}_{new} - \hat{\mathbf{z}}_{new}$$

where $\hat{\mathbf{z}}_{new} = \mathbf{P}_A \mathbf{t}_{A,\,new}$, $\mathbf{t}_{A,\,new}$ is the ($A \times 1$) vector of scores, and $\mathbf{P}_A$ is the ($q \times A$) matrix of loadings.

Monitoring $\mathbf{z}_{new}$ via T_A^2 will only detect whether or not the variation in the variables z in the plane of the first A PCs is greater than can be explained by common cause. However, this is not sufficient. Another chart to monitor the residuals is also necessary. By monitoring the residuals we can detect whether or not the noise in the system is similar to the one that existed during model development. A special event that was not included in the reference data used to develop the in-control PCA model may result in a change in the covariance structure of $\mathbf{Z}$. The squared prediction error (SPE_z) of the residuals of new observations is sensitive to this change:

$$\text{SPE}_z = \sum_{i=1}^{q}(z_{\text{new},i} - \hat{z}_{\text{new},i})^2 \qquad (9.21)$$

This statistic is also referred to as the Q-statistic (Jackson, 1991) or distance to the model. It represents the squared perpendicular distance of a new multivariate observation from the plane defined by the model. When the process is 'in-control', SPE_z should be within control limits developed from historical data; such limits can be computed using approximate results from the distribution of quadratic forms (Jackson, 1991; Nomikos and MacGregor, 1995). A very effective set of multivariate control charts is therefore a T_A^2 chart on the A dominant orthogonal PCs ($\mathbf{t}_1, \ldots, \mathbf{t}_A$) and an SPE_z chart.

When both $\mathbf{X}$ and $\mathbf{Y}$ historical data are available, the scores $\mathbf{t}$ in the X-space, the weights $\mathbf{w}$ and the loadings $\mathbf{p}$ are calculated from a PLS model between $\mathbf{X}$ and $\mathbf{Y}$. The standard deviation of the $\mathbf{t}$ scores is calculated to be used on the T_A^2-chart. For process monitoring, the new observations are available only for $\mathbf{X}$. New scores are calculated for the new observation $\mathbf{x}_{\text{new}}$ ($m \times 1$) as

$$t_{j,\text{new}} = \mathbf{w}_j^T \mathbf{x}_{j,\text{new}}$$

where $\mathbf{x}_{1,\text{new}} = \mathbf{x}_{\text{new}}$ and for $j > 1$, $\mathbf{x}_{j,\text{new}} = \mathbf{x}_{j-1,\text{new}} - t_{j-1,\text{new}} \mathbf{p}_{j-1}$.

Multivariate control is now achieved with a T_A^2-chart on the first A latent variables (equation (9.20)) and an SPE_x chart where

$$\text{SPE}_x = \sum_{i=1}^{m}(x_{\text{new},i} - \hat{x}_{\text{new},i})^2 \qquad (9.22)$$

where $\hat{\mathbf{x}}_{\text{new}} = \mathbf{P}_A \mathbf{t}_{A,\text{new}}$. As with the PCA method, the SPE_x plot will detect the occurrence of events that cause the process to move away from the hyperplane defined by the reference model. Control limits for the T_A^2 charts are chosen in the same manner as previously discussed, and the UCL on SPE_x is based on the chi-squared approximation.

The main concepts behind the development and use of multivariate SPC charts for monitoring continuous processes were laid out by Kresta et al. (1991), Wise et al. (1991), Wise and Ricker (1991) and Skagerberg et al. (1992). Several illustrations of the projection methods are also presented in these papers, along with the algorithms and details on estimating control limits. Other details of the construction and use of these charts and examples from applications in process industries can be found in Kourti and MacGregor (1996).

9.5.13 Fault diagnosis

In classical quality control, where only quality variables are monitored, it is up to process operators and engineers to try to diagnose an assignable cause for an out-of-control signal using their process knowledge and a one-at-a-time inspection of process variables.

Multivariate charts based on PLS or PCA provide a capability for diagnosing assignable causes. Diagnostic, or contribution, plots can be extracted from the underlying PLS or PCA model at the point where an event has been detected. The plot reveals the group of process variables making the greatest contributions to the deviations in SPE_x and the scores. Although these plots will not unequivocally diagnose the cause, they provide insight into possible causes and thereby greatly narrow the search. Details on the calculation of the contribution plots to detect variables responsible for an out-of-control signal on SPE_x, T_A^2 and on individual scores, are in Kourti and MacGregor (1996).

To summarise, multivariate process/product monitoring needs three charts as illustrated in Figure 9.6. The process is being monitored (on-line or off-line) with SPE_x and T_A^2 charts, constructed from measurements as they become available. Once a deviation from limits is detected, contribution plots point to the variable(s) that contribute numerically to the signal. T_A^2 checks that variables causing the main variation in the system are consistent with past operation. SPE_x checks that the

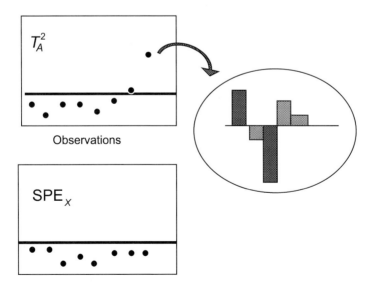

Figure 9.6 Process behaviour at a glance.

disturbances (noise in the system) remain within acceptable limits, consistent with past operation.

9.5.14 Multiway data

Up to now we have considered situations where **X** is a two-dimensional matrix, where one dimension is the variables and the other is the sample number. Such data are observations from continuous processes taken at different times, or data from batch processes where one row in the **X** matrix corresponds to one batch (i.e. per batch we have average temperature and pressure, total heat release, total reaction time).

However, we can consider situations where we have measurements for several process variables at multiple times within a batch and for many batches. Now the **X** matrix has three dimensions: time, batch and the process variable. The product quality measurements are usually made only at the end of a batch run, and thereby the **Y** dimension is usually two.

It is possible to conceive of data sets where **X** has four (spectral data (sample × emission × excitation × decay)) or more dimensions. The general class of high-dimension **X** and **Y** matrices is called multiway data. A very comprehensive work on multiway data analysis is that by Bro (1998).

When data are available in a historical database on many past batches, multivariate PCA and PLS models can be developed to establish on-line SPC charts for monitoring the progress of each new batch (Nomikos and MacGregor, 1994a,b, 1995). The interactions across the three dimensions create a complex problem for data analysis to which there are several possible solutions. Some of these solutions are compared for their applicability to batch process monitoring in Westerhuis *et al.* (1999). In these data sets, one dimension (time) is evolving during the progress of a batch, and the process variables take a nonlinear trajectory with time through the batch. In one of the approaches, the matrix is unfolded to a two-dimensional array, such that each row corresponds to a batch. In this case, mean centring of the variables effectively subtracts the trajectory, thus, converting a nonlinear problem to one that can be tackled with linear methods such as PCA and PLS (Nomikos and MacGregor, 1994a,b, 1995).

The set of multiway PCA and PLS charts is similar to those already described and provides the capability of detecting and diagnosing special events.

9.5.15 Multiblock data

There are very often multiple units in a process (e.g. each stage of a multistage synthesis can regarded as a process unit) and it could help operators to have a chart relating to each unit, rather than a chart for the whole process. Rather than building a model for each unit, it is possible to build a model for the full process that will take into account the interaction between units and their relative importance to the final product quality by weighting them differently. This is the approach of multiblock PLS (MB-PLS).

In the MB-PLS approach, large sets of process variables (**X**) are broken into meaningful blocks; usually each block corresponds to a process unit, or a section of a unit. Multiblock PLS is not simply a PLS between each **X** block and **Y**. The blocks are weighted in such a way that their combination is most predictive of **Y**. Multivariate monitoring charts for important subsections of the process, as well as for the entire process, can then be constructed and contribution plots are used for fault diagnosis as before (Kourti and MacGregor, 1996; Kourti et al., 1995).

As might be expected in multistage processes, there can be quite significant time delays between an event with a process variable in one unit and its effect on a product variable at the end of the process. These significantly affect the interaction and correlation structures of the process and need to be handled by lagged variables created from the original process variables. Data can be time-shifted to accommodate time delays between process units.

Several algorithms have been reported for multiblock modelling. It is suggested that the reader consults the work by Westerhuis et al. (1998), where all of the algorithms are presented and compared and a consensus on their use is derived.

9.5.16 *Issues in latent variable analysis and SPC*

Data validity checking
This stage should precede creating multivariate control charts. It helps us understand the behaviour of the process, detect outliers and find clusters of data. This stage should also precede any other empirical model building, no matter what type of regression model is used (MLR, PLS, neural nets). Chapter 2 describes the basic approaches.

Analysis of historical process databases with latent variables
In most industries, although massive amounts of process data are collected and stored in databases, very little analysis and interpretation is done. They are looked at when an event has occurred that disturbed production in a very costly manner. Of course, when a problem has occurred it is tedious to search among 400–600 variables individually to find out what went wrong in the process. By using PCA and PLS analysis,

plotting the latent variables and interpreting process movements using contribution plots, it is often possible to extract very useful information from historical databases, and to use this information to improve the process. Regions of acceptable operation can be detected and this information can be used to decide how best to set the control limits. If unexpected process disturbances are recorded in the database, then the analysis can help identify the variables involved.

There are many examples where historic data analysis has had benefit. They include:

- Identifying 4 out of 450 process variables responsible for a high-purity but low-yield product (MacGregor and Kourti, 1998).
- Finding new operating conditions for a low-density polyethylene plant (Moteki and Arai, 1986).
- Diagnosing systematic variations in a slurry-fed ceramic melter process (Wise *et al.*, 1991).
- The prediction of polymer properties from measured temperature profiles in a tubular low-density polyethylene process (Skagerberg *et al.*, 1992).
- Diagnosing reasons for poor control of Kappa number of a continuous digester in a pulp mill (Dayal *et al.*, 1994).
- The interpretation of the behaviour of mineral flotation and grinding circuits in a large mineral processing plant (Hodouin *et al.*, 1993).

Reference data set for SPC modelling
When dealing with empirical modelling (PCA or PLS), the data set upon which the model will be based must be chosen carefully to satisfy the needs of the intended application.

For inferential modelling and response surface modelling one needs data from designed experiments over a prescribed range of **X** and **Y** variables. Usually in this case a wide range of process conditions is considered to allow the choice of the optimal operating region. When the model is to be used for statistical process control, only a specific operating region is tackled. Historical process and/or product data corresponding to this operating region should be used. The objective is to model the good process behaviour in this operating region, and to test for any future deviations from this model. All the data should correspond to in-control operation and faults or disturbances are excluded from this model. If the preliminary analysis of the historical data in this region indicates clusters containing only a few points, or shows individual outliers, these data should not be discarded, but investigated. True outliers (measurement errors) should be discarded *if* they can be identified. If the small clusters reflect some real, unusual event that still produces acceptable product, then the data could be included in the model only if more data points in this region can be collected to establish a robust model. Otherwise these data should be left out during modelling, then tested with the model as if they were new data; you should identify what alarm signal they produce (direction of PCs) and store the information as a warning that the particular signal corresponds to this known 'rare situation' and not a bad product.

Finally, when we are exploring the process through historical data analysis, all the data should be used initially so that outliers are identified and discarded. Then the rest of the data should be used in the projection for the analysis of past behaviour.

Are linear models sufficient for SPC?
It has been argued that nonlinear latent variable models may be necessary to model batch or continuous processes correctly. This may be so if the models are created for inferential or predictive purposes and they cover a wide range of operating regions. Nonlinear versions of PLS are available. However, if the model is used for monitoring, linear models are usually a sufficient approximation to describe process fluctuations around the process target. Linear transformations of the raw data (using the logarithm or the inverse of a variable, etc.) may also be used to convert a nonlinear relation to a linear one. Finally, it should be emphasised that although the trajectories of the process variables are nonlinear in the batch processes, because of the way the data are mean-centred, the nonlinear trajectory is subtracted and process monitoring using linear models is possible.

Single or multiple models for quality properties (PLS1 or PLS2) for SPC?
One of the advantages of PLS over ordinary linear regression is that all the quality properties, **Y**, can be modelled together and related to **X** in a single model. When the quality properties are not correlated, it is customary to use models that relate **X** to each y variable separately. This approach is satisfactory, in general, if the model is just being used for calibration, inferential control or prediction. For monitoring purposes, however, since quality is a multivariate property, it is important to fit all the variables from the **Y** space in a single model in order to obtain a single low-dimensional monitoring space. The resulting model then describes how to keep all of the y variables in specification simultaneously rather than one at a time.

9.5.17 Other applications of multivariate charts

The principles presented so far are applicable to any scale of operation: laboratory, pilot-scale experiments/production or full-scale process units. In this section, we shall look at some applications of multivariate charting methods not related directly to controlling full-scale processes.

Applications of multivariate SPC in analytical laboratory methods

Standard or check samples (samples with known properties) are used to test analytical methods periodically to ensure that they are in control. For example, latex suspensions with known particle size are used to test the calibration in particle sizing methods, and samples with known concentration of an analyte to test gas chromatography. The results for these samples can be plotted on control charts to test that the value measured exhibits only common cause variation. Box *et al.* (1978, p. 559) give an example of how a Shewhart chart can be used for this purpose. When the response obtained from the standard is not a single number (i.e. instead of composition of one analyte we need composition of several analytes), one should consider multivariate charts.

In chromatographic systems, variables such as flow rate and pressure affect the performance of the system while, for example, column ageing or a contaminated detector can take the system out of control. Multivariate monitoring of a chromatographic system has been carried out using a check sample containing five analytes to test column performance (Nijhuis *et al.*, 1997). A T_A^2 chart and an SPE_x chart were used to monitor analyte peak area percentage of the five analytes. The results indicated that false alarms that would have occurred with univariate charts were avoided and points out of control owing to change in correlation could be detected (impossible with univariate charts—cf. Figure 9.4).

Establishing multivariate specification regions

All materials used in processes are bought or sold against specifications on certain properties. Up to now, these specifications have been set on a univariate basis, by setting upper and lower limits on each variable separately. This is equivalent to selling or buying a product by checking its properties against the square in Figure 9.4. Quality is a multivariate property and a material is of high quality only if it has the correct combination of *all* properties simultaneously. It has to be checked against the ellipse in Figure 9.4. Multivariate control charts are a simple way of identifying material that does not meet all of the requirements simultaneously. Rännar and Wikström (1998) present a multivariate procedure implemented at LKAB in Sweden, to monitor the properties of a fine powder sold as raw material in the steel industry. Material can now be sold showing that, in a multivariate sense, the properties are consistent.

To take this idea further, the customer may request that material meets certain multivariate specifications. Multivariate models can be established using process and product databases at both the producer and customer plants by tracking the lots of material through the customer plant to final product. As a result, proper quantitative multivariate specification regions can be defined and the producer's material can be sold by showing that its properties are multivariately consistent with specification.

The importance of this problem cannot be overstated. Companies that establish proper specification regions can potentially take large gains in market share. Furthermore, the multivariate specification regions provide the ultimate objective function for multivariate control systems (DeSmet, 1993).

9.6 Optimisation of processes

In this section we shall look at some of the optimisation approaches that are most commonly used in process industries and analytical laboratory environments. The data analysis methods involved with these procedures are usually regression based following some form of experimental design. For this reason, little detail is given here of the calculations and you should refer to the chapters on experimental design and linear regression as well as the earlier sections in this chapter. References are given for further reading where more complex calculations are involved.

9.6.1 General ideas

Optimisation problems are found at all levels in a company, from the research and development laboratories to process plant units; from the product testing methods to scheduling equipment and production. Optimisation is determining the set of values for a group of variables that will produce the desired optimum response for a chosen objective function, subject to various constraints.

Typical examples include the following.

- Optimising the performance of chromatography by identifying the combination of the mobile phase composition, temperature, etc. that maximises resolution between adjacent solutes under time and cost constraints.
- Conformational analysis of macromolecules to fit specific receptor shapes.

- The choice of polymerisation process conditions to produce a product with certain characteristics under the constraints of lowest cost and minimum emissions of pollutants to the environment.
- Scheduling the production of different grades of products in the same unit, in the most cost-effective sequence based on product demand, availability of storage space, availability of raw material, cost of startup and transition time.

In any of the above optimisation problems we have

- a goal we wish to achieve (optimality criterion);
- the manipulation of certain variables $(x_1, x_2, \ldots, x_j)$;
- constraints of equality or inequality on the variables and the goal;
- a procedure to solve the optimisation problem.

The optimality criterion is equivalent to minimising or maximising an objective function of the manipulated variables $f([x_1, x_2, \ldots, x_j]) = f(\mathbf{x})$. The manipulated variables $\mathbf{x}$ may appear in the literature with names such as input, control, design or independent variables or parameters. They may be continuous variables (e.g. temperature, flow rate, pressure) or discrete (emulsifier types; a pump may be on/off). $f(\mathbf{x})$ is called the response, the objective function or simply a function.

An extremum (maximum or minimum) can be either global (truly the highest or lowest function value within the range of the manipulated variable values available) or local (the highest or lowest in a subrange of the values and not on the boundary of that range). There can be more than one local optimum within the global range, but only one global optimum. As the maximum of a function is simply the minimum of its negative value $[\max(f) = \min(-f)]$, we shall treat optimisation problems as maximisation problems in this discussion.

Both the optimality criterion and the manipulated variables should be selected carefully. The answer to our optimisation problem will be a set of values $[x_1, x_2, \ldots, x_j]$ at which the optimum response occurs. However, we are interested not only in the optimal solution but also in its sensitivity to changes in the values of the parameters, that is how quickly the value of the response falls away as the value of one or more manipulated variables is changed by a small amount.

Sometimes, there may be detailed fundamental models describing the relationship between the manipulated variables, the objective function and the constraints. For example, the yield and the product quality in a chemical process can be expressed as a theoretical function of certain manipulated variables in the process (temperature, residence time, etc.). Such a function can be derived by taking into account fundamental laws and mechanisms such as mass and energy balances and the chemistry of the process (kinetics of main and side chemical reactions). Cost of raw materials, energy, clean up and prices of product will also be considered if cost constraints are to be included in the optimisation problem. The problem is that fundamental models are not often easily available. They are very tedious and time consuming to develop (in some plants often involving thousands of equations). In very many cases it is simply impossible to develop such models owing to lack of theoretical understanding, and when they are available they can be very difficult to solve. It is very frequently the case that empirical models have to be developed to describe the relations between objective function, the constraints and the manipulated variables. At other times the optimum is sought with no models at all (see simplex method below).

When mathematical models exist, there are numerous optimisation algorithms available for different levels of complexity of the optimisation problem (linear or nonlinear functions, continuity, with or without constraints, linear or nonlinear constraints, availability of first derivatives, storage requirements, etc.). Some of the algorithms need the evaluation of first derivatives of the function while others do not. Algorithms using the derivatives are somewhat more powerful, but not always enough to compensate for the additional calculations involved. Linear functions with linear constraints form a class of linear programming algorithms. Details of the optimisation techniques and algorithms can be found in several books (Edgar and Himmelbau, 1988; Fletcher, 1987; Press *et al.*, 1992).

Of course, without a fundamental model, pure computer experimentation (simulation) is not possible. In very many cases the reality faced by practitioners is that real-life experiments have to be performed. These are more costly and time consuming than computer experimentation and in many cases they have to be performed with the process operating while producing good-quality product.

Such procedures directly compare the function values measured experimentally at different points in the parameter space to find the set of $[x_1, x_2, \ldots, x_j]$ that results in the best operating conditions in that space (it may not be the global optimal though!). They may be either simultaneous, where the function values are measured and evaluated once for a predetermined group of manipulated variable sets (grid search or mapping the space), or sequential, where experimental results from one set of manipulated variable values directs the search to another set (simplex method). An approach that is perhaps intermediate between these two types is the use of empirical models based on experimental measurement of the objective function in the parameter space utilising experimental design. Again this approach can follow a simultaneous pattern (the first model is sufficient) or a sequential pattern (the model needs refining by more experiments).

348 DESIGN AND ANALYSIS IN CHEMICAL RESEARCH

Finally, the user should remember that when mathematical or empirical models are used (and not direct experimentation) it is the *model* that is being optimised, not the process itself. Therefore, one should spend time carefully choosing the procedure to be used, and judge the results taking into account all the assumptions and choices made on the way.

9.6.2 Optimise functions of many variables by changing one variable at a time

When several input variables affect the response, their effects should be studied by varying them simultaneously and not one at a time. This is because the one-at-a-time approach cannot take into account any interactions (correlations) between the variables. A common example of this is the effect of time and temperature on the yield of a reaction. Figure 9.7 shows contours of constant yield for a reaction as a function of time and temperature. Since the contour axes are not parallel to the coordinate axes, time and temperature are interacting.

Suppose we currently run the reaction at 130°C and 55 min. Had we tried experiments by varying one variable at a time, starting with $T = 130°C$ and varying the time we would have established that time = 62 min is best. Then, at that time, by varying the temperature only, we establish that 135°C is best. We are still far away from the true maximum

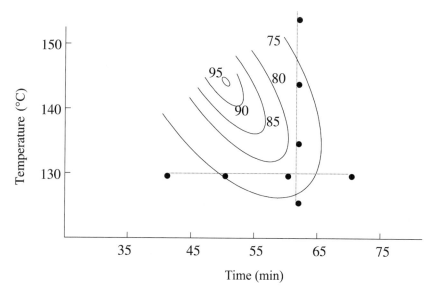

Figure 9.7 Searching one variable at a time.

of 145°C and 50 min, and no amount of one-at-a-time experimentation around this point will take us closer to the optimum. Varying one parameter at a time can work *only* if the parameters are independent of one another. It is left as an exercise to show that a different 'optimum' is obtained if temperature is varied first.

9.6.3 Multiresponse optimisation

This is the case where multiple objective functions have to be optimised simultaneously. The optimum value of one function may be at a different setting of the manipulated variables from the optimum of the other(s). A combination of the manipulated variables must therefore be found that satisfies several criteria. Multicriteria optimisation and multicriteria decision making (MCDM) are terms used for addressing these problems. A simple approach for multicriteria optimisation, when the manipulated variables are few, is to use overlay contour plots to determine the feasible region. If there are two objective functions to be maximised, the contours of the two functions can be plotted and a region where the contours overlap can be found by inspection.

Suppose we have two manipulated variables (temperature and time) and wish to maximise yield under the constraints that two quality properties (responses) A and B should remain within certain specifications. Separate contour plots of these properties against the two manipulated variables can be made. For the sake of simplicity we assume linear relationships between the properties and manipulated variables. If the two contour plots are overlaid on the contour plot for the yield (Figure 9.8), we can see the ranges of the variables where both specifications are met simultaneously while maximising yield. The shaded area is the window of possible operation, and we should operate near location Θ for maximum yield. Of course such a plot may show that all the requirements cannot be met, in which case the investigator might have to think of manipulating some other variable to achieve the desired goal.

Where many criteria must be met simultaneously, graphical methods are not appropriate and mathematics has to be used. Sometimes an overall function is determined by weighting the responses and converting the problem to univariate optimisation. A joint response measure from multiple responses can be constructed using fuzzy set theory (Otto, 1988). Similar approach is the desirability function approach, with membership functions given by Derringer and Suich (1980). The Pareto optimal method is also used (Keller *et al.*, 1991; Hendricks *et al.*, 1992) to address MDCM problems. Vanbel *et al.* (1995) discuss three MCDM methods in a work investigating the feasibility of developing rugged separations by

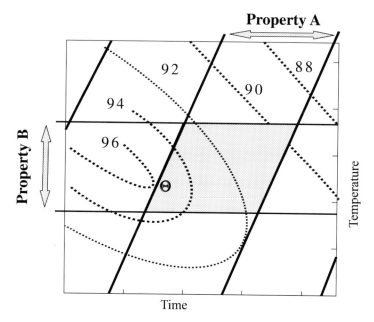

Figure 9.8 Contour overlay.

systematically varying the pH and solvent composition in chromatographic methods. Duiveland and Coenegracht (1995) present a multi-criteria steepest ascent method for a design space consisting of both mixture and process variables. Wienke *et al.* (1992, 1993) used multi-criteria target vector optimisation of analytical procedures with genetic algorithms as the searching agent.

9.6.4 *Procedure for optimisation with empirical models*

The procedure for optimisation using empirical models can be broken down into a few clear steps.

1. Define problem. Define response function(s). List all possible manipulated variables.

2. Select which manipulated variables and what range will be used for optimisation (screening):

    Fractional factorials
    Plackett–Burman
    D-optimal

3. Find region of optimum or optimum:

   Sequentially (response surface with steepest ascent or simplex)
   Simultaneously (grid search)

If satisfied with comparing function values only, stop here.

4. Response surface modelling (quadratic models) and optimisation:

   Experimental design (central composite designs, Box–Behken designs, D-optimal designs)
   Outlier detection (use PCA on response data, Y)
   Relate Y to X (MLR, PLS)
   Predict response Y in space of X
   Choose optimal conditions

For case studies where this procedure is illustrated, see Box *et al.* (1978) under response surface methods where linear regression was used for modelling, and Wold *et al.* (1989) where optimisation of an industrial process is discussed, with PLS models.

Problem definition
The importance of problem formulation cannot be overemphasised. At this point it is essential that the process or procedure to be optimised is analysed to define the variables involved, the responses to be measured and other specific characteristics of interest. A list of *all* the variables is made and the role of all of them is considered carefully. One should investigate the ranges of each variable. It is worth considering consulting people with more knowledge on the process regarding insights that may help. For example, it may be known that there are constraints that should be considered, or physical limitations to the range over which some variables may vary. If the system is very complex, it may be that the problem can be reformulated so that it can be solved with less effort. In any case, it is certain that time spent on this stage has enormous benefit in the simplification of the problem and the efficiency with which the solution is found.

Select manipulated variables and their range
This is a screening procedure to identify from the possible variables those that will be used for the optimisation and the likely ranges of values that should be explored. This way we restrict the searching space (which translates to number of experiments) for optimisation. Linear models with interactions are sufficient for screening and the designs most commonly chosen for this purpose are full or fractional factorials at two levels, Plackett–Burman and D-optimal designs (see Chapters 6 and 7). They are often augmented with a few centre points to allow evaluation of

reproducibility. Although these designs have the drawback of confounding of main effects with interactions, they are acceptable for this first screening process.

The variables with the highest influence on the response will be chosen.

Find region of optimum or the optimum
The objective now is to find the region that includes the optimum. This can be done with sequential or simultaneous methods, using the dominant manipulated variables found in the screening stage. The general principle can be illustrated with a one-variable optimisation.

We need to bracket the optimum (say, maximum). A maximum is known to be bracketed when there is an ordered triplet of values of the manipulated variable ($a < b < c$) such that the response $f(b)$ is larger than both $f(a)$ and $f(c)$. In this case we know that the function has a maximum in the interval (a, c) although b may not correspond to the maximum. To search for the maximum, choose a new point, x, between either ab or bc. Say x is between bc. If $f(b) > f(x)$, replace c with x. Repeat with points (a, b, x) and continue until the range on the variable is tolerably small. The position of x is arbitrary, but often the midpoint between the greater of ab or bc is chosen, although other fractions of the interval such as the Golden Section (0.381 97 into the larger of the two intervals measured from b) can be used and may give more efficient searching.

Sequential methods start with a few experiments and proceed sequentially in the direction of improvement by evaluating the results of experiments at each trial. Some of these approaches are purely experimental (the simplex method, Section 9.6.5) and others are a mixture of experiments and empirical modelling (response surface with steepest ascent, Section 9.6.5). The sequential methods creep towards favourable areas where the response is an improvement on the starting point. However, they do not guarantee that this is the area of the global optimum.

Having arrived at a favourable point, we can stop and accept the point as a sufficient improvement, we may repeat the process from a different starting point if we suspect a local optimum, or we may proceed with response surface modelling to study the response surface around that improved value.

With simultaneous methods, the space of the manipulated variables is searched either with a grid or using a set of designed experiments. All of the variables to be manipulated, and their levels, are chosen *a priori*. Data are collected at all the predetermined points and the optimum is decided either by directly comparing response values or by prediction using empirical models. At this point we can stop and accept the possibly crude estimate of optimum, or choose to proceed to the next step for a final design with response surface modelling around the points with the best performance.

With the simultaneous approach we gain an insight into the structure of the response surface, but we may miss the optimum if the range of the variables is not properly chosen. In that case, we may end up with more trials and the search becomes sequential. The approach has an advantage over some sequential methods that it can deal with discrete variables easily.

Response surface modelling for optimisation
At this stage we need to develop an empirical model relating the manipulated variables $\mathbf{X}$ to the response $\mathbf{Y}$ $(=f(\mathbf{X}))$ around the region of interest (which we think is near the optimum). We then use the model to predict $\mathbf{Y}$ at any point in this region and to study the behaviour of the response. We can also use the model as the mathematical function for optimisation under the constraints of the problem to predict the location of the optimum and the value of the response at the optimum.

The empirical model is usually a quadratic function of the manipulated variables and the response (see Chapter 7). In a limited region a quadratic function is a good approximation to different types of response surfaces: mound (minimum or maximum), saddle, stationary ridge or rising ridge (Box *et al.*, 1978).

To be able to form quadratic models, the experimental designs at this stage must be chosen properly. Factorial designs allow linear models with interactions but no quadratic terms, so the minimum or maximum cannot be found directly. Therefore we need composite designs, a combination of factorial or fractional factorial with star points. Central composite designs, Box–Behnken designs and D-optimal designs can provide enough data for such models.

After the data from the experiments/trials have been collected, and before proceeding to modelling, the data should be screened for outliers (wrong measurements, spoiled samples, etc.). Principal component analysis can be used to detect outliers in multiresponse data.

Regression methods (MLR, PLS) can be used to relate the response matrix $\mathbf{Y}$ to the matrix of the manipulated variables. Box *et al.* (1978) warn against the danger of trying to interpret an inadequately estimated response function. They suggest that before interpretation of the fitted surface the precision of the estimates should be considered.

The empirical models can be used to predict the response values at any point within the parameter space involved in the design (the experimental space). The shape of the response surface can be shown graphically for low-dimensional models and this aids the selection of optimum conditions as well as understanding of the influences of manipulated variables and their interactions. Specification charts can be constructed as shown in Figure 9.8. For high-dimensional models where the optimal point cannot be chosen visually, optimisation algorithms can be used to solve the empirical model for the optimum set of manipulated variables.

9.6.5 Sequential methods

Sequential optimisation methods are so widely used that some more detailed comments about the different types may be helpful.

Evolutionary operation (EVOP). It is very common that an industrial process is not operating at its optimal conditions even though it is producing product within specifications. This may be due to several factors:

- With the years, the process may have drifted away from its best operation.
- It may be required to deal with higher throughput.
- Environmental restrictions may have become tougher.
- The characteristics of a raw material may have changed.

Often, improved new operating conditions have to be found while the process is in full-scale operation producing satisfactory product. If mathematical models describing the process operations do not exist, on-line experimentation has to take place to improve the process. Evolutionary operation methods (EVOP), introduced by Box (1957), are used for this purpose. A sequential approach is used involving small step changes to the operating conditions determined by a simple experimental design. To keep things simple, it is usual to vary only two or three factors in any given phase of the investigation and to use 2^2 or 2^3 factorial designs, often with an added central point (see Chapter 6). The simplex method can be used as well.

Based on the results of the first design, decisions are made as to the direction in which and how much the variables will be varied and a second cycle follows with the conditions modified. The procedure continues, with each cycle suggesting improvements for the next. As only small changes are made in the levels of the process variables during EVOP, it is necessary to repeat the runs a number of times (using replicated designs) and average the observations to see the effects of the changes relative to the noise in the system.

Real-time optimisation (RTO) refers to the process of readjusting the set points of certain manipulated variables to reach a new optimal point during production. EVOP is a real-time optimisation using experimental/empirical approaches. The term RTO, however, has come to imply that a mathematical model of the process exists and that the current process performance is adjusted against an optimum predicted by the model.

Simplex method and modifications
The simplex method (not to be confused with the simplex method in linear programming) can be used either as an optimisation algorithm, when we have models for $f(\mathbf{x})$, or to dictate points for next search in an empirical experimental procedure. It was introduced as an EVOP method by Spendley *et al.* (1962). It is considered the most successful of the optimisation methods that merely compare function values $f(\mathbf{x})$ at several $\mathbf{x}$ in the parameter space.

A regular simplex is a set of $(n+1)$ equidistant points in an n-dimensional space, such as the equilateral triangle for $n=2$ and the tetrahedron for $n=3$. In optimisation, the vertices of the simplex correspond to separate sets of values of the manipulated variables $(\mathbf{x}_1, \mathbf{x}_2, \ldots, \mathbf{x}_{n+1})$. Associated with each vertex is the response of the process $(y_1, y_2, \ldots, y_{n+1})$. The current information kept in the method is the coordinates of the $n+1$ points and their corresponding function values.

Figure 9.9 illustrates the sequence of the procedure for the first 10 experiments as we approach the maximum, for the case when $n=2$. In the first iteration to maximise $f(\mathbf{x})$, we start with $n+1$ experiments (points 1, 2, 3 in Figure 9.9), and the vertex at which the function has the smallest value is identified. For the next iteration (or experiment) the coordinates of this vertex are reflected in the centroid of the other n vertices (point 4), thus, forming a new simplex. The response (function value) at this new vertex is compared with the remaining values from the first iteration. The procedure is repeated until no improvement in response is obtained.

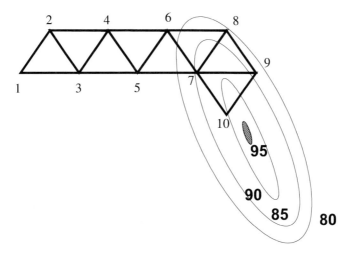

Figure 9.9 The simplex method.

There are many modifications and rules that have developed over the years to cater for a whole range of situations that create problems for the simplex approach. For example, after several iterations it might be that the newest vertex still has the smallest value and this could cause oscillation of the simplex without any improvement in response being seen; so the vertex with the next smallest value is reflected. Similarly, when a certain vertex has been in the current simplex for more than a fixed number of iterations, the simplex should be contracted by creating a simplex with vertices at the mid points of the edges of the current simplex. When the simplex is smaller than a prescribed tolerance, the routine (or experiment) is stopped.

It is also common to use irregular simplexes where distortions of the simplex are performed automatically in an attempt to take into account the local geometry of the function. With this approach (Nelder and Mead, 1965) the simplex expands in the direction of more favourable conditions and contracts if a move was in a direction of less favourable conditions (see Press *et al.*, 1988, for computer coding).

The simplex method handles one response at a time and really gives no insight of the response surface. It does move rapidly and efficiently towards the region of an optimum but, like all the sequential procedures, it may fall on local minima or maxima. Of course, if the cost and time allow, different starting conditions can be tried as well as starting with a very large simplex and letting it collapse onto an optimum. But we have to remember that the method started as an evolutionary operation method; as such it will certainly improve the performance of the process compared with the starting point.

Response surface methods and steepest ascent

Steepest ascent (descent in minimisation) is an approach to lead near to the optimum in situations where the region of interest is not known *a priori*. In this respect it is similar to the simplex method. An illustration of the procedure for optimisation for two variables is shown in Figure 9.10, where the starting points are the current operating conditions ($\diamond$).

A design is constructed (here a 2^2 factorial with a centre point) and a response surface model is determined. Based on this model, the direction of the largest increase in the response (steepest ascent) is determined. More experiments are performed along that direction, shown with $\oplus$, until the maximum is bracketed. At this stage, a conventional RSM design is chosen to locate the optimum closely. Tests for fitness of model and for curvature are done at each cycle where a surface model is obtained. Transformations for the variables or for the response may be tried in case of serious lack of fit (Box *et al.*, 1978).

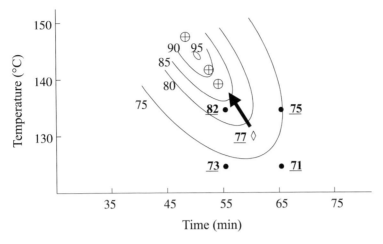

Figure 9.10 Method of steepest ascent.

Genetic algorithms and simulated annealing
Stochastic search methods based on simulated natural processes like simulated annealing and genetic algorithms have been widely used to solve high-dimensional optimisation problems. The two methods are compared in Davis (1987).

The increase in popularity of these methods is due to their many advantages when compared to other optimisation search approaches. Both methods search the space in parallel and therefore have a better chance of finding global optimum (mutations in genetic algorithms have a similar effect to the random steps in simulated annealing). They do not make any assumptions as to the linearity or nonlinearity of response function, nor do they require gradient information. This means that they can handle discontinuous, incomplete and noisy response functions and they can deal with discrete variables. However, they make many function evaluations, which makes them expensive in computing time when used to search for the optimum of a mathematically modelled function and prohibitive for search in an experimental space if measurements have to be made at each set of process variables to be evaluated. Therefore, they are used to solve complex optimisation cases (discontinuities, global optimum among local optima, mixed integer–real parameter values) when the response function can be evaluated at the search points using some fundamental knowledge.

Genetic algorithms
The first book about genetic algorithms (GA) was published as recently as 1989 by David Goldberg. His definition of genetic algorithms forms the first paragraph in many GA publications:

Genetic algorithms are search algorithms based on the mechanics of natural selection. They combine survival of the fittest among string creatures with a structured yet randomised information exchange to form a search algorithm with some of the innovative flair in human search. In every generation a new set of artificial creatures (strings) is created using bits and pieces of the fittest of the old; an occasional new part is tried for good measure. While randomised, genetic algorithms are no simple random walk. They efficiently exploit historical information to speculate on new search points with expected improved performance.

Genetic algorithms are different from the usual optimisation search procedures in that they work with a coding of the parameter set, not the parameters themselves (this feature makes them useful for a large variety of applications). They search from a population of points, not a single point; they use the objective function, not derivatives or auxiliary knowledge; and they use probabilistic transition rules, not deterministic rules.

The reader is referred to the first chapter of Goldberg (1989) for a 'Gentle Introduction to Genetic Algorithms'. A tutorial is also given in Davis (1991), where also the application of genetic algorithms to conformational analysis of DNA (Lucasius *et al.*, 1991) is reported. For a good introduction to the application of genetic algorithms in chemometrics, the reader is referred to Lucasius and Kateman (1993, 1994).

Simulated annealing
Simulated annealing is an analogy to thermodynamics in which solids are heated to melting and then slowly cooled to remove strain and crystal imperfections. A completely ordered pure crystal is thus formed, and this crystal is the state of minimum energy for the system. Slow cooling to allow ample time for redistribution of the molecules and atoms, as their mobility is lost, is essential to achieve this lowest energy state. All the optimisation algorithms we discussed earlier go for the rapid solution and they may find a local minimum or maximum; by analogy, with rapid cooling or quenching, metals would end in an amorphous or polycrystalline state or other metastable state.

The Boltzmann probability distribution $\text{Prob}(E) \sim \exp(-E/kT)$ expresses the idea that a system at thermal equilibrium at temperature T has its energy probabilistically distributed among all different energy states E; k is Boltzmann's constant. Even at low T, there is a chance, albeit very small, of a system being in a high-energy state. Therefore, there is a corresponding chance for the system to get out of a local energy minimum in favour of finding a better, more global one. For a minimisation problem the system sometimes goes uphill as well as downhill; but the lower the temperature the less likely is any significant uphill excursion. Offered a succession of options, the system is assumed to change its configuration from E_1 to E_2 with probability $p = \exp(-(E_2-E_1)/kT)$. For $E_2 < E_1$ this probability is >1 and in such case it is arbitrarily assigned $p = 1.0$, i.e. the system always takes such an option.

We illustrate the main points of the procedure with an example from Van Kampen and Buydens (1997) in which simulated annealing was used for the structure elucidation of a heptapeptide in the torsion angle space. Suppose the objective is to minimise a fitness function (FE). We start from randomly initialised torsion angles (**q**). A trajectory in the space is generated by making small modifications to these parameters; in the example their value increases or decreases with an amount ζ generated from a Cauchy distribution $\mathbf{q}_{new} = \mathbf{q}_{current} \pm \zeta$. If the evaluation of FE for the new set of parameters results in a decrease of its value, the parameter values are accepted, otherwise they are only accepted with probability $p_{accept} = \exp(-FE_{new} - FE_{old})/c_k)$. The control parameter c_k (analogue of temperature) is initialised such that most steps are accepted and the early stage resembles a random search. Then it is decreased following a schedule (annealing schedule) giving rise to lower acceptance probability. Subsequently a new trajectory through space is generated. By decreasing c_k, the chance of accepting detrimental steps decreases and forces the algorithm to a local search where only improvements are accepted. Because the whole of the variable space initially is explored repeatedly, the chances of a global optimum being found are high.

This technique has attracted a lot of attention as a suitable search engine for optimisation problems in which the space over which the objective function has to be minimised is a large, discrete configuration space, and for problems where a desired global extemum is hidden among many, poorer local extrema. Typical chemical applications are those where the number of combinations of variables is factorially large and they cannot be explored exhaustively by conventional methods. Finding the optimum three-dimensional structure of a heptapeptide and the identification of specific mass spectrum in the spectrum of a complex mixture are typical applications.

9.6.6 *Process optimisation using historical data*

Recently there has been a lot of interest in exploiting historical databases to derive empirical models (using tools such as neural networks regression or PLS) and use them for process optimisation. The idea is to use already available data rather than collecting new data through a design of experiments. The problem is that for process optimisation causal information must be extracted from the data, so that a change in the operating variables can be made that will lead to a better-quality product or higher productivity and profit. However, databases obtained from routine operation contain mostly noncausal information. Inconsistent data, range of variables limited by control, noncausal relations, spurious relations due to feedback control and dynamic relations are some of the problems the user will face using such happenstance data. These are discussed in detail in the section 'Hazards of fitting regression equations to happenstance data' by Box *et al.* (1978), where the advantage of experimental designs as a means of obtaining causal information is emphasised. In a humorous way, the authors warn young scientists that they need a strong character to resist the suggestion of their boss to use data from past plant operation every time they suggest performing designed experiments to collect data.

In spite of this, several authors have proposed approaches to optimisation and control based on interpolating historical bases. However, in all these cases success was based on making strong assumptions that allowed the database to be reorganised and causal information to be extracted. Jaeckle (1998) investigated an approach referred to as 'similarity optimisation' that combined multivariate statistical methods for reconstructing unmeasured disturbances and nearest-neighbour methods for finding similar conditions with better performance. However, it too was shown to fail for many of the same reasons. In general, it was concluded that one can only optimise the process if there exist manipulated variables that change independently of the disturbances and if disturbances are piecewise constant, a situation that would be rare in historical process operations.

9.6.7 *Product design with latent variables*

Given the reservations about the use of historical databases, one area where some success has been achieved is in identifying a range of process operating conditions for a new grade of product and in matching two different production plants to produce the same grade of product (Jaeckle and MacGregor, 1998). If fundamental models of the process exist, then these problems are easily handled as constrained optimisation problems. If not, optimisation procedures based on response surface methodology described earlier can be used. However, even before one performs experiments, there exists information within the historical database on past operating conditions for a range of existing product grades.

In this case the historical data used are selected from different grades and therefore contain information on variables for several levels of past operation (i.e. there is intentional variation in them and they are not happenstance data). The key element in this empirical model approach is the use of latent variable models that both reduce the space of **X** and **Y** to a lower-dimensional orthogonal set of latent variables and provide a model for **X** as well as **Y**. This is essential in providing solutions that are consistent with past operating policies. In this sense, principal component regression and PLS are acceptable approaches, while MLR, neural networks and reduced rank regression are not.

The methodology has been demonstrated via simulation using both linear and nonlinear approaches to the design of new grades of low-density polyethylene. It has also been applied to two industrial batch polymerisation processes and to a continuous industrial manufacturing process in Jaeckle (1998). The major limitation of this approach is that one is restricted to finding solutions within the space and bounds of the process space **X** defined by previously produced grades. There may indeed be equivalent or better conditions in other regions where the process has never been operated before, and hence where no data exists. Fundamental models or more experimentation would be needed if one hoped to find such novel conditions.

References

Box, G.E.P. (1957) Evolutionary operation. A method for increasing industrial productivity. *Applied Statistics*, **6** 81-101.

Box, G.E.P., Hunter, W.G. and Hunter, J.S. (1978) *Statistics for Experimenters. An Introduction to Design, Data Analysis and Model Building*, Wiley, New York. Wiley Series in Probability and Mathematical Statistics.

Bro, R. (1998) *Multi-way analysis in the food industry. Models, algorithms and applications*, PhD thesis, Royal Veterinary and Agricultural University, Denmark.

Crosier, R.B. (1988) Multivariate generalisations of cumulative sum quality-control schemes. *Technometrics*, **30** 291-303.

Dayal, B., MacGregor, J.F., Taylor, P.A., Kildaw, R. and Marcikic, S. (1994) Application of feed forward neural networks and partial least squares regression for modelling Kappa number in a continuous Kamyr digester. *Pulp and Paper Canada*, **95** (1) 26-32.

Davis, L. (ed.) (1987) *Genetic Algorithms and Simulated Annealing*, London, Pitman.

Davis, L. (ed.) (1991) *Handbook of Genetic Algorithms*, Van Nostrand Reinhold, New York.

Derringer, G. and Suich, R. (1980) *Journal of Quality Technology*, **12** 214.

DeSmet, J. (1993) *Development of multivariate specification limits using partial least squares regression*, MEng thesis, McMaster University, Hamilton, Ontario, Canada.

Duiveland, C.A.A. and Coenegracht, P.M.J. (1995) Multicriteria steepest ascent method for a design space consisting of both mixture and process variables. *Chemometrics and Intelligent Laboratory Systems*, **30** 23-36.

Edgar, T.F. and Himmelblau, D.M. (1988) *Optimisation of Chemical Processes*, McGraw-Hill, New York.

Fletcher, R. (1987) *Practical Methods of Optimisation*, Wiley, New York.

Geladi, P. (1989) Analysis of multi-way (multi-mode) data. *Chemometrics and Intelligent Laboratory Systems*, **7** 11-30.

Geladi, P. and Kowalski, B.R. (1986) Partial least-squares regression: a tutorial. *Analytica Chimica Acta*, **185** 1-17.

Goldberg, D.E. (1989) *Genetic Algorithms in Search, Optimisation and Machine Learning*, Addison-Wesley, Reading, MA.

Harris, T.J. and Ross, W.H. (1991) Statistical process control procedures for correlated observations. *Canadian Journal of Chemical Engineering*, **69** 139-148.

Healy, J.D. (1987) A note on multivariate Cusum procedures. *Technometrics*, **29** 409-412.

Hendricks, M.M.W.B., de Boer, J.H., Smilde, A.K. and Doorndos, D.A. (1992) *Chemometrics and Intelligent Laboratory Systems*, **16** 175.

Hodouin, D., MacGregor, J.F., Hou, M. and Franklin, M. (1993) Multivariate statistical analysis of mineral processing plant data. *CIM Bulletin of Mineral Processing*, **86** (975) 23-34.

Hunter, J.S. (1986) Exponentially weighted moving average. *Journal of Quality Technology*, **18** 203-210.

Jackson, J.E. (1991) *A User's Guide to Principal Components*, Wiley, New York.

Jaeckle, J.M. (1998) *Product and process improvement using latent variable methods*. PhD thesis, McMaster University, Hamilton, Ontario, Canada.

Jaeckle, J.M. and MacGregor, J.F. (1997) Product design through multivariate statistical analysis of process data. *AIChE Journal*, **44** 1105-1118.

Keller, H. R., Massart, D.L. and Brans, J.P. (1991) Multicriteria decision making: a case study. *Chemometrics and Intelligent Laboratory Systems*, **11** 175-189.

Kourti, T., Nomikos, P. and MacGregor, J.F. (1995) Analysis monitoring and fault diagnosis of batch processes using multiblock and multiway PLS. *Journal of Process Control*, **69** 35-47.

Kourti, T. and MacGregor, J.F. (1996) Multivariate SPC methods for process and product monitoring. *Journal of Quality Technology*, **28** 409-428.

Kresta, J., MacGregor, J.F. and Marlin, T.E. (1991) Multivariate statistical monitoring of process operating performance. *Canadian Journal of Chemical Engineering*, **69** 35-47.

Lowry, C.A., Woodall, W.H., Champ, C.W. and Rigdon, S.E. (1992) A multivariate exponentially weighted moving average control chart. *Technometrics*, **34** 46-53.

Lucasius, C.B. and Kateman, G. (1993) Understanding and using genetic algorithms. Part 1. Concepts, properties and context. *Chemometrics and Intelligent Laboratory Systems*, **19** 1-33.

Lucasius, C.B. and Kateman, G. (1994) Understanding and using genetic algorithms. Part 2. Representation, configuration and hybridization. *Chemometrics and Intelligent Laboratory Systems*, **25** 99-145.

Lucasius, C.B., Blommers, M.J.J., Buydens, L.M.C. and Kateman, G. (1991) A genetic algorithm for conformational analysis of DNA, in *Handbook of Genetic Algorithms*, (ed. L. Davis), Van Nostrand Reinhold, New York, pp. 251-281.

MacGregor, J.F. and Kourti, T. (1998). Multivariate statistical treatment of historical data for productivity and quality improvements, in *Foundation of Computer Aided Process Operations* (eds. J.F. Pekny and G.E. Blau). (AIChE Symposium Series No. 320, Vol. 94) Brice Carnahan, University of Michigan, Ann Arbor Press, pp. 31-41.

Marlin, T.E. (1995) *Process Control: Designing Processes and Control Systems Dynamic performance*, McGraw-Hill, New York.

Montgomery, D.C. (1991) *Introduction to Statistical Quality Control*, 2nd edn, Wiley, New York.

Montgomery, D.C. and Runger, G.C. (1994) *Applied Statistics and Probability for Engineers*, Wiley, New York.

Montgomery, D.C., Runger, G.C. and Hubele, N.F. (1998) *Engineering Statistics*, Wiley, New York.

Moteki, Y. and Arai, Y. (1986) Operation planning and quality design of a polymer process. *Proceedings of IFAC Symposium, DYCORD-86*, Bournemouth, UK, pp. 159-166.

Nelder, J.A. and Mead, R. (1965) A simplex method for function minimisation. *Computer Journal*, **7** 308-313.
Nijhuis, A., de Jong, S. and Vandeginste, B.G.M. (1997) Multivariate statistical process control in chromatography. *Chemometrics and Intelligent Laboratory Systems*, **38** 51-62.
Nomikos, P. and MacGregor, J.F. (1994) Monitoring of batch processes using multiway principal component analysis. *American Institute of Chemical Engineers Journal*, **40** 1361-1375.
Nomikos, P. and MacGregor, J.F. (1995a) Multivariate SPC charts for monitoring batch processes. *Technometrics*, **37** (1) 45-59.
Nomikos, P. and MacGregor, J.F. (1995b) Multiway partial least squares in monitoring batch processes. *Chemometrics and Intelligent Laboratory Systems*, **30** 97-108.
Otto, M. (1988) Fuzzy theory explained. *Chemometrics and Intelligent Laboratory Systems*, **4** 101-120.
Pignatiello, J.J. Jr. and Runger, G.C. (1990) Comparisons of multivariate Cusum charts. *Journal of Quality Technology*, **22** 173-186.
Press, W.H., Teukolsky, S.A., Vetterling, W.T. and Flannery, B.P. (1992) *Numerical Recipes in C. The Art of Scientific Computing*, 2nd edn., Cambridge University Press, Cambridge.
Rännar, S. and Wikström, C. (1998) Multivariate quality monitoring. Presented at the *2nd International Chemometrics Research Meeting, Veldhoven, The Netherlands, May 1998*.
Shewhart, W.A. (1931) *Economic Control of Quality of Manufactured Product*, Van Nostrand, Princeton, N. J.
Skagerberg, B., MacGregor, J.F. and Kiparissides, C. (1992) Multivariate data analysis applied to low-density polyethylene reactors. *Chemometrics and Intelligent Laboratory Systems*, **14** 341-356.
Spendley, G., Hext, G.R. and Himsworth, F.R. (1962) Sequential application of simplex designs in optimisation and evolutionary operation. *Technometrics*, **4** 44.
Stephanopoulos, G. (1984) *Chemical Process Control: An Introduction to Theory and Practice*, Prentice-Hall, New York.
Tracy, N.D., Young, J.C. and Mason, R.L. (1992) Multivariate control charts for individual observations. *Journal of Quality Technology*, **24** 88-95.
Vanbel, P.F., Tilquin, B.L. and Schoenmakers, P.J. (1995) Criteria for developing rugged high-performance chromatographic methods. *Journal of Chromatography*, **697** 3-16.
van Kampen, A.H.C. and Buydens, L.M.C. (1997) The ineffectiveness of recombination in a genetic algorithm for the structure elucidation of a heptapeptide in torsion angle space. A comparison to simulated annealing. *Chemometrics and Intelligent Laboratory Systems*, **36** 141-152.
Westerhuis, J.A., Kourti, T. and MacGregor, J.F. (1998) Analysis of multi-block and hierarchical PCA and PLS models. *Journal of Chemometrics*, **12** 301-321.
Westerhuis, J.A., Kourti, T. and MacGregor, J.F. (1999) Comparing alternative approaches for multivariate statistical analysis of batch process data. *Journal of Chemometrics*, **13**, 397–413.
Western Electric (1956) *Statistical Quality Control Handbook*, AT&T, Chicago, IL.
Wierda, S.J. (1994) Multivariate statistical process control—recent results and directions for future research. *Statistica Neerlandica*, **48** (2) 147-168.
Wienke, D., Lucasius, M. and Kateman, G. (1992) Multicriteria target vector optimisation of analytical procedures using a genetic algorithm. Part 1. Theory, numerical simulations and applications to atomic emission spectroscopy. *Analytica Chimica Acta*, **265** 211.
Wienke, D., Lucasius, C., Ehrlirch, M. and Kateman, G. (1993) Multicriteria target vector optimisation of analytical procedures using a genetic algorithm. Part 2. Polyoptimisation of the photometric calibration graph of dry glucose sensors for quantitative clinical analysis. *Analytica Chimica Acta*, **271** 253-268.
Wise, B.M. and Ricker, N.L. (1991) Recent advances in multivariate statistical process control improving robustness and sensitivity. *IFAC Symposium, ADCHEM'91, Toulouse, France*, Pergamon Press, Oxford, pp. 125-130.

Wise, B.M., Veltkamp, D.J., Ricker, N.L., Kowalski, B.R., Barnes, S. and Arakali, V. (1991) Application of multivariate statistical process control (MSPC) to the West Valley slurry-red ceramic melter process. Waste *Management '91 Proceedings, Tucson, Arizona.*

Wold, S. (1978) Cross-validatory estimation of the number of components in factor and principal components model. *Technometrics*, **20** (4) 397-405.

Wold, S., Esbensen, K. and Geladi, P. (1987). Principal component analysis. *Chemometrics and Intelligent Laboratory Systems*, **2** 37-52.

Wold, S., Carlson, R. and Skagerberg, B. (1989) Statistical optimisation as a means to reduce risks in industrial processes. *The Environmental Professional*, **11** 127-131.

10 Grouping data together—Cluster analysis and pattern recognition
W. Melssen

10.1 Introduction

Research chemists perform many quite complex experiments daily. These can yield (large) data sets that, in turn, contain much (hidden) chemical or physical information. Proper interpretation then can give insight into the underlying reaction mechanisms, molecular structure or properties, process dynamics, etc.

These data are usually continuously monitored, sampled at discrete moments, digitised and, finally, stored in a computer memory. Chemists then order the measured data in tables or, speaking more chemometrically, in matrices. In such a data matrix, each row represents an object (one experiment, one batch process, one infrared spectrum or one chromatogram, etc.). Each column in the matrix represents an individual variable (say, a wavelength or a retention time). Matrices can be heterogeneous; that is, variables are expressed in different units (for example, chromatographic retention time, molecular volume, concentration, ppm, calorific value, etc.). Homogeneous matrices contain variables that are all expressed in the same unit (for instance, Raman spectra, titrimetric curves, voltammetric time series or liquid chromatograms, etc.).

This chapter provides the reader with a number of chemometric visualisation, clustering and classification techniques that will unravel in a straightforward way valuable information contained in the matrix of measurements.

10.2 Where to go

10.2.1 Visualisation of data

To get a first glimpse of what kind of information has been gathered during the experiments, each researcher should look first at any structure in the collected data. When only a few (up to three) variables are measured, this visual inspection can be done in a straightforward manner (see also Chapter 2). For instance, when two variables per object are involved, an x–y plot can be constructed. This plot provides, in a direct way, some insight into the structure present in the collected chemical data. For instance, inspection of Figure 10.1 reveals that there are apparently two

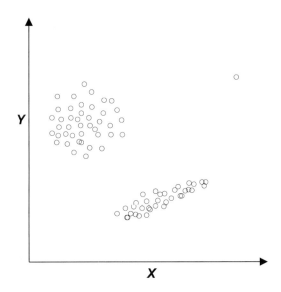

Figure 10.1 An x–y plot of a data set consisting of objects characterised by two variables.

groups (clusters) of objects and one data point that is isolated from the others. This point might be an outlier that it might be advisable to remove from the data set before, say, a classification model is built. However, it might be a valid point, in which case the model used to interpret the data might need to be modified to include the point.

When many variables are involved, such as an entire Raman spectrum per measured object, the experimenter has to use more advanced visualisation techniques from the chemometrics toolbox. This chapter describes the following visualisation and mapping techniques:

- Principal component analysis
- Nonlinear mapping
- Kohonen self-organising feature map neural networks
- Parallel coordinates

Each of these techniques will yield different insights into the structure of the collected multivariate data.

10.2.2 Similarity and finding clusters and groups in the data

After this preliminary inspection of the data, it might be worthwhile to apply techniques that group or cluster objects in an automated and robust way. For example, if one observes three clearly distinct groups of objects in, say, an x–y plot, one might suspect that these groups represent different classes of object properties. Obviously, the clustering of objects is based on similarities between the variable values of these objects. How to define and choose an appropriate similarity measure is discussed in Appendix B. How to represent and interpret in a simple way the outcome of a cluster analysis is outlined in the next section.

Three categories of clustering techniques are presented:

- Single, average, and complete linkage
- Ward's hierarchical clustering method
- Forgy's nonhierarchical clustering method

10.2.3 Using known groupings to predict membership of new samples

So far, we have assumed that no particular *a priori* knowledge is available about the class type or properties of the measured data. The previously mentioned data analysis methods do not use this information and, hence, belong to the class of unsupervised techniques. However, if preknowledge is available, more sophisticated and powerful techniques can be used to analyse and interpret the measured data. For example, if a chemist wants to examine, on the basis of near-infrared spectroscopy, whether various material types (classes) form separated spectral groups, and for a number of measured objects this material type information is known, then so-called supervised methods can be invoked. In general terms, the measured data as well as the known corresponding object class guide such supervised classification techniques. An advantage of supervised classification techniques is that they yield a prediction model that can be used for the classification of newly measured objects of which the class type is not yet known. In Section 10.5 three powerful and widely used classification techniques are discussed:

- Linear discriminant analysis
- Soft independent modelling of class analogy
- Artificial neural networks

10.2.4 Transforming data for better models

In many cases it is beneficial to perform a transformation of the raw data before they are subjected to a particular visualisation, clustering or classification technique, particularly when the underlying distribution of the data is non-Normal. For instance, when measuring the biological activity of a batch of synthesised macrobiomolecules, the resulting activity values are usually scaled in a logarithmic way. This transformation will to a large extent influence the outcome of, say, a selected classification technique. Appendix A elaborates a subset of very useful data transformation and scaling techniques. Those readers who are not familiar with the matrix representation of data sets are advised to read the introductory section of this appendix first. The effects of different transformations are shown by means of an example of a principal component analysis of the chromatographic interaction effects between chalcones and the composition of the mobile phase.

10.2.5 A brief route through the chapter

As stated, visualisation of the (experimental) data should be mandatory for any research chemist. Visual inspection quickly provides the first insight into what is contained in the measured data. For example, suppose an analytical chemist suspects an out-of-specification situation during a near-infrared (NIR) spectroscopic analysis of a set of synthetically produced yarns. It would be a very cumbersome task to compare the series of measured NIR spectra with a reference database containing thousands of standard spectra. Here, principal component analysis provides a fast and elegant solution to the problem. This same technique can be used to examine whether a series of chemical experiments yields multivariate data that appear as different clusters belonging to different experimental conditions. If principal component analysis fails to unravel sufficient information, two other mapping techniques might provide a satisfactory solution: nonlinear mapping and Kohonen neural networks. A third and easy to implement alternative is the representation of the data by means of parallel coordinates. Chemists who are interested in these topics and want to know how to group high-dimensional multivariate data in an automated way are referred to Section 10.3.

Section 10.4 deals with a number of techniques designed to build classification models in a supervised way. These classification models are based on a set of measured objects with known properties or class types. Once a classification model has been constructed and tested, the researcher will be able to forecast the (unknown) class or type of a newly measured object. Researchers having a chemical problem at hand that is roughly characterised by an underlying (local) linear relationship between the measured variables and the class property of the objects are recommended to read the subsections concerning the supervised classification techniques of linear discriminant analysis and soft independent modelling of class analogy. When these statistical linear classification techniques fail, a researcher might try nonlinear adaptive artificial neural networks.

10.3 Visualisation and mapping

This section outlines in a comprehensive way, three powerful visualisation and mapping techniques that are applicable to the examination of the structure in high-dimensional data sets: principal component analysis, nonlinear mapping, and Kohonen self-organising feature map neural networks. We shall touch briefly the use of parallel coordinates because this technique provides insight to the structure of the data in a totally different way from the other methods.

10.3.1 Principal component analysis

As with any other unsupervised technique, the input matrix (which is denoted by **X**) serves as the starting point for the analysis of the measurements. The general concept of principal component analysis (PCA) is to search for a coordinate transformation in such way that the new coordinate axes capture (or, more popularly, 'explain') in descending order of magnitude the variance present in constructed data matrix. Here, coordinate transformation means that the original (measured) variables are in some way weighted and combined in a linear fashion to form new variables in a new coordinate system. In chemometrics and statistics literature, these new variables are usually referred to as *features, canonical variates* or *latent variables*. The axes in the new coordinate system are called principal components (PCs). Mathematically, this coordinate transformation can be expressed as a decomposition of the data matrix **X** into three matrices, **U**, **L** and **V**,

$$\mathbf{X} = \mathbf{ULV}$$

In practice, this decomposition of the data matrix is performed by a singular value decomposition. To understand what PCA is and what it does with the data, it is not necessary to know the theoretical aspects of singular value decomposition, so we shall skip the theory at this point and refer the interested reader to the bibliography at the end of the chapter.

The matrix **U** contains the *scores* of the objects in the new coordinate system. The scores can be imagined as the projection of the original object vector (a spectrum or chromatogram) onto the new coordinate axes. The diagonal matrix **L** contains the *eigenvalues* of the respective principal components. These eigenvalues express the amount of variation, or variance, explained by each PC. Software used to calculate PCs generally presents the eigenvalues in decreasing size, with the first eigenvalue having the largest amount of explained variance and the last eigenvalue the smallest amount. The matrix **V** contains the actual transformation vectors that convert the original variables into the new variables. In PCA, these transformation vectors are called the *loading* vectors. As we shall see, these loading vectors provide information regarding the degree of correlation (or similarity) between the original variables.

In other words, the **U** matrix gives information about how the original objects (*not* the original variables) will appear in the space spanned by the new coordinate system, whereas the matrix **V** tells us how the original variables (*not* the original objects) are combined in a linear way to form the latent variables. Summarising this, **U** is an object-oriented matrix and **V** can be regarded as a variable-oriented matrix.

A graphical representation of how PCA operates on the original data is shown in Figure 10.2. For simplicity, we consider here that each object (represented by a filled dot) is characterised by just two variables. As can be seen, the measured objects form an elongated group of points in this two-dimensional data space. When we apply PCA, then the first principal component will be oriented along the direction that corresponds to the highest amount of variation in the data. This matches the elongation direction of the group.

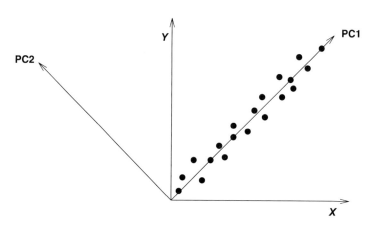

Figure 10.2 Principle of a PCA in a two-dimensional data space.

In this simple example, the first PC captures approximately 90% of the total variation in the data set. The second PC is by definition orthogonal to the first one and explains the remaining 10% of the variation. This small amount of variation might presumably be associated with, say, the measurement noise. Because all PCs are orthogonal to each other, in this example there are exactly two PCs. More generally, if one has an m-dimensional data space (thus, m variables per object) there are exactly m

PCs. However, and this an advantage of the PCA, usually a large number of PCs (having small eigenvalues) will contain just noise, so, the relevant (chemical) information is captured by the first few PCs. As a consequence, only these PCs have to be considered when examining the structure in the data.

Now the question arises of how many principal components should be taken into account. A simple and direct way is to construct a scree plot. In scree plots one depicts graphically the amount of explained variance per PC in descending order. As shown in Figure 10.3, the horizontal axis represents the PC number and the vertical axis represents the amount of variance captured by the PCs. Many sophisticated and statistically founded ways exist for an adequate interpretation of scree plots. For most research applications, however, it is sufficient to take into account only those PCs that have variances greater than the baseline level in the scree plot. Note that it is implicitly assumed that the remaining PCs represent the noise in the data. In this particular example, the scree plot shows there are four significant PCs out of the eight possible (this plot was generated from the magnetic resonance imaging (MRI) data described later in this section).

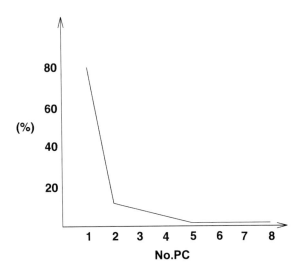

Figure 10.3 Scree plot corresponding to the PCA of the multivariate MRI images.

A study concerning the classification of human brain matter by means of multislice MRI (radiologists prefer the term 'tissue segmentation') provides a useful case study. Here, the term multislice means that several (in this particular case, seven) parallel brain images were measured as 2D planes across the brain.

Briefly, it is known from MRI studies that two parameters strongly influence the observed intensity contrast in MRI images. These key parameters are the pulse repetition time (PRT) and the pulse echo time (PET), respectively. During an experiment with a healthy volunteer, eight combinations of PRT and PET were used. In the experiments, each individual image slice consisted of 256×256 pixels. So for each multivariate pixel in the slice (which is related to the spatial location of a tiny volume in the brain of the volunteer) there are eight different intensity values corresponding to the respective PRT/PET combinations. The data is then arranged into a matrix with eight columns corresponding to the eight PET/PRT combinations and 65 536 rows, one for each pixel in an image. In this way, one complete multivariate MRI image slice is represented as a single matrix.

Now principal component analysis comes into play. The aim of applying PCA in this medical case was to investigate whether a relationship exists between the individual pixel vectors of PET/PRT values (which can be considered as a kind of spectrum or fingerprint) and the brain tissue type to which it corresponds. The principal component analysis was conducted on all 65 536 objects simultaneously. It appeared that 88% of the variance in the data could be explained by just the first two principal components. Hence, it is possible that a reliable visual inspection of this complex data can be made using an x–y plot of the object in just these two PCs. Figure 10.4 suggests there are different regions in the two-dimensional score plot. Each cluster in the plot corresponds to either the background pixels, the skull bone, fat or a particular tissue type. However, the dense cluster in the middle consists of three overlapping subclusters, which correspond to white matter, grey matter and cerebrospinal fluid. This indicates that for a proper separation of the latter tissue types one has to do something more. We will return to this topic later.

Figure 10.5 shows the loading plot for the first two PCs. As can be seen, there are four distinct groups of variables (combinations of PET and PRT). One contains the outlying PET/PRT combinations 20/550 and 20/3000. These variables have a major contribution to the PCA model. Next, two small clusters (80/3000, 55/3000, 120/3000) and (15/350, 15/550, 15/750) can be discerned. Since grouping of variables in a PCA loading plot indicates that the variables are strongly correlated, it is possible that some of these variables can be discarded in future experiments. The fourth cluster contains the variables 15/350, 15/750, 55/3000, and 120/3000 and, as these are close to the zero point of the plot, a negligible amount of information will be lost if they are skipped. Thus, only four (instead if eight) of the variables are necessary to preserve almost all of the relevant information.

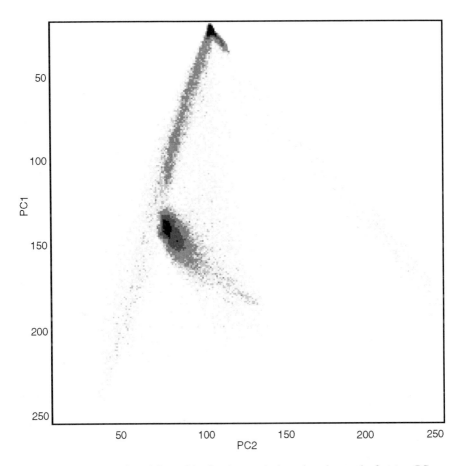

Figure 10.4 Score plot of the multivariate image pixels projected onto the first two PCs.

10.3.2 Nonlinear mapping (NLM)

Nonlinear mapping is a powerful dimension reduction technique. It forces high-dimensional data to fit into just two or three dimensions in a nonlinear way, thus allowing a quick global visual inspection of the high-dimensional data space. NLM does not focus on the variance in the data, as PCA does, but it tries to keep the distances between all objects in the two- or three-dimensional space the same as in the original data space. Hence, the technique uses two distance matrices: one for the objects in the original data space having distances (d_{ij}) and one for the mapped objects in the low-dimensional space (with distances d^*_{ij}).

Let us consider a mapping to only two dimensions. In this case, NLM searches for a two-dimensional plane in the original data space in such

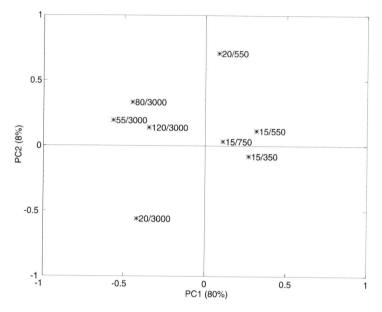

Figure 10.5 Loading plot of the eight PET/PRT combinations in the PC1–PC2 space.

way that a distance error function will be minimised. This distance error function, E, is defined by

$$E = \sum_i \sum_j \left(\frac{d_{ij} - d^*_{ij}}{d_{ij}} \right)^2$$

The general idea behind NLM can be envisaged as follows. Consider the old-fashioned stick-and-ball model for representing the spatial structure of a molecule. Suppose that the sticks that connect the atoms are flexible in length and mutual angles but do not bend. NLM squashes the three-dimensional molecular model between two solid parallel plates. As the parallel plates are pushed together, the sticks move and adjust to flatten the structure of the molecule in such a way as to best preserve the original structure (and thus, the distances between the atoms) in a two-dimensional representation.

A drawback of NLM is that it is a very computationally intensive technique. The process of searching for the best low-dimensional plane in a high-dimensional data space is a complex and time-consuming optimisation problem. Therefore, this technique is usually hyphenated with a powerful optimisation techniques such as genetic algorithms and simulated annealing.

In the following medical example, single pulse ^{1}H NMR spectroscopy is used to examine the chemical content (lactate, isoleucine, alanine, etc.) of the cyst fluid from 28 female patients with ovarian cysts. One of the aims of this research was to examine the relationship between the chemical fluid composition and the type of the ovarian tumour: benign or malign.

The peak areas of 54 metabolites were determined in the NMR spectrum of each cyst fluid. This data set of peak areas was used to characterise each ovarian fluid. The data set was examined by PCA and it appeared that a lot of variation was present in the data among all the variables, as indicated by the identification of 14 significant PCs. To get a full view of the structure in the data by simple graphical means, one has to inspect 91 ($=14 \times (14-1)/2$) PC score-plots.

As an alternative, the NLM technique was applied to force the 54 variables to fit in just two dimensions. The resulting x–y plot is depicted in Figure 10.6. Two clusters of data points can be discerned and, indeed, the left cluster corresponds to the benign group, whereas the other cluster contains exclusively malign tumours. Strikingly, the points indicated by a '✶' were so-called borderline patients for whom the exact tumour type (benign or malign) was not known at the time the data were analysed. Some months afterwards, a second analysis was conducted and these patients then appeared to have malign cysts. Moreover, the patient (located at position $(0, -1)$ in the x–y plot, although labelled as having a

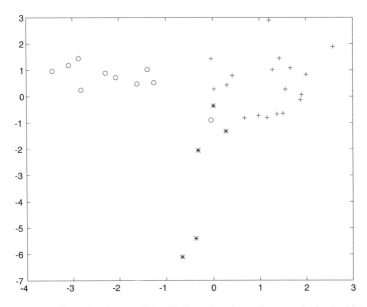

Figure 10.6 Two-dimensional map of the 54-dimensional cyst data set obtained with NLM.

benign cyst, developed a malign cyst after a while. This convincing data analysis result could not have been achieved by a straightforward PCA.

10.3.3 Kohonen self-organising feature map neural network

Kohonen self-organising feature map neural networks have analogies with NLM. They try to capture the structure (or more precisely, the topology) in the multidimensional space of the original data and embed this topology in the weight vectors of the processing units in the network. Many variants of Kohonen neural networks exist, but here we consider only a simple two-dimensional network. Basically, a Kohonen neural network consists of a rectangular grid with processing elements (units or neurons) mounted on the vertices of the grid lines (see Figure 10.7).

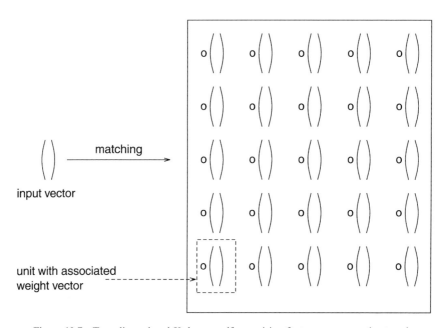

Figure 10.7 Two-dimensional Kohonen self-organising feature map neural network.

Each unit is equipped with a weight vector that contains as many elements as the number of variables characterising each data object. Without going into detail, the weight vectors of the units are adjusted in an iterative way by presenting examples (the data objects) to the Kohonen network. Suppose now that we present one data object, **x** (**x** is a vector of the object's values for each of the variables), to the Kohonen network. The distance between the weight vector and the object vector is determined for each of the units. The unit that has the smallest distance is assigned as the winning unit. Now, the weight vector of this winning unit is adjusted according to the following formula,

$$W_{ij}^{new} = W_{ij}^{old} + \alpha(x_i - W_{ij}^{old})$$

Here, W_{ij}^{new} represents the new weight vector of winning unit j for the object variable i, and W_{ij}^{old} the weight vector of the same unit and variable before it was adjusted.

Thus, the weight vector of the winning unit is slightly rotated (the amount of adjustment is controlled by the parameter α) towards the vector of the presented object and, hence, it becomes a little more similar to that object. The weight vectors of the units surrounding the winning unit (referred to as the neighbourhood of the winner) are updated in the same way.

Next, another object is presented to the network and the whole learning process is repeated. This adjustment process continues until the network converges (meaning that the weights do not change significantly). Usually, convergence occurs when all the data objects have been presented to the network many times (say, 100 up to 1000 times). By means of this specific learning strategy and the adjustment of the neighbouring units, the structure (topology) of the original data space is embedded in the weight vectors of the Kohonen neural network. This implies that each unit in the network, in fact, represents some specific local multivariate feature of the original data space. Here lies the crux of the Kohonen network. Local multivariate features in the data are usually directly related to specific properties or class types of the objects. The Kohonen map has clusters of units representing similar local multivariate features, and hence the specific classes or classes of objects, as depicted schematically in Figure 10.8.

In Figure 10.8, each column (1 to 4) represents the weight vector of a network unit with the magnitudes of the weights for each of the variables (a to g) shown by filled circles (large weights) and open circles (small weights). As can be observed, similar large and small values of the variables serve as connecting features between neighbouring units. So, while going from one unit to the other, the weight vectors are chained as in a game dominoes.

To demonstrate the power of a Kohonen neural network, we presented a database containing more than 3000 infrared spectra (each IR spectrum consists of 266 absorption values at wavelengths in the range of 400–4000 nm) to a rectangular 20×20 Kohonen neural network. After the

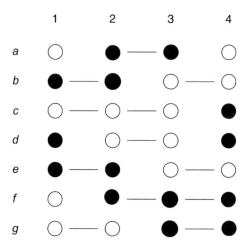

Figure 10.8 Chained multivariate features by a Kohonen neural network.

network had converged, various analyses were conducted on the weight vectors that were generated and the so-called Kohonen map (for details, the reader is referred to Melssen et al., 1993).

Figure 10.9 shows a particularly good visual presentation of complex data using a Kohonen map. The rectangle shows the borders of the network. However, instead of displaying the units themselves, the positions

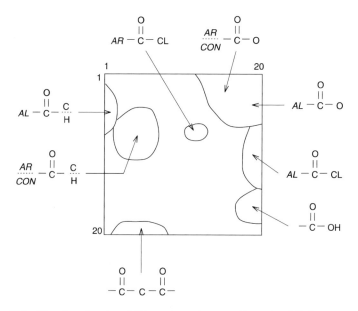

Figure 10.9 Mapping of carbonyl IR spectra onto a two-dimensional Kohonen network.

10.3.4 Parallel coordinates

Parallel coordinates were proposed by Inselburg in the early 1980s as a novel way to represent multivariate data. Unlike the previously described mapping techniques where two or three orthogonal axes are used to display the data, parallel coordinates depicts all of the data on parallel vertical axes that are equally spaced on a horizontal axis. Each vertical axis is scaled to display the values of just one variable in the data. Figure 10.10 shows an example of a 5-axis plot with a single object vector having the values $(-7, 4, 3, -2, 5)$. If there are more objects in the data set they are plotted in the same way.

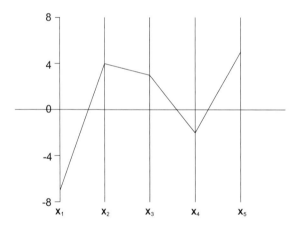

Figure 10.10 Representation of the object vector $(-7, 4, 3, -2, 5)$ in parallel coordinates.

Now consider a set of data points that form a straight line in a conventional two-dimensional plot (see Figure 10.11). In this case, the two variables are perfectly anticorrelated. Such a straight line appears in the two parallel coordinates representation as a star-like structure (each line in the star is the data from one object) with a crisp intersection point between both variables (Figure 10.11). This is one important property of parallel coordinates: a linear relationship between two variables shows up as parallel lines (in case of a positive correlation) or as lines crossing a single intersection point (if two variables are anticorrelated).

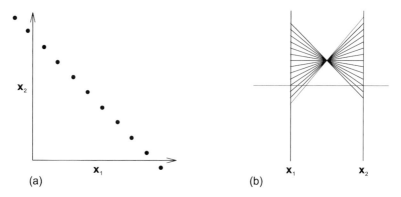

Figure 10.11 Representation of data objects characterised by two anticorrelated variables (a) in the original space and (b) in parallel coordinates.

Note that the patterns can be observed only if they exist between two adjacent variables in the plot. Hence, differently chosen orderings of the variables might produce substantially different representations in parallel coordinates. However, the correlation structure is preserved in the plot, irrespective of the ordering of the variables, and it can be followed from one variable pair to another pair elsewhere in the plot. (Some experimentation with the plotting of a small number of objects showing positive, negative and no correlation between, say, five variables is recommended to convince yourself of this!) In addition to the correlation structure, parallel coordinate plots also highlight similarity between objects. This is seen as objects having similar values on one or more axes.

A major advantage of this representation is that it facilitates the overview of the distribution of all variable values at once. Moreover, as the calculations involved in producing them are trivial, parallel coordinates can be used in an on-line fashion to monitor multivariate data in a dynamic system.

Let us return to the example of the multivariate MRI images. A map in the parallel coordinate space was made for a subset of pixels belonging to the categories of grey matter (GM) and cerebrospinal fluid (CSF). Figure 10.12 shows the result. As indicated by the abundance of parallel lines connecting the first variables (PET/PRT combinations) this representation suggests that the corresponding MRI images possess many positively correlated pixels. On the other hand, for the lines corresponding to the CSF, the intersection points between respectively variables 5 and 6 and variables 6 and 7 strongly suggest that the corresponding MRI images are anticorrelated. Moreover, this data representation by means of parallel coordinates demonstrates that GM and CSF are characterised

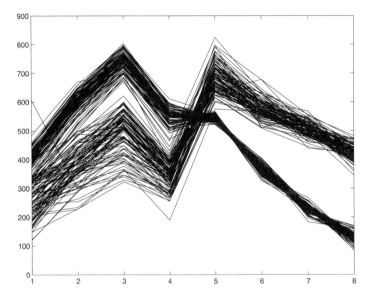

Figure 10.12 Representation of a subset of multivariate pixels corresponding to grey matter and cerebrospinal fluid in parallel coordinates.

by different multivariate MRI fingerprints. This latter conclusion could not have been drawn from a standard PCA and, moreover, the researcher retrieves direct information based on correlations between consecutive original variables.

10.4 Clustering of multivariate measurements

There is a wide variety of clustering algorithms available and it is impossible to discuss the whole gamut here; only the most representative ones are explained. We shall consider hierarchical as well as nonhierarchical clustering methods. Here, hierarchical implies that once some data points are assigned to a particular cluster, they will stay in that cluster always, whereas in nonhierarchical methods, previously clustered objects might be separated again and join different clusters during the clustering process.

The hierarchical single, average and complete linkage methods and Ward's method are considered. As an example of a nonhierarchical technique we shall discuss Forgy's clustering method. All of these methods have an agglomerative nature, meaning that they take as the initial situation all of the individual data points and then proceed to group together the data points one by one. Divisive methods, on the other hand, start with one initial cluster that encompasses all of the data points and then subdivide this cluster in two subclusters, and so on. Divisive methods also yield a hierarchical clustering and might present some computational advantages. However, the general principles of these divisive methods will provide no additional insights to those gained from agglomerative methods and they are therefore not considered in this section.

Each clustering method needs as a starting point the so-called *similarity matrix*. This matrix contains the mutual distances (or correlation coefficients) between all objects in the data set. For a more detailed discussion on similarity measures, the reader is referred to Appendix B.

Suppose that for five soil samples we measured four concentrations of heavy metals. These results are arranged in a measurement table (see Table 10.1). We can calculate for each pair of objects, say, the Euclidean distance between them. These distances are grouped into a new matrix to form the basic similarity matrix for this data set. The similarity matrix element $D(A, B)$, for example, denotes the distance between the objects A and B. Because the distance between A and B is equal to the distance between B and A, thus, $D(A, B) = D(B, A)$, the similarity matrix is symmetrical with respect to its diagonal and hence only the lower triangle has to be considered.

Note that the size of the matrix is determined by the number of objects and is independent of the number of variables; if there is a large number of objects, the similarity matrix will be very large. The similarity matrix, based on a Euclidean distance measure, is shown in Table 10.2.

Table 10.1 Heavy metal concentrations in arbitrary units in five soil samples

Sample	Metal 1	Metal 2	Metal 3	Metal 4
A	100	80	70	60
B	80	60	50	40
C	80	70	40	50
D	40	20	20	10
E	50	10	20	10

Table 10.2 The similarity matrix for the measurements given in Table 10.1. The numbers are the Euclidean distances between pairs of object, e.g. objects B and C have a distance value $D(B, C) = 17.3$

	A	B	C	D	E
A	0.0				
B	40.0	0.0			
C	38.7	17.3	0.0		
D	110.4	70.7	78.1	0.0	
E	111.4	72.1	80.6	14.1	0.0

10.4.1 Single, average and complete linkage

In the single, average and complete linkage methods, one searches for those two objects that are most similar. In terms of distances, one looks for the pair of different objects, X and Y, having the smallest value for $D(X,Y)$. In this example, $D(D, E) = 14.1$, is the smallest value.

Objects D and E are now joined to form a new object DE. Based on this observation, we start to construct a dendrogram (a reversed tree-like structure) in which we connect D and E at a branch length of 14.1. Now we have to recalculate the similarity matrix, because the new object DE replaces the old objects D and E and we need the distances of the other objects from this new one. To do this, the position of object DE has to be defined so that its distance from all other objects (A, B, C, etc.) can be calculated.

Suppose we want the distance of DE from object A. Three strategies are possible. In the average linkage clustering method, the mean of the distances $D(A, D)$ and $D(A, E)$ is taken, so

$$D(A, DE) = \frac{1}{2}(D(A, D) + D(A, E))$$

In general this average distance is not very different from the distance calculated between A and the mid-point between D and E. Both methods of calculating the average linkage are in use.

In the single linkage method, the new distance $D(A, DE)$ is equal to smallest distance, that is the minimum of $D(A, D)$ and $D(A, E)$, thus,

$$D(A, DE) = \min(D(A, D), D(A, E))$$

In the complete linkage method, it is the largest of both distances that is taken:

$$D(A, DE) = \max(D(A, D), D(A, E))$$

To continue this example, we shall use the average linkage method. The size of the new similarity matrix is reduced by one in both dimensions (rows and columns) to 4 rows by 4 columns, from the original 5×5 matrix. After recalculation of the respective distances, the new similarity matrix looks like Table 10.3. The smallest distance in this matrix is identified as BC, and the whole process is now repeated. This continues until all the objects are joined in the dendrogram. During this clustering process, object-connecting branches are added to the tree (with branch lengths corresponding to the recalculated distances).

The resulting dendrogram for this particular example is depicted in Figure 10.13. The dendrograms resulting when the single or complete

Table 10.3 The resulting similarity matrix after merging the objects D and E using the average linkage method

	A	B	C	DE
A	0.0			
B	40.0	0.0		
C	38.7	17.3	0.0	
DE	110.9	71.4	79.3	0.0

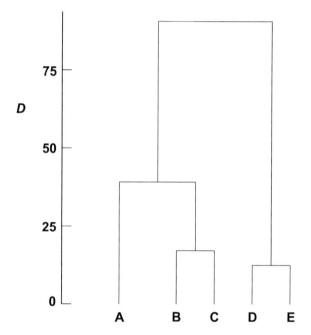

Figure 10.13 Dendrogram obtained with the average linkage clustering method for the data set given in Table 10.1.

linkage clustering methods are used would look slightly different. Which linkage method should be used in practice is highly problem dependent. Unfortunately, there exists no general recipe and one has to try possibly two or maybe even three of the linkage techniques in order to obtain the dendrogram giving the best possible clustering. As a rule of thumb, one should start with the average linkage method.

Consider the following real-world example. About 600 iron meteorites have been discovered on earth. These meteorites have been subjected to exhaustive inorganic analyses for their metal content. For most of them, assays for many metals (e.g. Ni, Ga, Ge, Ir and Au) are available. About 13 of these parameters are necessary for a proper clustering and classification of all meteorites. Such clustering and classification is important because astronomical experts are convinced that each group or class of meteorites originates from a different celestial body or region. Hence, an adequate clustering and classification allows a better understanding of the astronomical history of these meteorites. Thus, the problem is to find different groups among these 600 meteorites.

As stated, each object (meteorite) is characterised by 13 variables. When one wants to examine the relationship between all of these variables by means of, say, x–y plots, one has to construct and inspect $1/2 \times 13 \times 12 = 78$ such plots. This is a cumbersome and error-prone task and it is unlikely to show multivariate relationships. The average linkage clustering procedure using a Euclidean distance measure has been applied to the data. The result is shown in Figure 10.14, as a dendrogram. Only part of the full dendrogram is illustrated as the full one, with 600 end points, would spread over several pages of this book! We can see that meteorites m_1, m_3, m_4, m_7, m_8 and m_9 are very similar (linked at short distances), whereas they differ substantially from meteorites m_2, m_5, m_6, m_{10}, m_{11}, m_{13} and m_{14} (the two groups are linked at a much longer distance). Meteorite m_{12} appears to be very different from all the other meteorites and was classed as a 'loner' in the study.

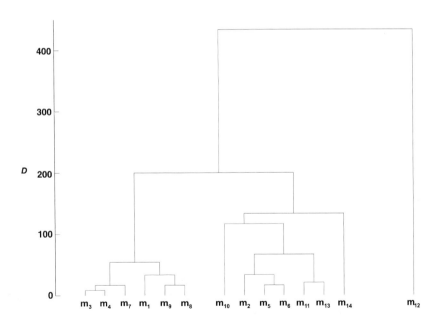

Figure 10.14 Dendrogram of the meteorite data set obtained with the average linkage clustering method.

10.4.2 Ward's clustering method

Another hierarchical clustering method is Ward's. This method is based on the criterion of a minimal increase in the error sum of squares (ESS). The ESS is a measure of the variability of the objects within the groups to which they have been assigned, averaged over all of the groups. Thus, if the groups contain very similar objects, the variability within the groups is low and, hence, ESS is low. If a new object is added to a group, and it is very similar to the objects already there, it will cause only a small change in the variability. If it is very dissimilar, then a much larger change in variability will be seen. Ward's method selects objects giving the smallest in ESS to add to a group.

In the univariate case, the ESS is defined by

$$\text{ESS} = \sum_{j=1}^{k}\left[\sum_{i=1}^{N_j} x_{ij}^2 - \frac{1}{N_j}\left(\sum_{i=1}^{N_j} x_{ij}\right)^2\right] = \sum_{j=1}^{K}\left[\sum_{i=1}^{N_j}(x_{ij}-\hat{x}_j)^2\right]$$

$$= \sum_{j=1}^{K}(N_j-1)\,\text{var}_j$$

In this formula, x_{ij} represents the ith object belonging to class j. K denotes the number of classes, N_j the number of objects in class j and var_j the variance of group j. The three forms of the equation are all equivalent ways of saying that ESS is the sum of variances within the groups. The left-hand form is computationally less sensitive to rounding errors than is the middle form.

As with any agglomerative technique, the individual objects serve as starting point for the clustering method. Thus, supposing one has N objects consisting of M variables, one starts with N clusters (each cluster corresponds to one object). Thus, $K=N$ and $N_j=1$. Then one calculates the ESS for each combination of pairs of objects (intuitively, one can compare this with the construction of the Euclidian average linkage similarity matrix).

The two objects causing the smallest increase in the ESS are joined (ESS has an initial value of zero when each object is its own cluster). As was the case with the object linkage methods, a dendrogram is composed, but now with the branch lengths equal to the ESS. Again, this process is repeated until all objects are joined together in the dendrogram. The number of clusters can be determined by looking for large changes (indicated by the long branch lengths in the dendrogram) in the ESS value.

The calculations for a simple example with five objects are summarised in Table 10.4, with the resulting dendrogram shown in Figure 10.15. The first row of the table gives the values of the five objects (we show only a

CLUSTER ANALYSIS AND PATTERN RECOGNITION 387

Table 10.4 Summary of the calculations for Ward's ESS clustering method

Stage					
1	A = 2	B = 5	C = 9	D = 10	E = 16
2	AB = 4.5	AC = 24.5	AD = 32.0	AE = 98.0	BC = 8.0
	BD = 12.5	BE = 60.5	CD = 0.5	CE = 24.5	DE = 18.0
3	AE + CD = 98.5	BE + CD = 61.0	CDA = 38.0	CDB = 14.0	CDE = 28.7
	AB + CD = 5.0				
4	ABCD = 41.0	ABE + CD = 109.2	CDE + AB = 33.2		
5	ABCDE = 113.2				

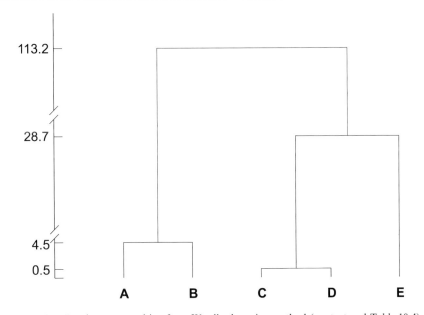

Figure 10.15 Dendrogram resulting from Ward's clustering method (see text and Table 10.4).

simple univariate example here—usually each object would have several values associated with it). The second row gives the ESS, for each possible combination of the objects that gives four clusters. Only the clusters containing two or more objects are listed. The set containing cluster CD has the lowest ESS, so objects C and D are linked and remain together throughout the rest of the calculation. Stage 3 gives the ESS values for all possible combinations of objects A, B, E and cluster CD that give three clusters. The set containing clusters AB and CD has the lowest ESS. At stage 4 the set containing clusters CDE and AB has the lowest ESS and at stage 5 all of the objects have been combined into a single cluster so the process stops.

A tasteful example of how well Ward's method works can be found in the examination of the concentrations of 20 amino acids in four different French wine species (Bourgogne/Beaujolais, Bourgogne/non-Beaujolais, Côte du Rhone, and Bordeaux). A total of 200 wines were analysed, yielding a data matrix of 200 objects and 20 variables (the 20 amino acid concentrations). The outcome of Ward's clustering method applied on this data set is shown in Figure 10.16.

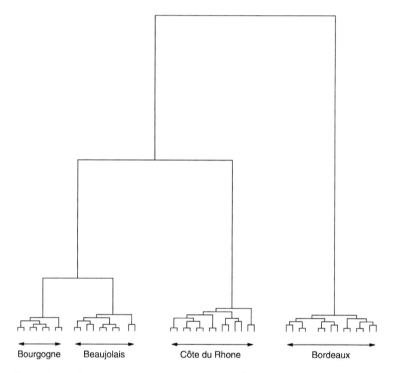

Figure 10.16 Grouping of four French wine species by Ward's clustering method.

The dendrogram shows that the four wine species are nicely clustered according to their amino acid content. Note that only a subset of the full data set is displayed for clarity. This strong correlation between the wine species and the amino acid content is related, among other things, to the geographical origin of each wine species (and thus, to the local chemical soil composition).

10.4.3 Forgy's clustering method

Let us now cluster the objects depicted in Figure 10.17 with a non-hierarchical clustering method. Instead of clustering data by sequentially

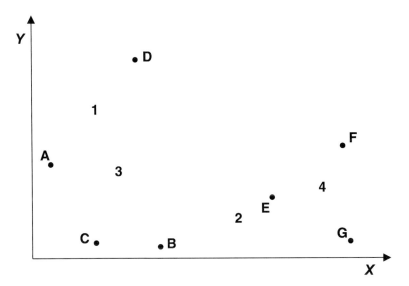

Figure 10.17 Clustering of objects according to Forgy's nonhierarchical method.

joining individual objects, Forgy's method determines directly a K clustering, meaning division of the available objects into K clusters in one step. Here, K is the key parameter, which has to be chosen by the chemist.

Figure 10.17 shows that for this data set, $K=2$ is the proper choice because there are two distinct clusters observable: ABCD and EFG. To obtain exactly two clusters in Forgy's method, one needs to select two initial *seed points* from the objects. Forgy's method then assigns each object to its nearest seed point. In this way, a provisional clustering is obtained.

Let us start with a bad first guess of the seed points for this data set by choosing objects A and B. In this two-dimensional example it is quite easy to see what is happening, but in reality it would be very difficult to spot that a bad guess had been made. Clearly, D is nearest to A, whereas C, E, F and G are closest to B. The initial clustering therefore is AD and BCEFG. For these clusters, one determines the centroid (the point corresponding to the mean of the variable values). The centroids of the clusters AD and BCEFG are respectively marked as 1 and 2 (see Figure 10.17). In the next step, each object is reassigned according to its respective distance from the centroids 1 and 2. This reassignment yields the new clustering ABCD and EFG. The whole procedure is then repeated. For ABCD and EFG two new centroids are calculated (3 and 4). Then, reassignment of the objects to these centroids yields again the clustering ABCD and EFG. Consequently, this clustering is considered as the definitive one.

Summarising this, Forgy's clustering method is based on five steps:
1. Select an appropriate number of initial centroids (seed points).
2. Select a suitable similarity measure.
3. Assign the objects to one of the centroids according to this similarity measure.
4. Determine the new centroid for each new cluster.
5. Repeat from step (3) until no change in the clustering occurs.

In general, as in the previous example, one uses high-dimensional data, so that it is (nearly) impossible to observe by eye the actual number of clusters. Here lies the power of nonhierarchical clustering methods. In hierarchical methods, previously clustered objects will remain forever in the same cluster—one is bound to the path paved by the initially chosen clustering and, in this particular example, this would be disastrous. During the nonhierarchical clustering process, however, clusters of objects might be partially divided and reassigned to other clusters at each iteration. This is certainly an advantage when dealing with data that cannot be visualised owing to the high number of uncorrelated variables involved.

A recent application of Forgy's nonhierarchical clustering method can be found in the challenging field of phylogenetics. Many functional proteins are believed to have descended from one specific common ancestor. The descendant proteins emerged by means of mutation of the ancestor protein. A tree-like structure that fixes the evolutionary history of the proteins is called a phylogeny or phylogenetic tree. Besides the topology of the tree, which describes the mutual coherence of the proteins, the branch lengths are important: they give insight into the evolutionary distance, that is how many mutations have taken place or how many years have elapsed since such genetic bifurcation occurred.

A phylogeny (star-like dendrogram) is shown in Figure 10.18 for a set of G-protein-coupled receptors (in phylogenetics this is a common alternative representation of a dendrogram). The membrane proteins play an important role in the cell's interaction with its environment. Small molecules outside the cell can bind reversibly to the receptor, causing chemical reactions inside the cell. The prediction of the properties of G-protein-coupled receptors can be facilitated if the phylogenetic tree is known.

Owing to the complexity of the data and the large number of variables (used to characterise each G-protein-coupled receptor), it was necessary to apply a nonhierarchical clustering method. As can be seen, this phylogeny shows nicely the grouping of the receptor types and the length of the genetic pathway (indicated by the length of the branches). For example, the H1 and H2 types are very old receptors and are quite similar to the common ancestor, as is indicated by the long branches joining near to the origin. In contrast, the Ok, Om and Od end-points correspond to

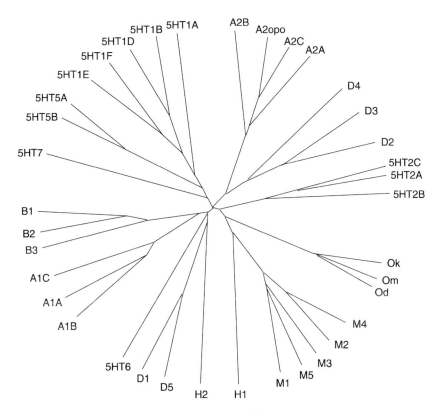

Figure 10.18 Star-like dendrogram representing the phylogenetic tree for the G-protein coupled receptor data set obtained with Forgy's clustering method.

relatively young proteins. The 5HT1 types (appearing in the upper left part of the dendrogram) show a more complex evolutionary history.

10.5 Classification

Up to this point we have considered exclusively unsupervised pattern recognition. In this section we shall deal with labelled data sets, that is data sets for which the class type or property of each object is known. This allows us to invoke more powerful methods that belong to the category of supervised pattern recognition techniques.

10.5.1 Linear discriminant analysis

Statistical linear discriminant analysis (LDA) is a widely applied classification technique. It is fast, provides some statistical information for assessing the significance of the results, and it is easy to use. The technique yields very good classification results if the relationship between the variable values of the objects and the corresponding class types is of a linear nature. LDA is a parametric classifier, meaning that it requires the clusters in the multivariate data space be characterised by a Normal distribution. Thus, each cluster is (in a statistical sense) fully described by a mean value (the centroid vector) and the standard deviation (spread of the variable values). In addition, the standard deviations of all the clusters should be more or less comparable for each dimension (variable).

The driving force behind LDA is the following mechanism. LDA searches for a line (usually referred to as the discriminant line or weight vector) onto which the data objects are projected in the multivariate space (see Figure 10.19). During this search process the within-cluster variation is minimised (yielding a narrow projection on the discriminant line for the clusters) and simultaneously, the between-cluster variation is maximised (which ensures a maximum distance between the projected clusters). In this way, an optimal classification of the individual clusters is achieved. In mathematical terms, LDA seeks for the weight values, w_i, of the linear discriminant function D of the variables x_i,

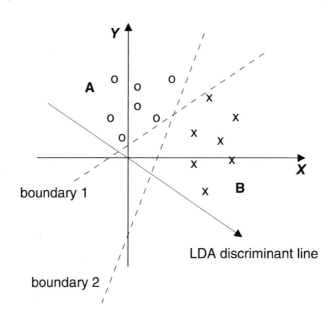

Figure 10.19 Decision lines obtained with LDA (solid line) and a MLF neural network (dashed lines).

$$D(\mathbf{x}, \mathbf{w}) = \sum_{i=1}^{M} x_i w_i$$

This equation is precisely the inner product of the discriminant line **w** and the variable vector **x** of an object, thus,

$$D(\mathbf{x}, \mathbf{w}) = \mathbf{x}' \cdot \mathbf{w}$$

where the prime denotes the transpose of the vector **x**.

In the case where just two clusters have to be classified, there is exactly one discriminant function required. If there are K classes expected in the data set, then up to $K-1$ discriminant functions are needed. This is because one discriminant line might be sufficient for the adequate classification of, say, three clusters.

An example of a simple discriminant function is illustrated in Figure 10.20. The objects of class A are projected onto the left side of the discriminant line, whereas the objects of class B are all projected onto the right side. In other words, if the projection of an object is on the right side of the discriminant line (or equivalently, if the function D has a positive value) then the object belongs to class B. A negative value of D indicates membership of class A. In practice, when more clusters are present, LDA combines several of these discriminant functions in order to determine the particular class membership of an unknown object.

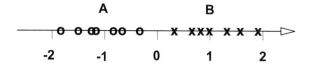

Figure 10.20 Projection of the objects shown in Figure 10.19 onto the discriminant line **w**.

It should be emphasised here, that LDA performs optimally if the variances of the clusters are similar for all dimensions (variables). Figure 10.21 shows two situations in which LDA certainly should not be applied. The left side of this figure shows two clusters having totally different spreads (dispersion) of objects. On the right side, two clusters are seen that possess different orientations. LDA fails to find the right discriminant lines in these situations because of either dispersion or direction in the data. In such cases, a local modelling technique (e.g. soft independent modelling of class analogy (SIMCA)) that takes into account the different cluster dispersions and directions has to be invoked to obtain a proper supervised classification model. Another approach would be to use an artificial neural network (ANN) that bases its decision lines on the

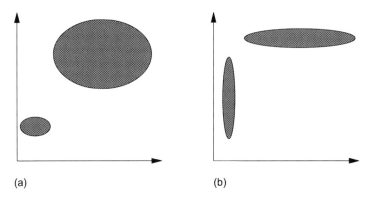

Figure 10.21 Two situations in which LDA should not be applied: (a) clusters with different variances; (b) clusters with different orientation.

class boundary instead of the class mean and dispersion. Both SIMCA and ANN will be discussed in the later sections.

Now let us consider an LDA-based real-world application that was developed during an EC project called AutoSort (De Groot *et al.* (1998)). The aim of this project was to design a reliable and robust classification system for the automated sorting of demolition waste on a large scale. Three main demolition fractions had to be separated by the waste sorting system: treated and untreated wood, plastics and stone. To achieve this, a number of representative demolition waste samples were collected and subjected to infrared reflectance analysis by means of a fast InGaAs diode-array spectrometer.

The spectra were arranged in a data matrix, **X**, with rows as objects and columns as wavelengths. For each object the material type (wood, plastic or stone) was known and stored in a vector, **Y**, that has the same number of rows as the **X** matrix. The input matrix **X** and the class membership vector **Y** were subsequently subjected to a LDA analysis. Five representative spectra of plastics (PE, PP, PS, PET, and PVC) are shown in Figure 10.22.

Application of LDA to the set of measured NIR spectra and corresponding material class types yields a clustering (and hence, a classification) in the space spanned by the two linear discriminant lines as depicted in Figure 10.23. One can compare this linear discriminant space to the space that is spanned by the two first principal component axes in a PCA. Three clusters (○, wood, ×, plastic, +, stone) are seen.

First consider the horizontal linear discriminant axis (the x-axis). Clearly, the projections of wood samples yield discriminant values below -2.0, whereas the other two fractions are projected on the right side of this value. Now the second discriminant line (the y-axis) becomes important. One can see that a number of objects (which are projected on

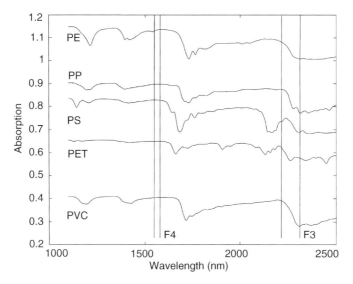

Figure 10.22 Mean NIR spectra of five plastic categories in the demolition waste data set.

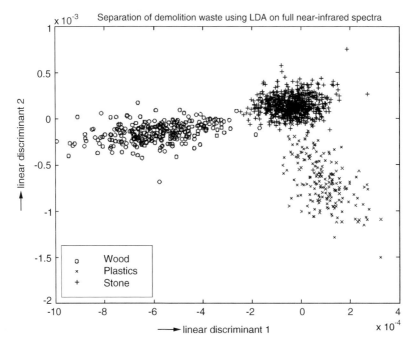

Figure 10.23 Projection of the wood, plastic and stone objects onto the first two linear discriminant lines.

the right side of the first discriminant line) have a projection value on the second line of -0.2 or higher. These objects belong to the stone category. The remaining objects are the plastic waste samples. This example illustrates in a convincing way that LDA can generate suitable functions from the infrared spectra of a waste samples to obtain a proper classification of samples.

Even better classifications can be obtained for some chemical problems if LDA is combined with a distance measurement criterion. The Euclidean and Mahalanobis distances are two commonly applied similarity measures.

Let us return to the example of the multivariate medical imaging problem. Again, the data matrix of PET/PRT combinations and pixels is used to form the data matrix **X**. A number of pixels was selected where the tissue type (fat, skull bone, white matter, grey matter and cerebrospinal fluid) was known. This knowledge was provided by an expert in the field of radioneurology and summarised in the **Y** vector for building the classification model. Figure 10.24 shows the result of LDA for these selected pixels. The left-hand plot shows the projection of the multivariate data points on the first two linear discriminant lines. Compared with the score plot obtained with the PCA analysis, a more fine-grained clustering is achieved. The tiny cluster on the left side corresponds to the background pixels, whereas the clusters with centroids at $(500, -100)$ and $(1500, -400)$ belong to fatty tissue and skull bone, respectively. The other elongated cluster contains pixels belonging to white matter, grey matter and cerebrospinal fluid.

Now the class type of all of the pixels not used in the construction of the LDA model is predicted using the LDA classification model, ignoring any classification information known about these pixels (in this case the LDA model used a Mahalanobis distance criterion). The result is shown in the right-hand plot of Figure 10.24. It is expected, of course, that the pixels should fall into the clusters obtained in the left hand plot. Apart from the rather noisy picture (notice for instance the edge of the skull), the classification is quite poor. The LDA model misclassified too many pixels in the multivariate image. Apart from other reasons, this suggests that the relationship between the multivariate pixel values and the tissue type could be nonlinear. To cope with such nonlinear relationships we need more powerful classification techniques such as artificial neural networks.

This example demonstrates very clearly the need to test models constructed with small data sets with new data.

10.5.2 Soft independent modelling of class analogy

Before we discuss a nonlinear classification technique, we first introduce another linear pattern recognition method that might circumvent certain

CLUSTER ANALYSIS AND PATTERN RECOGNITION 397

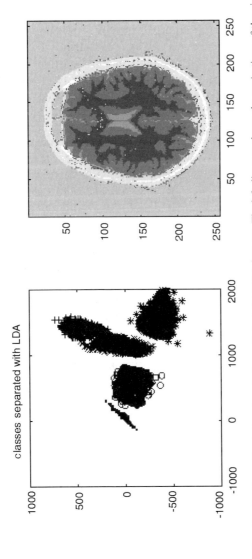

Figure 10.24 Results of a linear discriminant analysis applied on the MRI data set. The left diagram shows the projections of the pixels onto the first two discriminant lines. The right diagram depicts the resulting segmentation of the various brain tissues.

problems that can be encountered with linear discriminant analysis. As has been outlined, LDA searches for the optimal set of discriminant lines, based on the distribution of the whole data set in the multivariate space. Soft independent modelling of class analogy (SIMCA), in contrast, focuses locally (in the data space) on clustering properties of predefined subsets of the entire data set.

In a nutshell, the principle underlying SIMCA is as follows. First, a 'local' PCA is conducted on each cluster of objects (of which the class type is known). Based on the dimensionality and the variation within each cluster, one or more principal components are used to build a local description (in fact, a local regression model) of these clusters. The appropriate number of principal components can be determined from the corresponding scree plots.

Thus, SIMCA allows so-called *class-boxes* to be built that might have different dimensions for different classes. Consider, for example, the three-dimensional plot given in Figure 10.25. Two clusters (A and B) and one outlying data point (C) can be seen. Cluster A shows a large variation in both variables, whereas cluster B clearly exhibits a strong orientation in the direction of the y-axis. Based on pre-knowledge concerning the class type of the objects, in SIMCA a local PCA is conducted for both clusters. As one expects, cluster A is characterised by two significant principal components, whereas for cluster B precisely one PC is sufficient to capture the variance in the data points.

Boundary planes that enclose the class-box are defined according to the explained variance in each principal component. Following this, a decision model is built that determines in a statistical way the class membership of an unknown object. Formally, this statistical decision is based on a local regression model, which is built in the space spanned by the significant principal components of the class. For each class type, K, such a regression model is calculated and characterised by a mean and a variance (S_0^K). Then the distance of a new object is measured to the class-box boundaries is calculated (S_k^K).

Taking into account the number of degrees of freedom in the respective regression models, one can obtain a critical F-value for each class-box, for a given level of statistical significance (in this example a 5% level is used, indicated by $F^{0.05}$). In SIMCA, an unknown object belongs to class K if the inequality

$$\left(S_k^K\right)^2 < \left(S_0^K\right)^2 \cdot F^{0.05}$$

is satisfied.

Consider again the demolition-sorting example. The set of NIR spectra was subjected to a SIMCA analysis. Thus, for each demolition waste fraction, a local PCA was conducted first. By inspecting the resulting

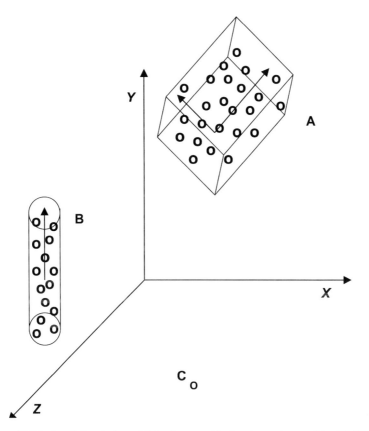

Figure 10.25 Local description of data clusters with class-boxes (see text) by SIMCA.

scree plots for the classes, it appeared that the variation in the wood class was modelled with three principal components. The plastic class required four significant PCs, while the variation in the stone fraction was modelled with two PCs. In this particular example we took a statistical significance level of 1%. Table 10.5 presents the performance of the SIMCA model for the whole set of demolition waste NIR spectra.

Table 10.5 Classification results of the SIMCA model for the three demolition waste fractions. Reading from left to right, 97% of the wood fraction is classified is wood and 3% of wood is classified as plastic. The bracketed numbers are from a LDA model

Class type	Percentage correct classification of the SIMCA model		
	Wood	Plastic	Stone
Wood	97 (94)	3 (5)	0 (1)
Plastic	2 (5)	96 (91)	2 (4)
Stone	3 (3)	3 (5)	94 (92)

As can be observed, the SIMCA model performs quite well. Moreover, for each of the waste fractions, the percentage of correct classification is better than the LDA performed on the same data set. Obviously, this improvement is due to the local, class-wise, modelling ability of the SIMCA classification technique.

10.5.3 Multilayer feed-forward neural networks

Since the early 1990s, artificial neural networks (ANNs) have been applied for solving complex problems in the field of chemistry. Owing to their adaptability to the measured data set and the capability of modelling nonlinear relationships, ANNs are very promising and powerful modelling and classification techniques. A large number of scientific studies in the last decade have shown neural networks outperforming many other function approximation and pattern recognition techniques, particularly where there may large uncontrolled variations in the data or there are nonlinear responses to the classification factors.

What are neural networks? We will restrict ourselves here to the so-called multilayer feed-forward (MLF) neural networks because these generic networks are most widely used in chemometrics. Basically, a multilayer feed-forward neural network consists of three layers of processing units (neurons).

The neurons in the first layer (usually referred to as input neurons) serve as a kind of flow-through unit (see Figure 10.26). There is usually one input neuron for each variable in the multivariate objects. They clamp the variable values and distribute them via the connections to all of the neurons in the core of the neural network. Thus, each neuron in the second (or hidden) layer receives a linearly weighted sum of all of the variable values of the presented object. The task of the hidden units is to make a nonlinear transformation of this weighted sum. In general, a sigmoidal function is chosen, but any function can be used. Then, each neuron in the third (output) layer collects a weighted sum of the output of all the hidden units. In mathematical terms, we can express the output of a neural network, here for unit m in the output layer, by

$$o_m = \psi \left[\sum_{j=1}^{L} \nu_{jm} \sigma \left(\sum_{i=1}^{M} x_i w_{ij} + w_{0j} \right) + \nu_{0m} \right]$$

where w_{ij} denotes the weight between unit i in the first layer and unit j in the second layer, and ν_{jm} is the weight between hidden unit j and output neuron m. The symbol $\sigma()$ refers to the nonlinear sigmoidal transfer function that is applied on the weighted input values in the hidden layer,

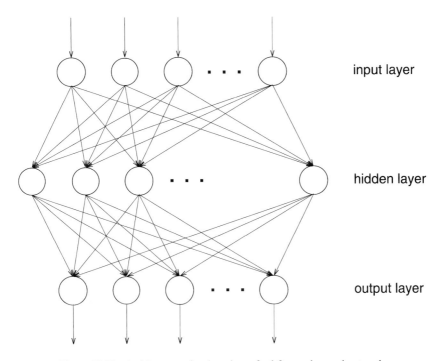

Figure 10.26 Architecture of a three-layer feed-forward neural network.

whereas $\psi(\ )$ is the transfer function of the output unit; this can be a linear function or a sigmoidal one. The terms w_{0j} and v_{0m} control the position of the transfer functions of the hidden and output neurons and are referred to as 'bias' or 'threshold'.

Summarising, a neural network first performs linear weighting of the input data, then it makes a nonlinear transformation in the hidden layer, and finally it combines the outputs of the hidden unit to form the input to the units in the last layer of the network. Depending on the task the network has to do, the output neurons are equipped with linear (quantitative function approximation problems) or sigmoidal (qualitative classification problems) transfer functions. Here, we shall consider only artificial neural networks suited for classification tasks, thus having output neurons equipped with a sigmoidal transfer function.

As with linear discriminant analysis, the weights of the connections between the units have to be determined according to a certain criterion. Instead of focusing on the variances within and between clusters of data points (as in LDA), a neural network adjusts its weights for object, by minimising the error between the predicted class (the output of MLF network, o_j) and the actual (known) class type of the object (d_j). The network output error is defined as

$$E = \frac{1}{2}\sum_j (d_j - o_j)^2$$

Based on this error function, one can show that the weights have to be adjusted in the following way,

$$\Delta w_{ji} = \alpha \delta_j o_i$$

where the error correction term δ_j depends on the type of layer. The weight of connection between i and j is adjusted by an amount of Δw_{ij}. For weights between units in the input and hidden layer δ_j is equal to

$$\delta_j = \sigma'(\text{net}_j) \sum_k \delta_k w_{kj}$$

The term net_j refers to the total input of unit j and $\sigma'()$ is the first derivative of the transfer function. Note that there is no direct way to define the output error of a hidden unit. Instead, this error is estimated by propagating the errors of the units in the output layer backwards via the weights between the hidden and output units. For this reason, such a neural network is usually referred to as a *back-propagation* neural network. The error correction term for the connections between the units in the hidden and output layer can simply be expressed as

$$\delta_j = \sigma'(\text{net}_j)(d_j - o_j)$$

Each object contained in the data set is presented repeatedly to the neural network together with its known class type. The weights between the neurons are gradually adjusted so that the network output error for the whole data set is minimised. Usually, the process of presenting examples to the neural network is called the training phase. The training of a neural network is stopped when, for instance, the total error (the accumulated error for all the objects in the data set) falls below a certain predefined threshold. Another criterion might be a maximum number of presentations (epochs) of all the objects in the training set to the artificial neural network.

Consider now again Figure 10.19. Apart from the discriminant line, obtained with the LDA, two dashed lines are drawn. These lines represent two possible solutions (decision lines) that can be obtained with a neural network. In fact, depending on the initial weights set for the network, a huge number of such solutions exist. As can be observed, the neural network focuses on the borders of the clusters, instead of on the centroids and variances. As a consequence, by including a sufficient number of hidden units, a neural network will even be able to classify a cluster surrounded entirely by another cluster possessing a different class type.

It should be emphasised here that the design of a neural network is not as straightforward as it might seem. Many parameters have to be determined, such as the learning rate and the number of units in the hidden layer. The latter parameter, especially, determines to a large extent the complexity of the model built by the neural network. If the number of hidden units is too small, the classification model will be too simple, resulting in unreliable performance of the network. Too many hidden units, on the other hand, usually result in an overtrained network (this situation can be compared to a polynomial fit of a too high order). In this case, the neural network will lose its generalisation capability, meaning that the network will not be able to predict the correct class type for an unknown object.

Another critical point of concern is the initial setting of the weight values. Hence, for one particular data set, many trial runs have to be conducted for various initial weight settings, learning rates and number of hidden units, respectively, in order to get a stable model.

To demonstrate what might be happening inside a neural network, we discuss the following classification problem. Here, objects are characterised by two variables (x and y). Objects belonging to the black K-like structure depicted at the right side of Figure 10.27 are encoded with a class type value of 1 (black), whereas objects outside the K are encoded with a 0 (white). For various combinations of variable values (covering the entire two-dimensional input space) we made a map of the outputs of the units in the hidden layer. As can be seen, each of the hidden units takes care of one of the borderlines of the K-shaped region (a dark area corresponds to an output equal to 1, whereas a white zone corresponds to 0). The output unit shown at the right side of the neural network combines these decision lines to establish the desired output pattern.

We shall now apply a three-layer feed-forward neural network-for the classification of the multivariate MRI images. Because many pitfalls can be encountered when one starts to use a neural network, we shall go into reasonable detail on how to construct a neural network-based classification model.

First of all, one has to assemble a database (preferably as large as possible) for training the neural network. In the MRI case we need to select (guided by an expert in the field of radioneurology) a sufficient number of pixels corresponding to the following tissue types:

- Background pixels including the skull bone
- Outer-skull tissue, squamous bone, muscles

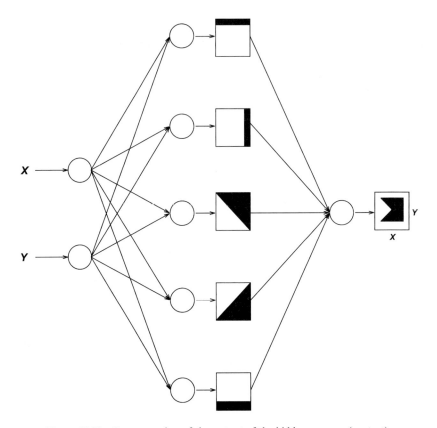

Figure 10.27 Representation of the output of the hidden neurons (see text).

- Fat
- Cerebrospinal fluid
- Grey brain matter
- White brain matter

The training data set contained approximately 3000 patterns (roughly 500 pixel vectors per type of tissue). For each MRI image, the single pixel intensity values were in the range 0 to 1024. Because neural networks cannot deal with such large input values, we first performed a range scaling transformation on the data set (see Appendix A), yielding for each variable a value in the range [0, 1].

Because of the number of variables (8), the network has eight input units. The tissue types were encoded as binary vectors of six elements; a value of 0 indicates that the particular tissue type is absent, whereas a value of 1 indicates its presence. For example, the vector (0, 0, 1, 0, 0, 0) identifies here the tissue 'fat' . Accordingly, the network had six output units equipped with sigmoidal transfer functions that ideally yield, for a well-trained neural network, output values close to 0 or 1.

The network was trained several times. As part of the study, the number of hidden units was varied between 3 and 9; five different initial weight settings were used (each time, the weight values were chosen randomly in the range $[-0.3, 0.3]$) and two learning rates were applied: 0.005 and 0.01, respectively. Thus, a total of 70 (that is, $7 \times 5 \times 2$) neural networks were trained to solve this particular brain segmentation problem.

It appears that the optimal number of hidden units is 7 with respect to the performance of the network. This neural network converged after some 4000 epochs. The number of misclassified pixels for the chosen training set was 1 (out of 3000 pixels!). Then, all of the pixels in the multivariate image that had not been used in training the network were submitted to the trained neural network to predict their tissue type. In Figure 10.28a the classification of the complete multivariate image of brain slice 7 is given. Compared with the LDA of this particular image, the ANN model yields a sharper, more detailed and, obviously, less noisy picture. For instance, one can observe that none of the background pixels is wrongly classified, as was the case with LDA.

Encouraged by these promising results, we applied the same ANN model to the six other brain slices recorded during the same multislice MRI measurement. Obviously, for the neighbouring slices 6 and 5, a still acceptable classification is achieved (see Figure 10.28b,c). However, when the fixed ANN model was applied to the multivariate images of slices 4 to 1, the classification performance deteriorated rapidly the farther away the slice was from slice 7. Outer-skull tissue, fat and white brain matter is still recognised well, but an increasing number of the cerebrospinal fluid and grey matter pixels are misclassified as being outer-skull material. An explanation for this might be that the average intensity of the individual MRI images varies as function of the slice level. To circumvent this problem, a more appropriate data transformation technique has to be applied to the data. Nevertheless, without a thorough optimisation of the data preprocessing procedure and the ANN parameters, the outcome of the brain segmentation by the three-layer feed-forward ANN is very good for at least half of the measured brain slices. For comparison, the LDA model seriously failed to predict the correct tissue types for all the other brain slices.

10.5.4 Validation of the classification model

So far, we have looked at the data set from the point of view of creating a model. In this context it is usually referred to as a training set. However,

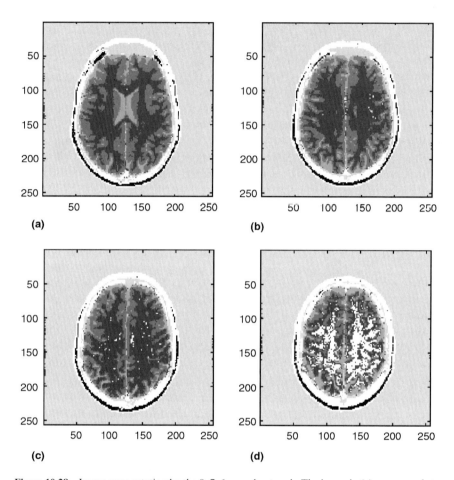

Figure 10.28 Image segmentation by the 8–7–6 neural network. The image in (a) corresponds to the segmentation of the multivariate image obtained for brain slice 7. The same ANN model was used for the segmentation of the multivariate images of slices 6 (b), 5 (c) and 3 (d).

when a linear discriminant analysis or a MLF neural network classification model has been built, it is necessary to test and validate it before it is used for practical applications.

The test of a classification model is based on a subset of the data set that has not been used to construct the classification model. This so-called test set must contain enough objects to provide a real test, and it must be representative of the application in which it will be used. In the waste-sorting problem, for example, the three waste fractions should be sufficiently covered by the objects in the test set. In fact, the same holds for the training set—if it does not contain a particular type of object, the model will not contain any information about it and, consequently, test samples will fail. Here, we assume explicitly that there are enough objects available to construct representative training and test sets. If this is not the case (for example in biochemical QSAR studies where only a few objects will in general be available) one has to rely to other test and validation strategies, like the leave-one-out method or other cross-validation procedures. These topics are discussed in depth in, for example, Massart *et al.* (1988).

With LDA and SIMCA, a classification model is constructed based on the training set. The performance of the model is then examined by means of the test set. This second step provides the researcher with an indication of how well the model is able to generalise. That is, the classifier will be able to predict reliably the class type of new (unknown) objects.

For neural networks, the test set can fulfil two functions. Apart from determining the generalisation capability of the network on new unseen data, the test set can be used to determine whether the network is sufficiently well trained and stop the training process. Although many variations on this theme are reported, they are all based on the following principle. By monitoring the behaviour of the output error of the network during the training phase, separately for the training and test sets, the minimum in the test set curve determines the point where the training of the network should be stopped (see Figure 10.29). However, it is strongly

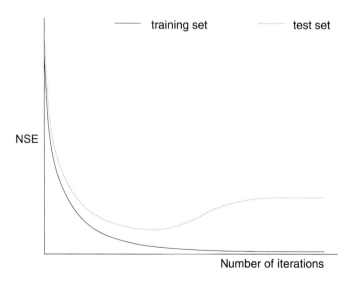

Figure 10.29 Output error (NSE) of a MLF neural network for both the training set (solid line) and test set (dotted line) as function of the iteration number.

advised that the test set be subdivided in two sets: one set that will be used to stop the training of the network, and the other for the actual determination of the performance of the neural network once sufficient training has been determined.

If the model is to be used under similar laboratory conditions to those of the model building and testing then, in general, these procedures for constructing a classification model will be sufficient. However, if the laboratory model has to be transferred to, say, an industrial plant, then it is mandatory to perform a second test (validation) of the model using samples taken from the target plant. The reason is simply that samples from the plant are exposed to different factors and environment from those created in a laboratory.

10.6 Appendix A. Transformation and scaling of the data

For the sake of simplicity, this appendix considers only scaling and transformations of the raw data that serve as input for a visualisation, clustering or classification technique. However, the same transformations can be applied to data destined for other types of calculation and to the output variables used in quantitative modelling.

As a starting point we take a number, N, of measured objects (say, spectra). Each spectrum consists of absorption values at M wavelengths and forms a row in the input matrix, $\mathbf{X}$. Thus, $\mathbf{X}$ is an $N \times M$ matrix containing N rows and M columns. An element of $\mathbf{X}$ is denoted by x_{ij}, where i is the row number and j is the column index number. Thus, in our example, x_{78}, represents the absorption value at wavelength 8 for spectrum 7.

In the following sections, this matrix will be subjected to various data transformation techniques. Note that data transformations are performed on the data by columns of the matrix $\mathbf{X}$ and not by rows. This implies that scaling or transformation is performed on the variables and not the objects. This kind of data scaling is frequently applied by chemometricians. Row-wise transformations (thus, scaling or transformation by object) are also possible but will not be considered here.

10.6.1 No transformation or scaling

This may seem to be rather superficial, but in some cases it might be sufficient to use the raw data, just as they were measured during an experiment. Unless there are clear reasons to perform a transformation immediately (such as the logarithmic scaling of biological activity values), visualisation should use the raw data in the first instance.

10.6.2 Range scaling

The aim of range scaling is to squash the range of values in a column of measured data between a minimum and maximum value. This can be expressed as

$$x_{ij}^* = \frac{x_{ij} - a_j}{b_j - a_j}$$

a_j and b_j become the lowest and highest values, respectively, that the transformed data could take. In this way the values in column j of the transformed matrix vary between $(\min_j - a_j)/(b_j - a_j)$ and $(\max_j - a_j)/(b_j - a_j)$, where $\min_j$ and $\max_j$ represent the lowest and highest values, respectively, of column j in the matrix $\mathbf{X}$.

A widely used choice for the scaling numbers a_j and b_j are the values $\min_j$ and $\max_j$ themselves. In this case, the transformed matrix contains exclusively values that are scaled between 0 and 1. Often such a transformation, usually referred to as *unit length scaling*, is performed if the data have to be presented to, for instance, artificial neural networks.

10.6.3 Mean centring

Another way to scale the measured data is by centring the mean of each column of the input matrix $\mathbf{X}$ to zero. The mean, μ_j, of a column is defined by

$$\mu_j = \frac{1}{N} \sum_{i=1}^{N} x_{ij}$$

and the mean centred matrix elements of $\mathbf{X}$ are calculated as

$$x_{ij}^* = x_{ij} - \mu_j$$

Such a type of transformation places the data for each variable around the origin or zero value in the multivariate space without influencing the original range or variation of each variable.

10.6.4 Autoscaling

An extension of mean centring is autoscaling. First, one has to calculate the standard deviation of the mean-centred data for each column. Then the mean-centred values are divided by the corresponding standard deviations.

In this way, each scaled variable covers approximately a value range between −3 and +3 (assuming a more or less Normal distribution of the values of each variable). As a consequence, after autoscaling, each variable will have the same weight when used as input to a visualisation, clustering or classification technique. Mean centring and autoscaling are usually applied if the data are subjected to a principal component analysis, where variables have different magnitudes of values and where variables of different types (e.g. temperature and pH) are used together. However, it is not usual to autoscale spectral data even though some absorbance values may be two orders of magnitude smaller than others.

10.6.5 Other transformations

Various other transformations and scaling options can be applied to the raw data. As mentioned in the Introduction, a logarithmic scaling of the data is preferable if one deals with biological activity values, because such activity values usually cover a very large range. The logarithmic transformation is often used with data where the underlying distribution of data values is non-Normal but the data processing technique has an assumption of Normality. Other transformations, such as taking the square root of values, may be appropriate for other types of distribution.

Logarithmic scaling means that small variations in small values will have a similar effect or impact to those of large variations in large values. In contrast to this, if one wants to focus just on large variations, taking the square, or a higher power, of the data values might provide a suitable data transformation.

10.6.6 Example

The effect of three types of data transformation (that is, no transformation at all, range scaling between 0 and 1, and autoscaling) will be illustrated with an example. In this example, the effect of modifications of the mobile phase composition on the behaviour of compounds on one particular stationary phase is investigated by means of PCA. The compounds chosen are the naturally occurring benzylideneacetophenone pigments,

$$R^1 - C_6H_5 - CH = CH - CO - C_6H_5 - R^2$$

These compounds are commonly known as chalcones. The chalcones occur in (E)-s-*cis* and (Z)-s-*cis* molecular conformations as described by Walczack and co-authors. These authors examined the relationship between the capacity factors, k' of the (E)-s-*cis* and (Z)-s-*cis* conformations and eight chromatographic systems with modified mobile phases (CH$_2$Cl$_2$, THF, dioxane, ethanol, propanol, octanol, DMSO, and DMF). The capacity factor is a measure of the ratio between the residence times of the chalcones in the mobile phase and in the stationary phase. It is influenced by, for example, the polarity of the solvent, the polarity of the chalcone and molecular size.

Principal component analysis offers a convenient means: first to reduce the dimensionality of the rather large data set and, second, to visualise the underlying relationship between the measured objects. In this example we shall consider only 23 (*E*)-s-*cis* molecular conformations taken from the whole data set.

Figure 10.30 shows the results of a principal component analysis when the raw data are used. About 95% of the variance in the data is captured by the first two principal components (Table 10.6). The loading plot shows that six out of eight modified mobile phases appear to lie on a straight line, whereas the other two (DMF and DMSO) are more or less located on a line orthogonal to it.

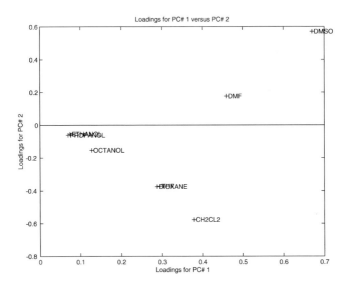

Figure 10.30 Loading plot of the chalcone data set. No data transformation is applied.

When range scaling (between 0 and 1) is applied to the data (Figure 10.31), the situation looks quite different. Roughly, there appear two clusters of mobile phases, that is (CH$_2$Cl$_2$, THF, dioxane, octanol) and (ethanol, propanol, DMSO, DMF). When autoscaling is applied to the

Table 10.6 Amount of variance explained by the first two PCs using different scaling of the chalcone data

Preprocessing	PC1(%)	PC2(%)	PC1 + PC2(%)
Raw data	92	3	95[a]
Mean centring	85	14	99
Autoscaling	84	15	99

[a] There is still 5% variance not accounted for by PC1 and PC2 due to the orientation of the first PC towards the mean of the data.

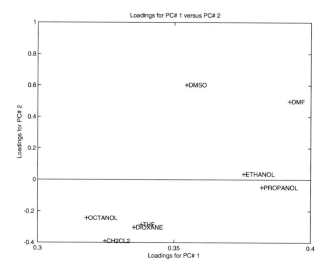

Figure 10.31 Loading plot for the chalcone data set after range scaling of the data.

raw data, the loading plot changes drastically (Figure 10.32). Now, six mobile phases appear as relatively compact clusters, whereas DMF and DMSO appear as two isolated points in this plot. And indeed, this could be expected from chemical and chromatographic considerations.

It should be remarked here, that the first two PCs of the scaled data explain approximately 99% of the variation in the data (Table 10.6).

Autoscaling (or alternatively, mean centring of the data) is widely applied as data preprocessing technique before a PCA is conducted. If no autoscaling or mean centring is applied to the data, the first PC often gets lost because it simply represents the mean of all data objects without providing any useful information about the finer structures in the data. This is shown in Table 10.6 by the considerably increased contribution of PC2 with the scaled data compared with the raw data.

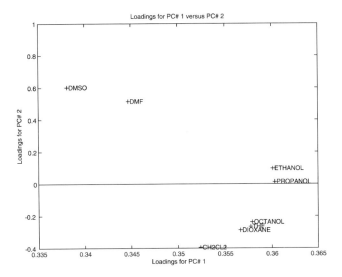

Figure 10.32 Loading plot for the chalcone data set after autoscaling of the data.

As this simple example demonstrates, the way the data are preprocessed (transformed or scaled) will substantially influence the outcome or performance of any particular clustering or classification (or function approximation) technique. Hence, it is strongly recommended that for a small range of data preprocessing techniques (including at least the raw data itself and autoscaling) is used to examine the performance of the chosen clustering or classification method.

10.7 Appendix B. Measures of (dis)similarity

To be able to group or cluster a number of objects, one first has to define a similarity measure. One good candidate will be 'distance'; another is 'correlation'. However, many mathematical definitions of distance exist. This section summarises the most common ones.

10.7.1 Minkowski distance

The Russian mathematician Minkowski defined the following generic equation for distance between continuous variables:

$$D(\mathbf{x}, \mathbf{y}) = k\sqrt{\sum_{i=1}^{M} |x_i - y_i|^k}$$

where x_i and y_i denote the value of variable i of the objects **x** and **y**, respectively. The integer number M is the total number of variables. Here, the parameter k plays a crucial role. If k is taken to be equal to 1, we obtain the Manhattan or city-block distance (see Figure 10.33). This measure is used when ordinal variables (ordinal means values expressed as ordered integer numbers) are considered. Thus, the Manhattan distance between the two dots in Figure 10.33 is equal to $A + B$.

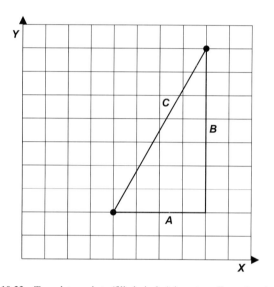

Figure 10.33 Two data points (filled circles) in a two-dimensional space.

When $k = 2$ one obtains the well-known Euclidean distance. The Euclidean distance measure is applied to continuous-valued variables (see Figure 10.33). In this simple example the Euclidean distance is equal to

$$C = \sqrt{A^2 + B^2}$$

An alternative mathematical expression for the Euclidean distance is given by the vector inner product,

$$D(\mathbf{x}, \mathbf{y}) = \sqrt{\mathbf{x} \cdot \mathbf{y}}$$

Higher integer values of k are seldom applied in chemometrics. However, in some cases it might be worthwhile to leave the paved way and apply higher-order distance measures. For example, a higher-order distance measure is recommended if the focus is on a large difference between two objects in a single variable, instead of rather small variations in all the variables (in other words, if one is interested in large local differences as opposed to small global differences).

10.7.2 Mahalanobis distance

A more sophisticated distance measure is given by the Mahalanobis distance. The Mahalanobis distance is defined by

$$D(\mathbf{x}, \mathbf{y}) = (\mathbf{x} - \mathbf{c}_0)' \mathbf{C}^{-1} (\mathbf{x} - \mathbf{c}_0)$$

It should be stressed here that the Mahalanobis distance is a measure of distance between a group of objects and a single object, $\mathbf{x}$.

The correlation structure of the group of objects forming the cluster is characterised by the variance–covariance matrix $\mathbf{C}$, and the centre of the cluster, $\mathbf{c}_0$, usually referred to as the cluster centroid. Note that, unlike the Euclidean distance, the distance is weighted by the inverse of the variance–covariance matrix $\mathbf{C}$. The Mahalanobis distance measure is very useful if the variances in the variables differ substantially. Consider, for example, Figure 10.34. Clearly, two distinct clusters are present. However, the spread in both directions is entirely different. When using the Euclidean distance, data point X would be closer to the centroid of cluster A than to the centroid of B. Intuitively, however, this data point matches cluster B much better. Indeed, if the Mahalanobis distance is applied, the point X will be assigned to cluster B, because this distance measure takes into account the different variances and covariances in both clusters. Putting this in other words, the actual distance is corrected for the correlation structure in the individual clusters.

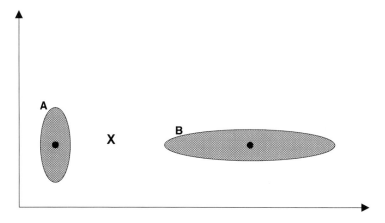

Figure 10.34 Two clusters exhibiting a different variability in both variables.

10.7.3 Correlation coefficient

Another way of defining similarity between two objects, **x** and **y**, is given by the correlation coefficient, $r(\mathbf{x}, \mathbf{y})$:

$$r(\mathbf{x}, \mathbf{y}) = \frac{\mathbf{x} \cdot \mathbf{y}}{\sqrt{\sum_{i=1}^{M} x_i^2 \sum_{i=1}^{M} y_i^2}}$$

Here, **x · y** again refers to the vector inner product, whereas M denotes the number of variables per object. The correlation coefficient is equal to the cosine of the angle between vectors **x** and **y**.

10.7.4 Which to use?

This section examines when to use which kind of similarity measure.

Consider the following example (Figure 10.35a). As can be seen, point X is much closer to cluster B than to A. When using the Minkowski or Mahalanobis distance, the centroid of cluster B is much closer to X than the centroid of cluster A (centroids are indicated by the two black dots). If one applies the correlation coefficient as the similarity measure, in this particular case $r(X,A)$ would be approximately equal to $r(X,B)$, because the angle between X and the centroid of A is identical to the angle between X and the centroid of B. Obviously, in this example the correlation coefficient should be rejected as a proper simularity measure.

In Figure 10.35b, the structure in the data is somewhat different. There are two elongated clusters present. When a distance measure is applied, object X would be closer to cluster A than to B. However, taking into account the structure (correlation) of the data, it is clear that point X belongs to cluster B. When the correlation coefficient is used as similarity measure, this object will be assigned to the proper cluster.

These two examples demonstrate that an appropriate choice for a distance measure must be based on the underlying structure or correlation in the data. This choice is not always straightforward but can be guided, for instance, by pre-knowledge of the problem at hand, by visual inspection of the multivariate data or by other (chemical) considerations.

As a final remark, the scaling or transformation that is used to pre-process the raw data will influence any distance-based similarity measure. The correlation coefficient is invariant to linear transformations or scaling of the data. However, nonlinear transformations of the data (like the logarithmic scaling of variable values) will influence the correlation coefficient as well.

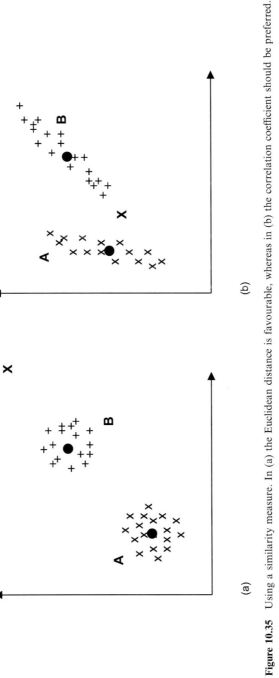

Figure 10.35 Using a similarity measure. In (a) the Euclidean distance is favourable, whereas in (b) the correlation coefficient should be preferred.

Bibliography

Principal component analysis

Dunteman, G.H. (1989) *Principal Components Analysis*, Sage Publications, Newbury Park, CA.
Cattell, R.B. (1966) The Scree test for the number of factors. *Multivariate Behavioral Research*, **1** 245.
de Jong, S. and Kiers, H.A.L. (1992) Principal covariates regression. *Chemometrics and Intelligent Laboratory Systems*, **14** 155.

Nonlinear mapping

Sammon J.W. Jr. (1969) A nonlinear mapping for data structure analysis. *IEEE Transactions on Computing*, **C-18** (5) 401.
Lerner, B., Guterman, H., Aladjem, M., Dinstein, I. and Romem, Y. (1998) On pattern classification with Sammon's nonlinear mapping. An experimental study. *Pattern Recognition*, **31** (4) 371.
Takahashi, Y., Miyashita, Y., Abe, H., Sasaki, S., Yotsui, Y. and Sano, M. (1980) A structure–biological activity study based on cluster analysis and the nonlinear mapping method of pattern recognition. *Analytica Chimica Acta*, **122** 241.

Kohonen self-organising feature map neural network

Kohonen, T. (1984) *Self-organization and associative memory*. Springer Verlag, New York.
Beckers, M.L.M., Melssen, W.J. and Buydens, L.M.C. (1997) A self-organizing feature map for clustering nucleic acids. Application to a data matrix containing A-DNA and B-DNA nucleotides. *Computers in Chemistry*, **21** (6) 377.
Melssen, W.J., Smits, J.R.M., Rolf, G.H. and Kateman, G. (1993) Two-dimensional mapping of IR spectra using a parallel implemented self-organising feature map. *Chemometrics and Intelligent Laboratory Systems*, **18** 195.

Parallel coordinates

Brodlie, K., Carpenter, L., Earnshaw, R. *et al.* (1992) *Scientific Visualization Techniques and Applications*, Springer-Verlag, New York.
Inselburg, A. and Dimsdale, B. (1990) Parallel coordinates: a tool for visualizing multi-dimensional geometry. *Proceedings of the First IEEE Conference on Visualization*, p. 361.
Wegman, E.J. (1990) Hyperdimensional data analysis using parallel coordinates. *Journal of the American Statistical Association*, **411** (85) 664.

Cluster analysis

Ward, J.H. Jr. (1936) Hierarchical grouping to optimize an objective function. *Journal of the American Statistical Association*, **58** 236.
Forgy, E.W. (1965) Cluster analysis of multivariate data: efficiency versus interpretability of classifications. *Biometrics*, **21** 768.
Massart, D.L., Kaufman, L. and Esbensen, K.H. (1982) Hierarchical and nonhierarchical clustering strategy and the application to classification of iron meteorites according to their trace element patterns. *Analytical Chemistry*, **54** 911.

Reijmers, T.H., Wehrens, R., Daeyaert, F.D., Lewi, P.J. and Buydens, L.M.C. (1998) Using genetic algorithms for the construction of phylogenetic trees: Application to G-protein coupled receptors. *BioSystems*, in press.

Linear discriminant analysis

Lachenbruch, O.A. (1975) *Discriminant Analysis*, Haffner Press, New York.

Coomans, D., Broeckaert, I., Jonckheer, M., Blockx, P. and Massart, D.L. (1978) The application of linear discriminant analysis in the diagnoses of thyroid diseases. *Analytica Chimica Acta*, **103** 409.

van den Broek, W.H.A.M., Wienke, D. *et al.* (1997) Application of a spectroscopic infrared focal plane array sensor for on-line identification of plastic waste. *Applied Spectroscopy*, **51**(6) 856.

Soft independent modelling of class analogy

Kvalheim, O.M., Oygard, K. and Grahl-Nielsen, O. (1983) SIMCA multivariate data analysis of blue mussel components in environmental pollution studies. *Analytica Chimica Acta*, **150** 145.

Sjostrom, M. and Wold, S. (1980) SIMCA: A pattern recognition method based on principal components models, in *Pattern Recognition in Practice* (eds. E.S. Gelsema and L.N. Kanal), North-Holland, Amsterdam, p. 351.

Wold, S., Johansson, E., Jellum, E., Bjornson, I. and Nesbakken, R. (1981) Application of SIMCA multivariate analysis to the classification of gas chromatographic profiles of human brain tissues. *Analytica Chimica Acta Computer Techniques and Optimization*, **133** 251.

MLF neural networks

Zupan, J. and Gasteiger, J. (1991) Neural networks: a new method for solving chemical problems or just a passing phase? *Analytica Chimica Acta*, **248** 1.

Rumelhart, D.E. and McClelland, J.L. (1986) *Parallel Distributed Processing, Explorations in the Microstructure of Cognition*, Vols 1 and 2, MIT Press, London.

Smits, J.R.M., Melssen, W.J., Buydens, L.M.C. and Kateman, G. (1994) Using artificial neural networks for solving chemical problems. Part I. Multi-layer feed-forward networks. *Chemometrics and Intelligent Laboratory Systems*, **22** 165.

Melssen, W.J., Witjes, H., Postma, G.J. and Buydens, L.M.C. (1997) Using neural networks for the analysis and interpretation of multivariate medical images. *Proceedings of the International EUFIT Conference, 1997, Aachen, Germany*, Vol 1, p. 369.

Transformation and scaling of the data

Massart, D.L., Vandeginste, B.G.M., Buydens, L.M.C., de Jong, S., Lewi, P.J. and Smeyers-Verbeke, J. (1998) *Handbook of Chemometrics and Qualimetrics: Part B*, Elsevier, Amsterdam.

Stein, R. (1993) Preprocessing data for neural networks. *AI Expert*, 32.

Measures of (dis)similarity

Fishback, W. (1969) *Projective and Euclidean geometry*, Wiley, New York.
Mahalanobis, P.C. (1936) On the generalized distance in statistics. *Proceedings of the National Institute of Science* (*India*), **12** 49.

General

Anderson, T.W. (1984) *Introduction to Multivariate Statistical Analysis*, 2nd edn, Wiley, New York.
Massart, D.L., Vandeginste, B.G.M., Buydens, L.M.C., de Jong, S., Lewi, P.J. and Smeyers-Verbeke, J. (1997) *Handbook of Chemometrics and Qualimetrics: Part A*. Elsevier, Amsterdam.
Brereton, R.G. (1990) *Chemometrics, Applications of Mathematics and Statistics to Laboratory Systems*. Ellis Horwood, New York.
Varmuza, K. (1980) Pattern recognition in analytical chemistry. *Analytica Chimica Acta*, **122** 227.
Frank, I.E. and Todeschini, R. (1994) *The Data Analysis Handbook*, Elsevier, Amsterdam.
Aarts, E.H.L. and Korst, J.H.M. (1989) *Simulated Annealing and Boltzmann Machines. A Stochastic Approach to Combinatorial Optimization and Neural Computing*, Wiley, Chichester.

11 Linear regression
R. Tranter

11.1 Introduction

This chapter is about the basics of generating quantitative and predictive models linking causes of change to their responses. We have seen in other chapters the tools that can be used to describe variability in data, and how the variability can be broken down into components related to particular causes. Here, we are concerned with changes in a variable that are greater than the expected noise level of random variation and that occur in a predictable way in response to a change in another variable.

Chemical theories often produce equations linking two or more variables together. For example, reaction kinetics links the concentration of a product, or reactant, to time and/or the concentration of other re-agents; in spectroscopy the absorbance of a sample at a given wavelength is related to concentration, path length and molecular properties.

Such equations are functional or theoretical models of the system. They often contain parameters or constants whose values either are completely unknown or which may be prescribed from a theoretical basis. In both situations, the ability to estimate values of the parameters from experimental data provides a good check on the theory generating the equation.

On the other hand, there may be no existing theoretical model to generate an equation. We then want to generate an empirical function describing the relationship, usually to predict future values of responses from expected values of other variables. The empirical function is often chosen to be a simple linear relationship between response and factor(s), but it is not restricted to this and any arbitrary function that adequately describes the relationship may be chosen. Such situations frequently occur in calibrating an instrument's response to an analyte, or in exploring the relationship between, say, molecular polarisability and functional group size in a molecule.

This chapter describes the principles of linear and nonlinear regression modelling. Examples are given to illustrate several situations in which the technique can be applied and how the results of the modelling can be interpreted.

11.2 Where to go

Information needed	Go to Section(s)
Does it matter which way round x and y are?	11.4.3, 11.6, 11.7.12
What is the benefit of having replicate measurements?	11.7.8
How many measurements should I make?	11.8.1
I think the yield of the reaction depends on several factors. Do I have to make separate models for each?	11.8
Does there have to be a cause/effect relation before I can use regression?	11.7.10
The scatter plot of my data shows a curve. Can I use linear regression?	11.8.2
Should I leave out the intercept term because the theory says it is not needed?	11.7.13
How do I know if the fitted equation is a good one?	11.7.6, 11.7.7, 11.7.8
Why are there different lines when I use $x = ay + b$ instead of $y = ax + b$?	11.6
How reliable is a prediction from the equation?	11.7.9
The software I use gives tables of numbers as well as the slope and intercept. What do they mean?	11.7.7
How can I fit a complex equation to my data?	11.9

11.3 Some terminology

Before getting into the detail of the subject, it is worth noting some of terms that are used in linear regression.

Cause, predictor, factor, regressor, independent variable
These may be regarded as synonyms. They usually form the x variables in the regression equations. The use of *cause* and *independent variable* implies a specific property or relationship that may not actually exist, so their general use should be avoided. In this chapter I shall use *predictor* or *factor* unless I am describing an established cause–effect relation.

Response, effect, dependent variable
These may be regarded as synonyms. They usually form the y variables in the regression equations. The use of *effect* or *dependent variable* implies a

specific property or relationship that may not actually exist, so their general use should be avoided. In this chapter I shall use *response* unless I am describing an established cause–effect relation.

Model
A model is the equation relating response and predictors. It includes *parameters* whose values need to be adjusted by regression to give the best fit of the model to the data.

Linear regression
Linear regression is used to estimate the parameters in the equation that best fits (models) the relationship between the response and one or more predictors, and to test whether a significant relationship exists.

Linear least squares
Linear least squares is one of the computational methods used to carry out linear regression. It is based on minimising the sum of squares of the differences between the model equation and the data values. Other methods use, for example, median or absolute differences.

Correlation
Correlation is a measure of the degree of association between variables. It is measured on a scale of ± 1 with either of these values indicating perfect association and a value of zero indicating no association. The existence of correlation does not imply the existence of a cause–effect relation between the variables.

Notation
Throughout this chapter I use a common convention for distinguishing between different types of variables and parameters. The true values of parameters in models are denoted by Greek letters such as β_0 for an intercept and β_1 for a slope. The estimates of these parameters derived by calculation from experimental data are denoted by the corresponding Roman letters, e.g. b_0 and b_1 for the intercept and slope, respectively.

Individual values of variables are denoted by lower-case letters in italic font. So, x represents individual values of variable X in general while x_i refers to the ith X value in a sequence of values.

If a complete vector or matrix of values is referred to it will be in upper case and bold font. So, **X** represents a complete vector of X values.

Where an average has been calculated it will be denoted by a horizontal bar over the symbol, as in $\bar{y}$. If a value has been calculated from a regression equation and is thus a predicted value from the model, it will be denoted by a 'hat' over the symbol, e.g. $\hat{y}$.

11.4 Cause and effect

11.4.1 General

We are very used to looking at the real world and saying that a particular event was caused by another event; the sun shone so it became warm, the tap opened and water came out, the solution of lead chloride cooled and crystals formed. In these, the sun, the open tap and the cooling solution are identified by us as the causes and warmth, water and crystals are the effects. In every day conversation we are very good at making these relations from what is really quite sparse data. However, if we wish to understand the relationships and to quantify them, we must be much more precise about causes and effects. The three examples are actually complex combinations of multiple cause-and-effect relations and a good predictive understanding requires that these be disentangled.

A simpler example to use here is the dissolution of a metal oxide in a stirred reactor, as illustrated in Figure 11.1. There are many variables that can be measured:

- Rate of stirring
- Concentration of the dissolving agent (e.g. nitric acid)

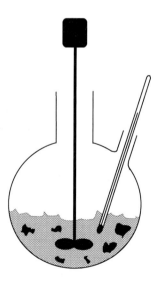

Figure 11.1 Process for studying the dissolution of a metal oxide in a stirred reactor.

- Compactness of the metal oxide (fine powder, lumps, sintered mass)
- Temperature of the dissolving agent
- The colour of the solution
- The amount of metal ion in solution.

These variables can be grouped into two classes. The first three fall into the class of causes and the last two into the class of effects. Why?

Consider the amount of metal ion in solution. Some simple experiments show that increasing the stirring rate, increasing the concentration of the dissolving acid and increasing the fineness of the metal oxide all increase the amount of metal ion in the solution at a given time interval from the start of the dissolution. However, if a metal ion solution is put into the reactor at the start of the experiment, the stirring rate, acid concentration and oxide fineness do not adjust themselves to match the ion concentration. Thus, ion concentration is an effect and the others are the causes, and similar arguments can be made about the colour of the solution.

The temperature of the dissolving acid is an interesting variable as it can be both a cause and an effect. Simple experiments again show that increasing the stirring speed, the acid concentration and the fineness of the oxide all increase the temperature of the acid; temperature is an effect. However, temperature of the acid can be controlled externally and deliberately changed. If the acid is cooled to $0°C$ or heated to $90°C$ then the amount of metal ion in solution, at a given time after the start of the dissolution, will change; temperature is a cause. How the temperature of the acid is treated in the data analysis then depends on the information required and on what is controlled in the measurements.

This particular example is quite simple in that the cause-and-effect relations are easily determined and, although each may involve some fairly complex relations in the underlying physics and chemistry, the observed relations at the macro level are firm.

This is not always the case, as seen in the classic example of the stork population in Hamelin many years go. Observations over many years recorded an increase in the number of storks nesting on chimneys in the town and in the number of babies born to the people of Hamelin (Figure 11.2). The obvious interpretation of these observations is that the number of storks is the cause of the number of babies born, in line with at least one theory of human reproduction!

Reality is, of course, quite different. Over the period of the observations people had moved to Hamelin from other districts. Many new houses with chimneys were built to accommodate them, so the local stork population found new nesting sites and rapidly occupied them. Human social dynamics being what it is, the number of babies born in the town each year also increased. Thus, the real situation is that both the nesting stork

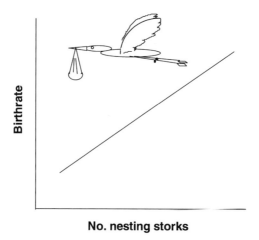

Figure 11.2 Correlation between birth rate and the number of nesting storks.

population and the birth rate are effects and that their cause is the increased human population in the town.

The significance of this example is that apparently 'obvious' cause–effect relations may be quite misleading and that the true relationships are very different. The true situation can be discovered only by questioning and by a careful consideration of all available information.

11.4.2 Relationships between variables

A cause and effect description implies that there is a functional relation between the variables. Thus, we might write the dependence of metal ion concentration on stirrer speed as

$$concentration = \mathbf{f}(stirrer\ speed)$$

The precise form of this relation is not yet known and its discovery is the purpose of the quantitative investigation. The usual assumption is that the function is simple, continuous and monotonic such as found with linear, quadratic or exponential functions. There may be theoretical justification for a function of this type as well. However, not all functions are continuous or monotonic and consideration must be given to these. It is quite possible, for example, that the ion concentration increases only above a critical value of stirrer speed and concentration hardly changes below this speed, as shown in Figure 11.3

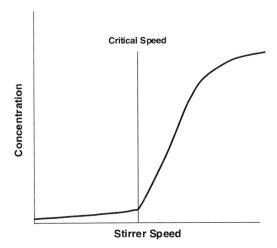

Figure 11.3 Relation between concentration and stirrer speed, showing nonlinearity and the effect of a critical stirrer speed.

Measurements above the critical speed will allow a model of the relation to be built, but it is quite clearly nonsense to use that model to extrapolate below the critical speed. Failure to detect the presence of the critical speed, either through a theoretical model or by experimentation, will lead to false understanding of the relationships between the variables and to the possibility of making dangerous predictions.

The oxide dissolution example illustrates the dependence of a variable on several causes. Thus,

$$concentration = \mathbf{f1}(stirrer\ speed)$$
$$concentration = \mathbf{f2}(acid\ concentration)$$
$$concentration = \mathbf{f3}(oxide\ compactness)$$
$$concentration = \mathbf{f4}(acid\ temperature)$$

Thus, *concentration* is a function of four factors and the factors act independently of each other. For example, no matter what acid concentration is used, the metal ion concentration will vary in precisely the same way with stirrer speed. Overall, *concentration* is then described by the sum of the individual functions:

$$concentration = \mathbf{f1}(stirrer\ speed) + \mathbf{f2}(acid\ concentration)$$
$$+ \mathbf{f3}(oxide\ compactness) + \mathbf{f4}(acid\ temperature)$$

Alternatively, there may be relations between the factors and there may be *interactions* between them. In this example, stirrer speed and oxide compactness may have an interaction that means that *concentration* increases more rapidly than would be expected by the independent model. The appropriate functional description is then

$$concentration = \mathbf{g}(stirrer\ speed,\ oxide\ compactness)$$
$$+ \mathbf{f2}(acid\ concentration) + \mathbf{f4}(acid\ temperature)$$

The function **g**( ) could have the form

$$b_0 + b_1 \times stirrer\ speed + b_2 \times oxide\ compactness$$
$$+ b_3 \times stirrer\ speed \times oxide\ compactness$$

where the last term in the function describes the interaction between *stirrer speed* and *oxide compactness*. Clearly, more complex functions are possible, involving quadratics, inverses, exponentials, etc., but the principle is no different.

The interaction of *stirrer speed* and *oxide compactness* described here is quite different from saying that the two are related. There is no fundamental function that describes stirrer speed in terms of oxide compactness, or vice versa. There is an accidental relationship between them that describes how one must vary with the other if a constant ion concentration is to be achieved, but it does not describe a cause–effect relation as either factor could be the response in the function and it would make no difference to the interpretation, or use, of the function. This type of accidental function is, of course, similar to the relation between birth rate and number of nesting storks.

11.5 Correlation, covariance, r and R^2

The simplest way to get an understanding of correlation is to look at some examples. Four small data sets are given in Table 11.1. The corresponding scatter plots for these data sets are shown in Figure 11.4. In data set A high values of x are always associated with high values of y, and low values of x are always associated with low values of y. This type of association is described as a high correlation. (Notice we could equally have said the high values of y are always associated with high values of x—x and y are still highly correlated). In data set B there seems to be no pattern of association between x and y values and in this case we should

Table 11.1 Four data sets to illustrate different forms of correlation

A		B		C		D	
13.66	14.64	10.63	10.62	9.89	7.60	13.89	7.91
12.53	12.93	8.32	9.27	10.06	12.92	10.09	11.79
12.33	11.60	9.62	8.80	9.81	12.73	8.27	14.58
12.79	13.50	8.33	13.34	13.43	14.65	12.50	8.75
10.17	10.15	8.13	8.99	12.08	12.94	8.28	13.11
9.89	9.95	11.88	13.81	13.35	15.03	13.57	7.78
13.74	13.25	8.05	14.00	11.79	8.96	9.06	13.88
10.55	10.10	9.06	11.02	9.58	8.98	12.43	10.11

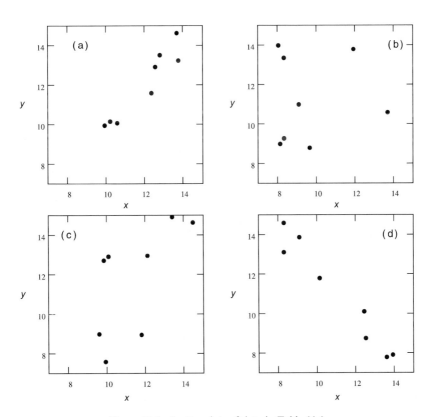

Figure 11.4 Scatter plots of data in Table 11.1.

say that there is no correlation between x and y. Data set C shows low correlation between x and y—high values of x are mostly associated with high values of y, but not always. Finally, in data set D we have the inverse situation to A: low values of x are associated with high values of y and vice versa. This is high negative correlation.

Correlation is measured by the *correlation coefficient*, which is usually given the symbol r and is defined as

$$r = \frac{\sum(y_i - \bar{y})(x_i - \bar{x})}{\sqrt{\left(\sum(y_i - \bar{y})^2\right)\left(\sum(x_i - \bar{x})^2\right)}}$$

where the summations are over all data pairs x_i, y_i and $\bar{x}$ and $\bar{y}$ are the average values of the x_i and y_i, respectively.

The denominator of this equation can be expressed as the product of the standard deviations of the x and y values:

$$s_x = \sqrt{\frac{\sum(x - \bar{x})^2}{n - 1}} \qquad s_y = \sqrt{\frac{\sum(y - \bar{y})^2}{n - 1}}$$

$$r = \frac{\sum(y_i - \bar{y})(x_i - \bar{x})}{(n - 1)s_x s_y}$$

The term $\left[\sum(y_i - \bar{y})(x_i - \bar{x})\right]/(n - 1)$ is called the covariance of x and y, or, cov_{xy}. Thus,

$$r = \frac{\text{cov}_{xy}}{s_x s_y}$$

The correlation coefficient is the covariance normalised by the standard deviations in the two variables. It is quite easily shown that if x and y are perfectly correlated then r will have a value of ± 1 depending on whether the correlation is positive or negative. The covariances and correlation coefficients for the data sets in Table 11.1 have been calculated and are given in Table 11.2.

Table 11.2 Covariance and correlation values for the data in Table 11.1

Data set	Covariance	Correlation
A	2.3218	0.9473
B	0.2896	0.0732
C	2.8604	0.6368
D	−5.4589	−0.9740

Both correlation coefficient and covariance are measures of the degree of association between the variables. Covariance is expressed in terms of the original units of the variables and so can take a wide range of values depending on the magnitudes of the variables. Correlation coefficient is a dimensionless number and it is always in the range -1 to $+1$. This makes it much easier to compare degrees of association between different pairs of variables.

Another correlation coefficient is R^2. This is a measure of the degree of association between measured values of a variable and the values predicted by the regression model. It is calculated in a similar way as r, but it involves only the responses and their values predicted by the model.

$$R^2 = \frac{\sum(\hat{y}_i - \bar{y})^2}{\sum(y_i - \bar{y})^2} = 1 - \frac{\sum(y_i - \hat{y}_i)^2}{\sum(y_i - \bar{y})^2}$$

R^2 tells us something about how well the model fits the data; it measures the amount of variability in the data that is accounted for by the model. Thus, a value of R^2 close to 1.0 indicates that the model accounts for nearly all of the variability seen in the y values—the model is a good description of the data. A value close to zero indicates that either the model is not a good description of the data or that the variation in data is essentially random. In the case where there is only a single predictor in the model, r^2 and R^2 are numerically equal.

Because R^2 is a measure of the association of the y values and the model, its value depends on the structure of the model. If the model is given greater flexibility by having many parameters that can be adjusted to fit it to the data, then a good fit will always be obtained; ultimately the model will fit to the noise in the data as well. Thus, increasing the number of parameters in the model automatically increases R^2. This means that R^2 presents an over optimistic view of the reliability of the model. To counter this, an adjustment may be made to R^2 using a factor that decreases as the number of parameters in the model increases. Thus,

$$R^2_{adjusted} = 1 - \frac{\sum(y_i - \hat{y}_i)^2}{\sum(y_i - \bar{y})^2} \cdot \frac{(n-1)}{(n-p)}$$

where n is the number of data points and p is the number of parameters in the model. Possible problems with the use of R^2 and $R^2_{adjusted}$ are discussed in Draper and Smith (1981) and Ryan (1997).

It is important to interpret a statistically significant correlation coefficient with care. It does not prove that changes in the predictor are the cause of changes in the response; there may be a third variable simultaneously influencing both. Similarly, it does not necessarily demonstrate that the relationship between the variables is a straight line, only that there is a significant slope if a straight line is fitted. Conversely, a nonsignificant or even zero correlation does not necessarily show that there is no association or dependence between variables, only that there is no significant slope over the range considered. For example, data sets can be constructed that clearly follow a quadratic curve on the scatter plot but for which the calculated linear correlation is zero, and a single observation with very high x and y measurements added to a data set with zero correlation may cause the correlation to become highly significant.

11.6 Regression

Regression is the process of calculating the line that best describes the correlation between variables. Historically it applied to data where replication of data points was impossible and, possibly, the cause-and-effect relationship was absent or ambiguous. The relationship between the height of a father and his son(s) is a classic example:

- A father has only one height that can be measured more or less accurately but it cannot be changed.
- The cause-and-effect relation between the father's and son's heights is not clear and there is no theoretical way of describing it.
- A son's height is also fixed and a father might have several sons all with different fixed heights.
- It makes equal sense to regress the son's height onto the father's or the father's onto the son's. Each is a satisfactory description of the correlation.
- The variable that is regressed onto is assumed to have zero error in its values, while the variable that is regressed is assumed to contain all of the error.

In contrast, very clear cause–effect relations are seen in laboratories, for example:

- A change in concentration of analyte causes a change in response such as UV absorbance.
- The relationship can be described theoretically (Beer–Lambert law in this example).
- True, independent replicates of the cause (concentration) can be used to assess the error in the response (absorbance).
- A theoretical model is fitted to the data rather than generating an empirical line to describe a correlation. (Note the order of the words here—we never try to fit the data to a model, do we!).
- The regression really only makes sense in one direction (absorbance on concentration) because of the cause–effect relationship.

Regression is quite simply described pictorially. If we have just a single response, y, and a single predictor, x, then the data will be a number of data pairs x_i, y_i as shown in Figure 11.5. When we are regressing y onto x, we focus on differences of the y values from the regression line, as shown by the solid lines in the figure. When we regress x onto y, we focus on the differences of the x values from the regression line as shown by the dashed lines. These different foci result in different regression lines, but both pass through the point that is the average of the x values and the y values.

Regression is the process of adjusting the position of the line, by adjusting the values of the two parameters of the model, to give the best representation of the data. There are several methods for determining the best representation, of which *least squares* fitting is the most commonly used.

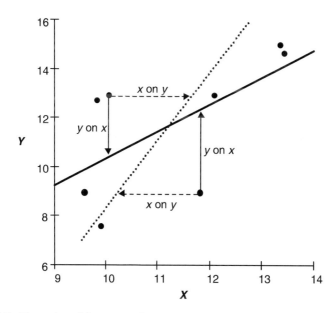

Figure 11.5 Illustration of the process of regression. The solid lines are regression of y on x. The dashed lines are regression of x on y.

11.7 Simple linear regression

11.7.1 Linear models

The simplest useful model has a linear functional relation between x and y and is also linear in the parameters of the model. The parameters are the intercept (or offset of y for a zero value of x) and the slope of the line through the data,

$$y = \beta_0 + \beta_1 x + \varepsilon$$

where y is the response that has been measured; x is the predictor; β_0 is the intercept of the line on the y axis; β_1 is the slope of the line; ε is the error associated with the measurements. This equation describes a linear model in x. If the equation involved an x^2 term then we would be trying to fit a quadratic model (a parabola) to the data; this would be a nonlinear model in x.

However, this is not the meaning of the word 'linear' in linear regression or linear least squares. It relates to the β_0 and β_1 of the equation, and these parameters appear only as linear terms; nonlinear terms do not appear. So, the model

$$y = \beta_0 + \beta_1 x + \beta_2 x^2 + \beta_3 e^{-x} + \varepsilon$$

would be treated by linear regression, as all the βs are linear terms, even though the predictor, x, appears as nonlinear terms.

Linear models may involve more than one predictor, for example temperature, time and flow rate, with a single response variable such as peak height or weight. Such systems will again have equations that are linear in β and they are described as multiple linear regression models (see Section 11.8).

11.7.2 Linear least squares

There are several ways of fitting a model to the data, each using a different criterion for giving the 'best' fit of the model. By far the most common method is the method of least squares and this is the one we shall look at here. The other methods have generally been designed to tackle specific problems with the data. So, the least median squares method, for example, is robust towards outliers in the data and it would be chosen when the experiment is known to generate occasional unusual results that cannot be stopped by redesigning the experiment (see Chapter 5), while partial least squares (PLS) handles large numbers of correlated predictors (see Chapter 12).

The process of least squares can be described pictorially. Figure 11.6 is a plot of three data pairs and shows three lines that could be fitted to them. Let us focus on line A first. We can measure the distance of each data point from the line, parallel to the axis of the response variable. These are marked as d_1, d_2 and d_3 and are called the residuals between the data values and the fitted line. The closeness of the fit of the line to the data can be measured by the sum of the squares of the residuals:

$$D_A = d_1^2 + d_2^2 + d_3^2 = \sum d_i^2$$

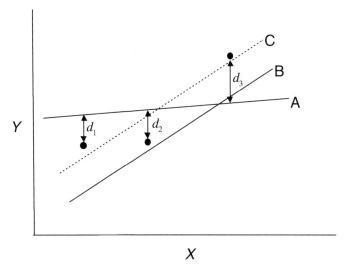

Figure 11.6 Illustration of the process of linear least squares.

If line A is rotated towards line C about its intersection with C we can see that the values of d_1, d_2 and d_3 will decrease. Thus, the value of D_A will decrease and approach the value of D_C. Looking now at line B, we can see that if we move it towards line C, without any rotation, then its value D_B will decrease towards D_C.

We can say that line C is a better fit to the data than lines A or B. The line that best fits the data is the one with the smallest value of D—if there were a perfect fit then D would be zero. Thus, the term linear least squares means that we are fitting a linear model to the data and that we are using the criterion of the least sum of squares of the residuals as the best fit criterion.

The choice of linear least squares instead of, say, least sum of absolute differences derives from its property of minimising the variance between the data values and the fitted line. This property is particularly useful where Normal distributions are expected for x, y and the error values.

This description has completely skipped over the theoretical derivation of the procedure and detailed discussion of many properties, but these are fully covered in the references.

11.7.3 *Assumptions*

It is useful to summarise at this point the main assumptions contained in linear regression and linear least squares.

- The true relationship is a straight line $y = \beta_0 + \beta_1 x$.
- The response variable, y, has errors that are Normally distributed.
- There is no bias, i.e. the errors have zero *average*.
- The magnitude of the error is independent of the magnitude of y and of the magnitude of x (i.e. the standard deviation of the measurement of y does not vary with y, nor with x).
- The predictor variable, x, can be regarded as not subject to error, i.e. there is no bias (systematic error) or random error in the measurement of x. In practice, the methods can be applied if the values of the predictor are fixed or predetermined by the experimenter or if any variability in the values of the predictor are small relative to the variability in the response (see also Section 11.7.12).

Gross deviations from these assumptions will require the use of a more complex analysis and expert help should probably be obtained.

11.7.4 Checking the results

Perhaps the first and most important item to examine is the plot of the residuals. If the model is good, and the data have a Normal distribution of errors at each data point, then the residuals will have random Normal distribution about zero. The residuals of the fit are calculated as

$$r_i = y_i - \hat{y}_i = y_i - b_0 - b_1 x_i$$

Plots of residuals against the predictor values are an essential tool for assessing the goodness of the model. Figure 11.7 shows some plots illustrating four common problems. In plot (a) there is a residual with an extreme value compared with the other residuals. This indicates an outlier data point that may be due to an incorrect response or predictor value. As such points can exert a strong influence on the regression equation, they need to be investigated to see whether they are truly discordant and can be corrected. Blocks of positive and negative values are seen in plot (b). This indicates that the actual relationship between the two variables is nonlinear and that a different model may be more appropriate. In plot (c) the values of the residuals get progressively larger as the value of the predictor increases, indicating that the error variance in the response is not constant. This might indicate a problem with the measurement system but it could be an inherent property of the data. In the latter case a transformation of the data to give constant variance or the use of weighted regression (Section 11.7.11) may be appropriate. Finally, the

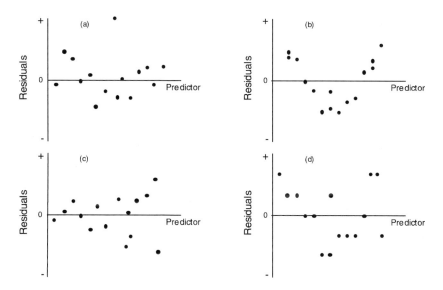

Figure 11.7 Examples of residuals plots. In these plots the residual has been plotted against the predictor value.

effects of other variables or of rounding often appear as bands of residuals as shown in plot (d).

Also useful are plots of residuals against the response or the predicted response. A Normal probability plot of the residuals can be used to examine the assumption of Normally distributed errors.

11.7.5 An example

To illustrate some of the points already made, we shall use a simulated example based around the dissolution of the metal oxide experiment described in Section 11.4.1. The conditions for this set of experiments are that the temperature of the dissolving acid was brought to a set temperature and a fixed quantity of sieved metal oxide was then added. The stirrer was started and after a fixed time period the amount of metal ion in solution was determined. The experiment was carried out at each of the temperature set points using fresh aliquots of acid and metal oxide. The stirrer speed was kept constant for all of the experiments. After going through all of the temperatures, repeat experiments were made at three of the temperatures.

Table 11.3 gives the temperature and metal ion concentration data and Figure 11.8a shows a scatter plot of the data together with the regression

Table 11.3 Simulated temperature–metal ion concentration data for the dissolution experiment described in Section 11.4.1

Temperature (°C)	Concentration (M)
10	0.0211
10	0.0174
15	0.0329
20	0.0354
25	0.0462
30	0.0488
35	0.0788
35	0.0675
40	0.0818
45	0.1054
50	0.1251
55	0.1588
60	0.1829
60	0.1782
65	0.2101

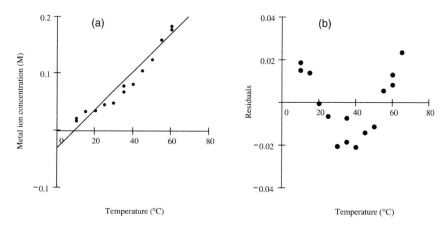

Figure 11.8 (a) Scatter plot of the metal ion concentration data with temperature with the linear regression line. (b) Residuals between the measured responses and those predicted by the regression line.

line. The scatter plot suggests that the response may not be a linear function of temperature, though it is not clear whether this is due to a critical temperature at about 30°C where the amount of metal ion increases more rapidly than below this temperature. This view is supported by the residuals plot (Figure 11.8b) where there are quite clear blocks of positive and negative values.

LINEAR REGRESSION

The slope and intercept of the regression line are

$$\text{Slope} = 0.003\,349$$
$$\text{Intercept} = -0.031\,209$$

The negative intercept of course has no physical meaning (the chemistry suggests that we should have a small positive intercept as some oxide should dissolve even at 0°C) and is a consequence of the apparent non-linearity of response.

The correlation coefficient, R^2 and R^2_{adjusted} values are

$$r = 0.971\,534$$
$$R^2 = 0.943\,878$$
$$R^2_{\text{adjusted}} = 0.939\,561$$

As we expect, the degree of association between the metal ion concentration and temperature is high and the model accounts for a large amount (94.4%) of the variation in the data, even though the linear model may not be appropriate.

We shall return to this example in the later sections.

11.7.6 Standard errors of estimates

The slope and intercept of the line through the data are readily calculated. However, it is very important that we also estimate the errors in these values and hence the confidence level we have in them.

Overall error
The overall error associated with the regression line is estimated from the residuals between the y values and the corresponding values calculated from the fitted line,

$$s^2 = \frac{\sum (y_i - \hat{y}_i)^2}{n - 2}$$

where y_i is the observed response at x_i; $\hat{y}_i$ is the response calculated from the regression line at x_i; n is the number of data pairs used in the calculation of the line. s^2 is an estimate of the *variance about the regression* (i.e. about the fitted line). It is used in the calculation of other estimates of error.

Standard error of parameters
The standard errors of the slope and intercept values are calculated quite simply from the standard deviation about the regression. They are

particularly useful as they allow us to calculate confidence intervals for the values or carry out hypothesis testing of the estimates with expected values or with other estimates (see Chapter 4).

The standard error of the slope is

$$se(b_0) = \frac{s}{\sqrt{\sum(x_i - \bar{x})^2}}$$

and the standard error of the intercept is

$$se(b_1) = s\sqrt{\frac{\sum(x_i^2)}{n\sum(x_i - \bar{x})^2}}$$

where s is the standard deviation about the regression; x_i is the ith x value; $\bar{x}$ is the average of the x values; n is the number of data pairs used in the calculation.

The 95% confidence intervals about the slope and intercept, using the two-sided t-value with $n-2$ degrees of freedom, are

$$\text{Slope} \quad b_1 \pm t_{0.05, n-2} \cdot se(b_1)$$
$$\text{Intercept} \quad b_0 \pm t_{0.05, n-2} \cdot se(b_0)$$

Alternatively, a t-test can be carried out to compare the calculated slope or intercept with some expected value such as a specification limit or a theoretical value. A common test value for a slope is 1.000 (particularly in calibration experiments) and that for an intercept is 0.000.

If we return to the metal oxide example, we can calculate the standard errors and 95% confidence intervals on the intercept and slope as shown in Table 11.4. We can now see that the intercept has a significantly negative value as its confidence interval is completely negative. In fact we have to go to the 99.5% confidence interval before the upper interval value becomes positive. Similarly, the slope has a highly significant positive value.

Table 11.4 Standard errors and 95% confidence intervals of the slope and intercept of the metal oxide dissolution data

	Value	Standard error	95% confidence interval	
Intercept	−0.031209	0.009315	−0.015133	−0.011085
Slope	0.003349	0.000226	0.002859	0.003838

11.7.7 The ANOVA table

Having obtained the values of the model parameters that give the best representation of the data, the next stage is to assess how well the chosen model actually fits the data.

The line calculated from the data must go through the point that is the average of the x and y values (a consequence of the least squares criterion). We can now look at the way in which the individual data points vary about the average point and the calculated line in order to quantify the significance of the regression.

Figure 11.9 gives a pictorial representation of the process. It shows a measured point x_1, y_1 (one of n used to calculate the regression line in the diagram) and its relation to the average of the data points $\bar{x}, \bar{y}$ and to the point predicted by the regression line, $x_1, \hat{y}$.

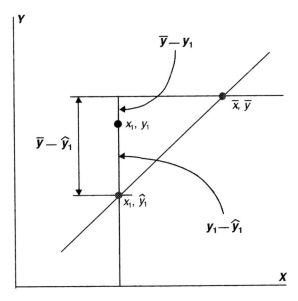

Figure 11.9 Pictorial representation of the relationships between errors in a linear regression model.

The relations between the y values are shown in the diagram and are summarised by the sum

$$(\bar{y} - \hat{y}_i) = (\bar{y} - y_i) + (y_i - \hat{y}_i)$$

Similar relations hold for all the data points. Some simple algebra (given in all texts on regression analysis) provides the sum of squares equation:

$$\sum_i (\bar{y} - y_i)^2 = \sum_i (y_i - \hat{y}_i)^2 + \sum_i (\bar{y} - \hat{y}_i)^2$$

The three sums in this equation have acquired specific names, which are used frequently in the ANOVA of regression:

$\sum_i (\bar{y} - y_i)^2$ Sum of squares about the mean
$\sum_i (y_i - \hat{y}_i)^2$ Sum of squares *about* the regression
$\sum_i (\bar{y} - \hat{y}_i)^2$ Sum of squares *due* to the regression.

If the fit of the model to the data is perfect, then the sum of squares about the regression will be zero. If the fit is good, but not perfect, then we might expect the sum of squares about the regression to be small compared with the sum of squares due to the regression and this forms the basis of a significance test for the regression as shown in the ANOVA table (Table 11.5).

Table 11.5 Basic form of the ANOVA table for a regression analysis

Source	Sum of squares	df	Mean squares	F-ratios
Due to regression	$\sum (\hat{y}_i - \bar{y})^2$	1	$\sigma^2 + \beta_1^2 \sum (x_i - \bar{x})^2$	Due/About
About regression	$\sum (y_i - \hat{y}_i)^2$	n−2	σ^2	
Total	$\sum (y_i - \bar{y})^2$	n−1		

The significance of the regression with respect to the measurement error is assessed by the F-test using the ratio of the due-to-regression mean square to the about-regression mean square. A significant ratio means that the amount of variation in the data due to the underlying model is greater than could be explained by the variation expected as a result of the measurement noise. In other words, the model and its calculated parameters are plausible. The ANOVA table for fitting a model to the metal oxide data (Table 11.3) is given in Table 11.6.

Table 11.6 ANOVA table for the simple linear regression of the metal oxide dissolution data of Table 11.3

Source	Sum of squares	df	Mean squares	F-ratios	p
Due to regression	0.05427470	1	0.05427470	218.6	< 0.0001
About regression	0.00322711	13	2.48239×10^{-4}		
Total	0.05750181	14			

The F-ratio of 218.6 shows that the fit of the model to the data is very highly significant, supporting the evidence we have already gained from the plot of the regression line and the confidence intervals on the model parameters. The column labelled p in this table is one that often appears in ANOVA tables produced by statistical packages. It is the significance value and represents the probability of getting an F-ratio at least as large as the one calculated by a random chance. The value of <0.0001 says that there is less than 1 chance in 10 000 of getting such a large value. The confidence level and significance level are related in that the confidence level $= 100 \times (1-p)$ (see also Chapter 4).

11.7.8 Lack of fit

The quantity estimated by the about-regression mean square is the variance of the actual y values about the calculated line. This is an estimate of the variance in the measurement of the y values *providing the model used to calculate the line is correct*. If the model is not correct, the variance about the line contains both the measurement variance and an extra quantity due to the deviation of the model. Thus, if the measurement variance is known, a measure of the correctness of the model can be obtained. An estimate of the measurement variance is obtained from true replicate measurements in the experiment (true here means that the measurements are not simply repeats on the same sample); this is called the *pure error*. The difference between the pure error and the about-regression sum of squares is termed the *lack of fit*.

Pure error is calculated from the sum of squares of the replicate values about the mean of the replicates at each x value where there are replicates,

$$\text{Pure error} = \sum_{i}^{k} \sum_{j}^{m_i} (y_{ij} - \bar{y}_i)^2, \quad \text{with} \quad \sum_{i}^{k} (m_i) - k \text{ degrees of freedom}$$

(k is the number of x values at which replicates were measured and m_i is the number of replicates at the ith x value).

The ANOVA can be modified to include the lack of fit and pure error components as shown in Table 11.7. This layout of the table is quite conventional and shows that the about-regression sum of squares has been split into two parts, the lack of fit and pure error (you can verify this by adding together the appropriate sums of squares and degrees of freedom).

If the lack of fit is significant by its F-test (ratio of lack of fit mean square to pure error mean square), the model being used is wrong for the

Table 11.7 The regression ANOVA table including lack of fit and pure error

Source	Sum of squares	df	Quantity estimated by mean square	F-ratio tests
Due to regression	$\sum(\bar{y}-\hat{y}_i)^2$	1	$\sigma^2 + \beta_1^2 \sum(x_i - \bar{x})^2$	Due/About
About regression	$\sum(y_i - \hat{y}_i)^2$	$n-2$	σ^2	
Lack of fit	$\sum(y_i - \hat{y}_i)^2 - \sum_i^k \sum_j^{m_i}(y_{ij} - \bar{y}_i)^2$	$n - 2 - \sum m_j - k$		LoF/PE
Pure error	$\sum_i^k \sum_j^{m_i}(y_{ij} - \bar{y}_i)^2$	$\sum m_j - k$	σ_m^2	
Total	$\sum(\bar{y} - y_i)^2$	$n-1$		

data and there is no point in trying to get any further information about the model from the data. Instead, a better model should be sought and/or the design of the experiments generating the data should be questioned.

As replicate data are available in the metal oxide data, the lack of fit can be calculated. Table 11.8 is the ANOVA table revised to include lack of fit. As it is significant, the linear model is not a good descriptor of the data, confirming what we have already suspected from the plot of data. At this point we should try to find a better model, but we shall defer doing this until Section 11.8.

Table 11.8 ANOVA table for the metal oxide dissolution data including lack of fit and pure error estimates

Source	Sum of squares	df	Mean squares	F-ratios	p
Due to regression	0.05427470	1	0.05427470	218.6	< 0.0001
About regression	0.00322711	13	2.48239×10^{-4}		
Lack of Fit	0.00314538	10	0.00031454	11.54	0.0342
Pure Error	0.00008174	3	0.00002725		
Total	0.05750181	14			

11.7.9 Confidence and prediction intervals

Confidence and prediction intervals are calculated for the regression equation in most statistical packages. Graphs are usually available showing the confidence interval and the prediction interval for the response over a range for the predictor. Usually 95% intervals are given. The equations for the intervals are given in the appendix to this chapter.

The confidence interval around the predicted values of the measured responses shows the band within which the true linear relationship is likely to fall. We should expect 95% of repeat experiments to give lines that fall inside this confidence interval, assuming a 95% t value was chosen (see also Chapter 4). Thus, a narrow interval implies a well-defined model.

The confidence intervals for the metal oxide linear model are shown in Figure 11.10a. The interval is quite narrow, indicating again quite a good model, but the pattern of data around the interval also suggest that the metal ion concentration may not be a linear function of temperature.

If a prediction is to be made for a new value of y at a given x value, then the actual measurement variance has to be included with the regression variance in the estimate in order to calculate the prediction interval. We should expect that 95% of new measurements at the given

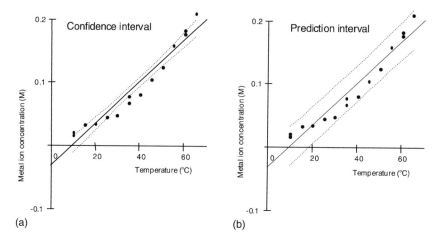

Figure 11.10 Confidence (a) and prediction (b) intervals for the metal oxide dissolution data.

predictor value to be within this prediction interval. The prediction interval for the metal oxide data is shown in Figure 11.10b. The prediction interval is wider than the confidence interval and, although it is the interval for future measurements, we can see that it includes the data (actually 95% of it) used to calculate it.

There is a point that has to be borne in mind about the prediction interval given here. It applies to the y values as they are used in the regression calculations. Thus if y_i is the average of m replicate measurements at x_i, then the prediction interval also applies to the average of m replicates and not to the individual values used to calculate the average. If a prediction interval is needed for the individual values, then a modified equation is required to account for the averaging. The appropriate equation is given in the appendix.

These equations should not be used if there is lack of fit as this means the model is not suitable for the data. I have not followed this advice here as I want to use a single example to illustrate the various calculation and plots—in a real study I should have already looked at different models, as we shall do in Section 11.8.

11.7.10 Calibration lines and their use

An essential requirement of measurement systems is to relate the signal generated by the measurement sensor to the property we are interested in. For example:

- A glass electrode generates a potential difference, measured in volts, relative to a reference electrode and this has to be related to the dimensionless pH scale.
- The length of a column of mercury in a glass capillary tube, measured in millimetres, has to be related to the Celsius temperature scale.
- The current, measured in amps, applied to the electromagnetic sweep coils of a mass spectrometer has to be related to the atomic mass scale in daltons.

Theoretical models may exist for the relationships, but the practical parameter values in the models have to be determined by experiment to account for imprecision and bias within the measurement system. Establishing the parameter values is the process of *calibration*. Once the linear equation relating y and x is established, it can be used to estimate an x value from a measured y value.

Let us use the mercury thermometer as an example. The first step is to establish the empirical linear relationship between the length of the mercury column, y, and temperature, x. In practice, this is done by immersing the bulb of the thermometer in fluids at different, well-controlled and precisely known temperatures. In principle, measurements at only two temperatures are needed to establish the linear calibration line. However, it is usual to use 3–5 temperatures as a check on the linear response over the temperature range to be calibrated.

Linear least squares is used to calculate the regression line of mercury length on temperature:

$$length = b_0 + b_1 \times temperature$$

To use the thermometer, its bulb is immersed in the fluid whose temperature is needed, the length of the mercury column is measured and the temperature is calculated using the equation,

$$temperature = \frac{length - b_0}{b_1}$$

The confidence interval in the calculated *temperature* can be estimated from the confidence interval of *length*. Figure 11.11 shows the process graphically. The usual plot of y (*length*) against x (*temperature*) with the confidence intervals of y is drawn. If the mercury length corresponding to the unknown temperature is y_0, the horizontal line corresponding to y_0 is drawn so that it crosses the confidence intervals. The temperatures corresponding to where the y_0 line crosses the confidence interval lines give the confidence range for the calculated temperature. In general the

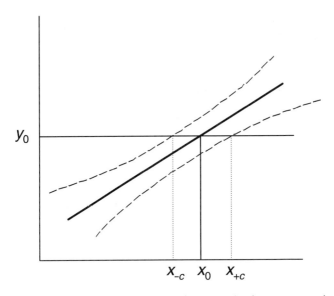

Figure 11.11 Estimating the confidence range for an x value from a measured y value.

interval will not be symmetrical about the calculated x value. This process can, of course, be carried out algebraically.

There are some situations when this process can give unexpected or counterintuitive results (see Draper and Smith, 1981).

The question may be asked, *Why not simply regress the temperatures onto the lengths, i.e. regress x on y?* This is called inverse regression and it will generate a calibration equation,

$$temperature = b_0^* + b_1^* \times length$$

The main reason for not adopting this approach is that the data in this form do not meet one of the assumptions of linear least squares regression: the predictors (now *length*) have a significant amount of error associated with them, while the response values (now *temperature*) have very little error (see Ryan, 1997). However, it is possible to handle such situations computationally and calibrations using this approach are made. In practice, most calibration exercises result in small errors in both x and y values, so the practical difference between reverse calculation and inverse regression is small.

11.7.11 Weighted linear regression

We have seen that outliers in data can have quite an influential effect on a linear regression calculation. We may see that some groups of responses are much more variable or uncertain than others. In both situations we may be concerned about the influence such effects may have on the linear regression model. A weighted linear regression model will reduce these effects.

The basic approach of weighted least squares is to transform the residuals, r_i, to new values, r_i^*, in such a way that the expected variance of y values at each x value is constant:

$$r_i^* = w_i \cdot r_i = w_i(y_i - b_0 - b_1 x_i)$$

The simplest weighting factors, w_i, are the inverses of the response variances at each x value.

The effect of weighting is that responses with large variance have a small weight and so influence the regression line less than the responses with larger weights. In this way, the effects of uncertain measurements or of outliers are reduced:

$$w_i = \frac{1}{\sigma_{x_i}^2}$$

Unfortunately, it is not often that these variances are known and estimates have to be made from the available data. If there are insufficient replicates in the data to get good estimates of variances, some form of iterative procedure may be required to estimate both the weights and the regression parameters. Draper and Smith (1981) and Ryan (1997) discuss the issues of weighting extensively.

An alternative that may be considered is the use of robust or non-parametric methods (see Chapter 5).

11.7.12 Errors in the x values

An assumption in linear regression is that the x values are error free. The assumption is generally reasonable in that the experimenter either sets the values with a controlled system or measures them with a precise method. However, occasions do occur when the assumption does not hold; for example, there is an unnoticed error in setting a specific x value or an imprecise method is used. What effect does this have?

The references in the bibliography should be consulted for a more detailed discussion on this topic. I shall present here a simple graphical interpretation to illustrate the main consequences, using Figure 11.12. The regression line (the solid line) was calculated using the data point labelled a (together with all of the other points not shown). This point is at its correct x value, i.e. there is no error in x. The point labelled b represents the situation where the same y value has been measured but

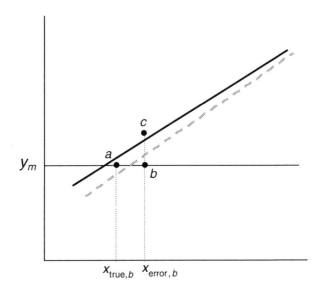

Figure 11.12 Illustration of the effect of error in the x values.

now there is an error in x—the correct x value is the same as for point a but it has been recorded as the value at b. It can be seen that one effect is to alter the position of the regression line slightly (shown by the dashed line). How much it alters the regression line depends on the magnitude of the error and its direction.

The other effect is to increase the apparent variability of y. Point b is the y value at an incorrect x value. Point c is the y value we might expect for that x value if there is no error in x. The two y values are, of course, different. If the effect over many data points is considered, then the overall effect is to increase the *apparent* variability in y, if it is assumed that x is error free. The consequence is widened confidence ranges on the parameter estimates.

11.7.13 What about a zero intercept?

The standard model for linear regression includes an intercept term, β_0. That is, we are assuming that at an x value of 0, y will not necessarily be zero. There are often very good grounds for believing that y should be 0 at $x=0$. For example, in spectroscopy the Beer–Lambert law relates the absorbance of a sample at a given wavelength to the amount of material in the light path:

$$A_\lambda = \alpha_\lambda l c$$

where A_λ is the absorbance at wavelength λ; α_λ is the specific absorbtivity of the material at wavelength λ; l is the thickness of the sample; c is the concentration or amount of absorbing species. Using the notation for models that we have used so far, this is

$$y = \beta_1 x + \varepsilon$$

i.e. there is no intercept term. Using this model will force the regression line to go through the point (0,0). The question then arises, *Should I use the equation as the model to fit my data?* The answer, in general, is *no*! This answer has nothing to do with casting doubt on the validity of the model. It has everything to do with the problems of making accurate and precise measurements.

There are two common situations that we shall look at where using the no-intercept model definitely should be avoided.

Bias in the measurements
Figure 11.13a illustrates the problem. A series of measurements has been made, for example, of solution absorbance at different analyte concentrations in solution. Unfortunately, the solvent (water) has been contaminated by the experimenter inadvertently dipping his finger in it (this is true—if you have access to a UV spectrophotometer try it!), giving a small background absorbance at the wavelength used for the sample

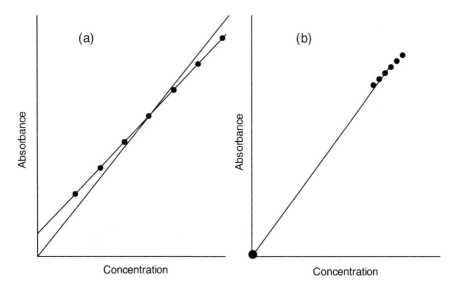

Figure 11.13 (a) Effect of measurement bias on the regression model with no intercept. (b) Effect of using the regression model with no intercept with a distant range of data.

measurement. Although the measurements are very precise and repeatable, the best line through the points quite clearly does not go through (0,0). Forcing the model to do so will give a poor representation of the data as illustrated.

As an aside to this particular problem, it is never wise to *assume* that the response at a predictor value of 0 is also 0 and to include the point $x=0$, $y=0$ with the measured values. Even if the model with intercept is used, some unusual results may arise. It is much better to make the experiment and measure the response at zero predictor. In the spectroscopic example above this would have given a positive reading owing to the contamination of the water.

Zero is a long way away
The first example had x values that were sufficiently close to $x=0$ that it could have been reasonably included in the data. The other situation is where the x values are quite distant from zero. This is typified by another spectroscopic example, where the calibration of a sample absorbance is required over a concentration range of 95–105% of the target amount of the analyte in the sample. Good practice requires the concentration range to be extended to about 90–110% to allow for samples that are close to the specification limit. Again, theory says that the calibration should go through the point (0,0). Figure 11.13b shows the effect of including this point in the regression calculations. The data set has effectively been reduced to a two-point set: (0,0) and $(\bar{x}, \bar{y})$. The resulting calibration line is not a good representation of the data in the concentration range of interest. In this case, even making an experimental measurement at zero concentration will not help as the data set will remain a two-point set.

Any model forced to include the (0,0) point will not account for any biases, nonlinearities or other effects in the required range. The metal oxide dissolution example shows a similar problem. A linear regression model in the 10–25°C temperature range will be quite different from the model using temperatures in the 50–70°C range. This is the situation where a local regression model in the target concentration range is best.

11.8 Multiple linear regression

So far we have looked at situations where there is only one predictor in the model equation. Now, we shall look at including several predictors. There will also be a change of notation used in the equations. Multiple regression equations are most easily and economically described using matrix notation rather than the algebraic form used so far. The two forms are exactly equivalent, but the matrix form is much more compact and one set of equations is applicable to any number of predictors, including the single-predictor cases already used. If you are not already familiar with using matrices, you may find the notation a little strange. However, do not be put off reading the following sections—the equations are supplementary to the descriptions. If you want to see how to use the matrix equations, Draper and Smith (1981) give quite a complete account and Tranter and Davies (1993) give a detailed spectroscopic example.

The whole subject of multiple linear regression (MLR) is vast and well studied. It is not possible, nor is it appropriate, to try to explain here this large amount of material—the references should be consulted for this. Instead I shall look at some quite simple examples to show some of the power and pitfalls of the approach. I shall also be establishing some of the basic links to other methods such as principal components regression (PCR) and partial least squares (PLS), which are discussed more thoroughly in Chapter 12.

11.8.1 Defining the model

The general MLR model is

$$y_i = \beta_0 x_{i,0} + \beta_1 x_{i,1} + \beta_2 x_{i,2} + \cdots + \beta_n x_{i,n} + \varepsilon_1$$

or

$$\mathbf{Y} = \mathbf{X}\boldsymbol{\beta} + \boldsymbol{\varepsilon}$$

The least squares solution of this equation for the parameters is

$$\boldsymbol{\beta} = (\mathbf{X}^T\mathbf{X})^{-1}\mathbf{X}^T\mathbf{Y}$$

This equation appears so frequently in regression literature that it is worth taking some time to understand its structure (terms such as $\mathbf{X}^T\mathbf{X}$ generate sums of squares).

As written above, the model appears to have no intercept. However, if all of the $x_{i,0}$ values are made equal to 1 then β_0 takes its familiar role as the intercept representing the value of y when all of the predictors are at 0.

A common form of MLR model has a single predictor but includes nonlinear terms in the predictor. For example, the metal oxide dissolution example we have been working with so far shows evidence of nonlinear response of metal ion concentration with temperature (see Section 11.7.5). A model to test this could be

$$concentration = \beta_0 + \beta_1 \times temperature + \beta_2 \times temperature^2 + \varepsilon$$

or

$$y = \beta_0 + \beta_1 x_1 + \beta_2 x_1^2 + \varepsilon$$

or

$$\mathbf{Y} = \mathbf{X}\boldsymbol{\beta} + \boldsymbol{\varepsilon}$$

Clearly, other models could be generated using higher powers of temperature, exponentials, logarithms, inverses, etc. All that is necessary is to ensure that the appropriate column of matrix **X** has the correct numerical values derived from the corresponding function.

Alternatively, we might want to include other predictors in the model together with their nonlinear terms and interaction terms between them. The metal oxide example was originally described in terms of several causes (Section 11.4.1) and a model that includes stirrer speed as well as temperature is

$$\begin{aligned} concentration = {}& \beta_0 + \beta_1 \times temperature + \beta_2 \times stirrer \\ & + \beta_3 \times temperature \times stirrer + \beta_4 \times temperature^2 \\ & + \beta_5 \times stirrer^2 + \varepsilon \end{aligned}$$

or

$$y = \beta_0 + \beta_1 x_1 + \beta_2 x_2 + \beta_3 x_1 x_2 + \beta_4 x_1^2 + \beta_5 x_2^2 + \varepsilon$$

or

$$\mathbf{Y} = \mathbf{X}\boldsymbol{\beta} + \boldsymbol{\varepsilon}$$

Just these two examples show the freedom there is to generate models of any complexity. But with such freedom comes danger. A model with too many terms, particularly high-order polynomial models, will actually model the noise in the data as well as the effects we are interested in. This situation is described as overfitting and it is a common problem. The best protection is to use models with the smallest number of terms that gives a reasonable representation of the data. Of course, reasonable in this context is not defined! Knowledge of the underlying chemistry and physics is of great help here, but there are some statistical tools such as stepwise regression (Section 11.8.3) and cross-validation (Chapter 12) that can help select suitable terms from a large possible set.

Three other dangers are colinearity of terms in the model, correlation of terms and not having sufficient data to get reasonable error estimates to quantify confidence in the model. The first of these, colinearity, is a mathematical property whereby a linear combination of two or more of the terms in the model is a constant. The effect is to make the calculation unstable to the point where it cannot be carried out. The difficulty with this is that the existence of colinearity in the model may not be obvious until the calculations are attempted, and then it may be difficult to identify just where it lies. The likelihood of colinearity increases as the number of terms in the model increases and it is unlikely to be a major problem for models with half a dozen or so terms.

Correlation is different from colinearity. As already described, correlation is the degree of association of variables. A linear regression model can be calculated when two predictors in the model are correlated (note that we are talking about correlation between two of the x variables here, not correlation between the x variables and y). The problem comes in interpreting the model and deciding whether either, or both, of the factors is important in the model and whether the calculated b values are actually the correct ones. This problem is similar to the one of confounding discussed in Chapters 6 and 7 and the solution is the same; use experiment design to ensure that important factors are not correlated. Knowledge of the chemistry and physics of the system may help to identify a plausible solution, but it is probably best to plan a small series of experiments where the factors are not correlated to obtain a definitive answer.

The third danger is not having enough data. A model that contains k terms needs $k+1$ measurements to define a unique solution, providing those measurements occupy unique positions in the measurement space. Three measurements are needed for a two-factor model, for example, but those measurements should form a triangle in the measurement space and not a straight line. Although $k+1$ measurements give a unique solution they will not give error estimates on the parameters or information about lack of fit. For these, more measurements than $k+1$ are needed, including some that are replicates. In addition to this, the number of levels at which each factor is used needs to be considered. A factor used in a quadratic term needs to have at least three levels in order to be determined. Clearly, the complexity of a model, with its ability to fit the data, needs to be balanced with the complexity and cost of generating suitable data.

11.8.2 *Interpreting the results*

We shall continue with the metal oxide dissolution example to look at modelling with a simple polynomial and with two different factors.

A simple polynomial model

All of the simple linear regression analysis so far (Section 11.7) has suggested that there might be a nonlinear response of metal ion concentration with temperature. We shall now carry out the analysis using a model incorporating a quadratic temperature term in the model,

$$y = \beta_0 + \beta_1 x_1 + \beta_2 x_1^2 + \varepsilon$$

where x_1 is the temperature. Table 11.9 gives the results of the calculations.

Table 11.9 Results of fitting a quadratic polynomial in temperature to the metal oxide dissolution data

Source	Sum of squares	df	Mean square	F-ratio	Prob. > F
Due to regression	0.05721824	2	0.0286091	1210.69	<0.0001
Temperature	0.00002756	1	0.0000276	1.17	0.3014
Temperature^2	0.00294355	1	0.0029436	124.57	<0.0001
About regression	0.00028356	12	0.0000238		
Lack of Fit	0.00020183	9	0.0000224	0.82	0.6408
Pure Error	0.00008174	3	0.0000272		
Total	0.05750181	14			

Term	Estimate	Std error	t-ratio	Prob. > |t|
Intercept	0.0208202	0.005476	3.80	0.0025
Temperature	−0.000367	0.000340	−1.08	0.3014
Temperature^2	0.0000505	0.000005	11.16	<0.0001
R^2	0.995069			
R^2_{adj}	0.994247			

The first part of the table is the ANOVA table. Its structure is very similar to Table 11.8 (the ANOVA table for the single-factor model) only now the due-to-regression line is followed by its two components: *temperature* and *temperature*^{2}. The regression is very highly significant with an F-ratio of 1210.69 (the single-factor model was 218.6) and the lack of fit is not at all significant whereas with the single-factor model it was. Of the two factors, the linear term *temperature* is not a significant contributor to the model, while the quadratic term *temperature*^{2} is highly significant. All of these indicate that the quadratic model is a far better representation of the data than the simple linear model.

The second part of the table gives the R^2 value and the estimates of the model parameters. The R^2 is very high at 0.995 (0.944 for the simple linear model), again indicating a very good fit of the model to the data. The parameter estimate for the quadratic temperature term is very significantly different from zero, while that for the linear term is not significantly different from zero. Finally, it should be noticed that the intercept has a positive value and is significantly different from zero, in line with what we should expect for the system, unlike the intercept for the simple linear model, which was negative. The residuals plot is shown in Figure 11.14 and now has no obvious features.

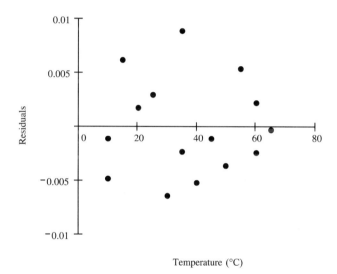

Figure 11.14 Residuals plot for the quadratic model of metal oxide dissolution with temperature.

This example also poses one of the vexed questions in MLR. The *temperature* term is not a significant contributor to the model but *temperature*^{2} is. So, should *temperature* be removed from the model? There is a logic that says no, *temperature* should be retained because *temperature*^{2} depends on it and its absence would be illogical. Indeed, some statistics packages warn you if you try to fit a model with such a term missing. However, there may be theoretical grounds for believing that, in this case, metal oxide dissolution is functionally dependent on *temperature*^{2} but not on *temperature*, so leaving out *temperature* is correct. Practice falls somewhere between these two extremes. The data should be modelled with *temperature* in. If it is found that the term is not significant and there is justification for it not being needed, then it should be removed and the regression recalculated. In this way we check for unusual effects in the data (see the problem of zero intercept, Section 11.7.13) that could seriously affect the model.

A more complex model with two factors
Table 11.10 gives metal oxide dissolution data for different combinations of temperature and stirrer speed. (Note that these data form what is called an all-combinations experiment design—there are 5 temperatures and 3 stirrer speeds and $5 \times 3 = 15$ experiments, excluding the replicates. This is not a particularly efficient design, but the experimenter had not read Chapters 6 and 7.)

Table 11.10 Simulated data for the effect of temperature and stirrer speed on metal oxide dissolution

Temperature (°C)	Stirrer (rpm)	Concentration (M)
10	50	0.0116
10	150	0.0468
10	250	0.0704
25	50	0.0328
25	150	0.0680
25	250	0.1253
40	50	0.0651
40	150	0.1143
40	250	0.1794
55	50	0.0791
55	150	0.1475
55	250	0.2056
70	50	0.0914
70	150	0.1936
70	250	0.2696
40	50	0.0577
40	150	0.1250
40	250	0.1738

The model we shall try to fit to the data is a full quadratic model in the predictors,

$$y = \beta_0 + \beta_1 x_1 + \beta_2 x_1^2 + \beta_3 x_2 + \beta_4 x_2^2 + \beta_5 x_1 x_2 + \varepsilon$$

where, x_1 is *temperature* and x_2 is *stirrer speed*.
Table 11.11 gives the results of the regression. In the ANOVA table, we now see that the due-to-regression sum of squares is split into five components, corresponding to the five predictor terms. This shows that *temperature*^{2} and *stirrer*^{2} are not significant contributors to the model.

Table 11.11 The regression results for the data in Table 11.10

Source	df	Sum of squares	Mean square	F-ratio	Prob. > F
Due to regression	5	0.080322	0.016064	260.4	<0.0001
Temperature	1	0.000496	0.000496	8.04	0.0150
Stirrer	1	0.000291	0.000291	4.72	0.0506
Temperature * Stirrer	1	0.003721	0.003721	60.3	<0.0001
Temperature^2	1	0.000026	0.000026	0.41	0.5322
Stirrer^2	1	0.000023	0.000023	0.37	0.5552
About regression	12	0.000740	0.000062		
Lack of fit	9	0.000640	0.000071	2.13	0.2896
Pure error	3	0.000100	0.000033		
Total	17	0.081063			

Term	Estimate	Std error	t-ratio	Prob. > |t|
Intercept	−0.018568	0.012361	−1.50	0.1589
Temperature	0.0012356	0.000436	2.83	0.0150
Stirrer	0.0002798	0.000129	2.17	0.0506
Temperature * stirrer	0.0000091	0.000001	7.77	<0.0001
Temperature^2	−0.000 003	0.000005	−0.64	0.5322
Stirrer^2	-2.383×10^{-7}	3.927×10^{-7}	−0.61	0.5552
R^2	0.990868			
R^2_{adj}	0.987062			

Their parameter estimates are also not significantly different from zero. The lack of fit is not significant, indicating that the model is a good representation of the data.

Although we have no justification based on dissolution mechanism, we shall remove the nonsignificant terms from the model and recalculate the regression. The results are in Table 11.12. We see that there are small changes to the *temperature* and *stirrer* parameter values, the lack of fit is slightly less significant and the amount of variability accounted for by the model has hardly changed. This is a more parsimonious model and is preferred over the more complex model, particularly as the model statistics have improved.

11.8.3 Stepwise regression

As we have seen, one of the big problems with MLR is identifying which factors are important to a model when there may be no theoretical or other basis for including them. This is the problem of factor or variable selection and it affects the principal component based methods as well as MLR (see also Chapter 12).

A method that has been used extensively in MLR for some considerable time is stepwise regression. The basis of the method is first

Table 11.12 The regression results for the data in Table 11.10 after removing the nonsignificant terms from the model in Table 11.11

Source	df	Sum of squares	Mean square	F-ratio	Prob. $> F$
Due to regression	3	0.080274	0.026758	475.1	<0.0001
Temperature	1	0.001501	0.001501	26.65	0.0001
Stirrer	1	0.000988	0.000988	17.55	0.0009
Temperature * Stirrer	1	0.003721	0.003721	66.06	<0.0001
About regression	14	0.000789	0.000056		
Lack of fit	11	0.000688	0.000063	1.8713	0.3318
Pure error	3	0.000100	0.000033		
Total	17	0.081063			

Term	Estimate	Std error	t-ratio	Prob. $> \|t\|$
Intercept	−0.01098	0.008491	−1.29	0.2169
Temperature	0.0009864	0.000191	5.16	0.0001
Stirrer	0.0002083	0.00005	4.19	0.0009
Temperature * Stirrer	0.0000091	0.000001	8.13	<0.0001
R^2	0.990272			
R^2_{adj}	0.988188			

to identify all of the possible predictors (linear, nonlinear and interaction) for a model and to then use a directed search method to select a small subset that describes the data adequately. Note that I use the word 'adequately' here rather than 'best' or 'perfectly'—we have seen already that by simply increasing the number of terms in a model we can get a better fit of the model to the data.

An approach that is implemented in many statistics packages uses an F-test to determine whether a term in the model has a significant impact on the amount of variability in the data that is accounted for. The process can start with either with no terms in the model or with all of the terms in the model. In the former case, a term is added to the model if it makes a significant change to the amount of variability that is explained (F to add is significant). In the latter, terms are removed if their removal does not significantly change the amount of variability accounted for. In both cases a search is made for the most significant contributors. Both of these unidirectional processes are heavily influenced by the first few terms put into or taken from the model. To combat this, a bidirectional process can be used. In this, the effect is examined of removing any of the terms from the model when a new term is added as well as the effect of adding any of the available terms when one is removed from the model.

Although the bidirectional process generally gives a more appropriate selection of terms than either of the unidirectional processes, it is by no means perfect and the model cannot be described as optimal. The basic reason for this is that this form of directed search explores only a small part of the space spanned by all possible models. Other search strategies, including those based upon genetic algorithms, aim to get round these limitations.

However, despite its limitations, the 'standard' stepwise regression is useful and can help reduce the size of a potential model considerably.

11.8.4 Principal components regression and partial least squares

Principal components regression (PCR) and partial least squares (or projection onto latent structures, PLS) are discussed fully in Chapter 12. They are introduced here as an extension of the MLR method.

One of the problems with MLR is that predictors in the model may be correlated with each other, particularly when there are many terms in the equation, such as when using complete spectra as predictors. A possible solution to this problem is to generate the principal components of the predictors. The PCs are guaranteed to be orthogonal to each other and as they are formed from linear combinations of the original variables, they can be used as predictors in an MLR process. This is principal components regression.

The orthogonality property means that the model is particularly simple, with no interaction terms,

$$\mathbf{y} = \beta_0 + \beta_1 \mathbf{p}_1 + \beta_2 \mathbf{p}_2 + \cdots + \beta_n \mathbf{p}_n + \varepsilon$$
$$\mathbf{Y} = \mathbf{P}\boldsymbol{\beta} + \boldsymbol{\varepsilon}$$

The orthogonality also means that selecting which of the PCs to keep in the model is a much simpler process than with the original variables. The downside of PCR is interpreting the regression model in terms of the original variables. Where MLR minimises the variance of the model, PCR minimises the correlation.

PLS is also a PC-based technique. However, instead of the simple PCs of the predictors, PLS generates PCs while taking account of the variability in the responses. Thus, PLS generates its principal components (which are different from those given by PCA) and a regression model in one step. The benefit is that a more resilient model is generated that is optimised to explain the variability in the response. A further benefit of PLS is that it readily handles problems involving multiple responses, generating a model that is a good simultaneous fit to all of the responses. PLS minimises the covariance of the predictors and responses.

11.9 Nonlinear regression

There is a large class of regression problems where the model necessarily includes terms that are nonlinear in the model parameters as well as the predictors. Very good examples of this class are the kinetic equations describing chemical reactions. For example, a first-order reaction for the conversion of compound A into compound B is described by the general first-order kinetic equation,

$$[A]_t = [A]_\infty - ([A]_\infty - [A]_0)e^{-kt}$$

where $[A]_t$ is the concentration of compound A at time t after the start of the reaction, $[A]_0$ and $[A]_\infty$ are the concentrations at the start of the reaction ($t=0$) and the end of the reaction (infinite time), respectively, and k is the reaction rate constant. The parameters in this equation are $[A]_0$, $[A]_\infty$ and k. In general, the initial and final concentrations of A will be known, which leaves k to be determined. Other more complex reaction schemes may have several nonlinear parameters. The following sections use this kinetic example to explore how a nonlinear parameter may be determined.

11.9.1 Linearisation by data transformation

The traditional way of handling this problem is to apply some form of linearising transformation so that the parameter(s) of interest form a linear equation. For convenience I shall rewrite the kinetic equation using the simpler notation and symbols that I have used in other regression equations so that it is easier to follow the various changes:

$$y_i = y_\infty - (y_\infty - y_0)e^{-\beta_1 x_i}$$

A simple algebraic rearrangement of the equation followed by taking logarithms gives

$$\log_e(y_i - y_\infty) = \log_e(y_0 - y_\infty) - \beta_1 x_i$$

or

$$y_i^* = \beta_0 - \beta_1 x_i$$

This is a familiar linear equation that has been used for decades by chemists to estimate rate constants of reactions. However, there are problems with its use. The first is that it is dependent on a good estimate of y_∞. A poor estimate causes severe curvature in the plot and so affects the estimate of rate constant. This is illustrated in Figure 11.15a. Fortunately, most simple first-order reactions have zero concentration of compound A (theoretically and practically) at the end of the reaction, but unfortunately, most pseudo-first-order reactions do not.

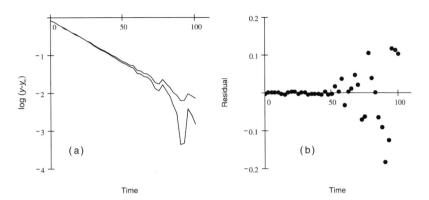

Figure 11.15 (a) Effect on y^* of not having a good estimate of y_∞. The upper line uses the true value of y_∞ and for the lower line y_∞ is 5% larger. (b) The residuals plot following linear regression of y^* with x; y^* was calculated with the true value of y_∞.

(Pseudo-first-order reactions are reactions that mechanistically are second order or higher but, because the reactant being monitored is at a very much lower concentration than the others, the reaction appears to be first order). The second problem is that the distribution of errors for y^* is not Normal. The measurement errors for y are expected to have a Normal distribution, but the logarithmic transformation changes this. This is shown in the residuals plot in Figure 11.15b.

A weighted regression model could be used to mitigate these two effects.

11.9.2 Linearisation by series approximations

Section 11.10.1 shows how the simple linear least squares equations for the parameter estimates are developed through solving a set of normal equations derived from the least squares criterion,

$$S = \sum \left(y_i - (\beta_0 - \beta_1 x_i)\right)^2 = \text{minimum}$$

We can follow exactly the same process for nonlinear models and set the least squares criterion as

$$S = \sum \left(y_i - \mathbf{f}(\beta, x_i)\right)^2 = \text{minimum}$$

where $\mathbf{f}(\beta, x)$ is the nonlinear function involving β and x. It is quite common in nonlinear regression to include a weight as well. As with the simple linear case, the normal equations are derived by the partial differentiation of S with respect to the β. The problem is that the normal equations also contain nonlinear terms and their direct solution is, in general nontrivial. A generally applied resolution of this problem is to substitute $b^0 + \Delta b$ for β where b^0 is some guess at the value of β and Δb is the unknown deviation of the guess from the correct value. The normal equations are expanded by a Taylor series in Δb and truncated after the linear term in the series. In this way the equations are converted into linear equations in Δb, which are then solved (Wentworth, 1965).

Because of the truncation of the Taylor series, it is unlikely that the estimate of Δb will be exactly that required to make b^0 equal to β, so the b^0 is updated with the value of Δb and the new values are used to calculate another estimate of Δb. This process of iteration is continued until the change in Δb is sufficiently small for us to say that the process has converged on the best estimates of β.

Having got the estimates of β, it is then possible to calculate the usual statistics in order to assess the fit of the model to the data.

There are many potential problems with the process. Two major ones are that the partial derivatives with respect to the parameters must exist and that the process is an iterative one. The first is not often a problem with chemical systems, where exponential functions are common. However, the second frequently gives problems. These stem from the linearisation during the series expansion, which can cause the iteration steps to be too large and oscillate around the minimum S, or to be too small and with slow convergence and even getting stuck in a local minimum caused by noise in the data. There are many computational fixes to these problems, but you need to be aware that not all of them work in all circumstances and nonlinear regression remains something of an art.

11.9.3 Residuals surface methods

An alternative approach to nonlinear least squares is to work directly with the sum of squared residuals equation and to treat S as a function of the b. In this way the problems of deriving and solving the normal equations are avoided. To illustrate the approach, let us return to the simplest linear equation,

$$y_i = \beta_1 x_i + \varepsilon_i$$

The sum of squares of residuals can be calculated and the equation rearranged as follows:

$$\begin{aligned} S &= \sum (y_i - b_1 x_i)^2 \\ &= \sum y_i^2 - 2b_1 \sum x_i y_i + b_1^2 \sum x_i^2 \\ &= a_0 + a_1 b_1 + a_2 b_1^2 \end{aligned}$$

We see that S is a quadratic function of b_1 and it will have a minimum value when $b_1 = \beta_1$ (see Figure 11.16, curve a). If we make some guess at b_1, i.e. b_1^0, we can calculate a value for S. We can then make a small adjustment to b_1^0, i.e. $b_1^0 + \Delta b_1$ and recalculate S. If the new value of S is lower than the first value, then we know that we should increase b_1^0 to move towards the minimum value of S. This process can be carried iteratively, but note that we are not trying to determine the a parameters in the equation for S; we are working purely empirically by finding the values of b_1 that reduce the value of S.

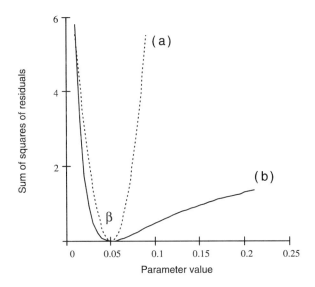

Figure 11.16 Curve (a): the sum of squares of residuals as a function of a linear model parameter b. Curve (b): the sum of squares of residuals as a function of a nonlinear model parameter b (exponential).

Exactly the same process can be applied when working with nonlinear expressions in many b. The problem is that now the response curve is a surface and it generally does not have a quadratic shape. This is often ignored and a quadratic shape is assumed. The assumption is quite reasonable in the region of the minimum S (i.e. b_i is not far from β_i), but it can be quite unreasonable farther away. The consequence is that if the initial guesses of b_i are not good, the corrections may not find a clear or rapid descent towards the minimum. This is illustrated above in Figure 11.16, curve b.

Again there are many ways of carrying out the process. The steepest descent method makes a small local model of the response surface, identifies the direction of steepest descent and moves all of the parameter values a small amount in that direction before repeating the calculation. The pit mapping approach (Sillen and Ingri, 1964) creates a local model using a star experiment design of the parameters. A quadratic surface is assumed and the model used to predict the location of the minimum. The parameters are moved to that location and the process is repeated.

The value of these approaches is that they are not dependent on solving complex equations and they can work with arbitrarily complex functions, even those that contain discontinuities. An example of a complex chemical kinetic problem (proton quantum-mechanical tunnelling in an aqueous solution) solved using the pit mapping technique is given in Bell and Tranter (1974).

11.9.4 Neural networks

Neural networks have been applied very successfully to the creation of models to fit data for predictive purposes (Wasserman, 1989). Their value is in situations where there are many predictors, the predictors are correlated, and the functional relationship between the predictors and response is not known or is complex. Neural networks can easily model simple linear regression, but the computational penalty in training the network is too high to make it worthwhile.

The basic structure of the neural network for quantitative modelling is the feed-forward, back-propagation network with one hidden layer (Gemperline et al., 1990), Figure 11.17. There are as many inputs to the neural network as there are predictors. The output layer can have one or more nodes (a node is the computing element of the structure that sums the weights into it and applies the transfer function; it is shown as a circle in Figure 11.17) depending on the application. When a simple numerical value is required, only a single node is used. The number of nodes in the hidden layer is determined empirically (usually) and it controls the complexity of relationship between predictors and response that can be modelled.

It is usual in networks of this sort to reserve one input node and one hidden layer node to have a simple constant input. These are the bias nodes and they serve to model the intercept in a regression equation.

Each node in the network has several inputs from the nodes in the previous layer. Weights are applied to each input and the sum of the weighted inputs is multiplied by a transfer function to give the output of the node. The choice of transfer function is wide, but the most common choice is the sigmoid function,

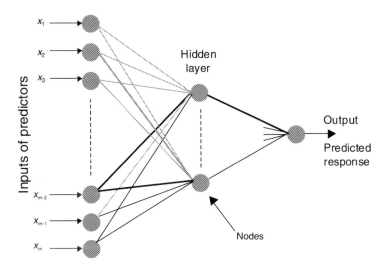

Figure 11.17 Schematic of a feed-forward, back-propagation neural network.

$$\frac{1}{1+e^{-x/\theta}}$$

where x is the weighted sum of inputs and θ is the gain, which controls the steepness of the sigmoid shape. This function has several useful properties for neural networks, but the one of interest here is that it has both an approximately linear region near the centre of the function and two highly nonlinear regions to each side. It is this property that allows the network to model nonlinear systems.

In operation, the weights of the node inputs are set to random values and one set of predictors from a calibration set of data is presented to the input layer. These values feed forward though the network to generate a predicted response at the output. The predicted value is compared with the measured value and the difference is then propagated backwards through the network, adjusting the weights so as to minimise the difference. The size of the correction is controlled through the so-called learning parameter. Another set of predictors is presented and the error signal they generate with their response is used to make another adjustment to the weights. This process is repeated many times (often several thousand) until the weights in the network have reached a stable state. The thickness of the lines between nodes in Figure 11.17 indicates the magnitude of the weights of the connections.

If a new set of predictors is presented to the inputs, the predicted response will be a good estimate of the true value. At no point has the functional relationship between the predictors and response been entered into the network. An empirical relationship has been calculated and is encoded in the weights of the node inputs.

11.9.5 Genetic algorithms

A genetic algorithm is a search method that minimises (or maximises) some predefined cost function by altering the parameters of the function to find the 'best' set. In this respect, a genetic algorithm is very like the residuals surface mapping techniques for nonlinear regression. The similarity is increased when the least squares criterion is used as the cost function. Note that here, unlike in the neural network approach, we do need to know the relationship between predictors and response in order to define the least squares function.

In operation, several sets of parameters are generated with random values. The value of the cost function is calculated for each set. The sets that give the poorest (highest in this case) values are discarded (death in genetic algorithm terms) and replaced by new randomly generated sets. The number of sets discarded is one of the controls in the genetic algorithm. A proportion of the sets that gave the best cost function values is taken and blocks of their parameters are swapped between the sets (reproduction). Finally, a small number of parameter values in some of the best sets are randomly changed (mutation). The cost functions are recalculated and the processes of death, reproduction and mutation are repeated. At each stage the parameter set with the lowest cost function value is saved. The process is stopped when it appears that there is no, or little, improvement in the cost function.

The similarities with the steepest descent and pit mapping techniques are apparent, but there are advantages to the genetic algorithm approach. The random sets of the parameters can cover the whole of the parameter space and as a consequence there is a better chance both of finding the global minimum and of not getting stuck on some relatively flat plain in the response surface. The reproduction steps work with the parameters closest to the minimum of the cost function and so explore this region of the response surface with a good chance of moving to a lower value. The mutations are like wild shots. They may, quite by chance, hit a region of the response surface with a lower value than anything found up to that point and so move the search out of a local minimum into a new and better region.

An extension of the process is to search for the best combination of predictor terms from a large number of possibles. The presence or absence of a term in the model can be coded as a binary switch—if it is 'on' the term is used in the function, if it is 'off' the term is not used.

LINEAR REGRESSION

The genetic algorithm manipulates the switches to search for the best combination of terms. It is possible to combine the search for the best combination of term and the search for parameter values into one process.

Once the best set of parameters has been found, it is then possible to calculate the model statistics and assess the significance of the parameters.

11.10 Appendix: Summary of equations for linear regression

11.10.1 Algebraic equations for simple linear regression

The simple linear regression model is

$$y = \beta_0 + \beta_1 x + \varepsilon$$

The least squares algorithm minimises the sum of squares of the residuals, S, between the measured and predicted responses,

$$S = \sum (y_i - \beta_0 - \beta_1 x_i)^2$$

As S is a function of the β, S is a minimum when its first derivative with respect to the β is zero. The partial derivatives of S with the β are

$$\frac{\partial S}{\partial \beta_0} = -2 \sum (y_i - \beta_0 - \beta_1 x_i) = 0$$

$$\frac{\partial S}{\partial \beta_1} = -2 \sum x_i (y_i - \beta_0 - \beta_1 x_i) = 0$$

These are the normal equations. By expanding the summations, a pair of simple linear equations is obtained and these can be solved simultaneously for β_0 and β_1 to give

$$b_1 = \frac{n \sum (xy) - (\sum x)(\sum y)}{n \sum (x^2) - (\sum x)^2}$$

$$b_0 = \frac{(\sum x^2)(\sum y) - (\sum x)(\sum xy)}{n \sum (x^2) - (\sum x)^2}$$

or

$$b_0 = \bar{y} - b_1 \bar{x}$$

The mean sum of squares about the regression, s^2, is the estimate of variance of the responses about the regression line:

$$s^2 = \frac{\sum(y_i - \hat{y}_i)^2}{n - 2}$$

This is used to estimate the standard errors of the b:

$$se(b_1) = \frac{s}{\sqrt{\sum(x_i - \bar{x})^2}}$$

$$se(b_0) = s\sqrt{\frac{\sum x_i^2}{n\sum(x_i - \bar{x})^2}}$$

The correlation coefficient between variables is

$$r = \frac{\text{Cov}(x, y)}{s_x s_y} \quad \text{where} \quad \text{Cov}(x, y) = \frac{\sum(x_i - \bar{x})(y_i - \bar{y})}{n - 2}$$

The correlation between the response and its predicted values is

$$R^2 = 1 - \frac{\sum(y_i - \hat{y}_i)^2}{\sum(y_i - \bar{y})^2} \quad \text{with an adjusted value of}$$

$$R^2_{\text{adjusted}} = 1 - \frac{\sum(y_i - \hat{y}_i)^2}{\sum(y_i - \bar{y})^2} \frac{(n - 1)}{(n - p)}$$

The pure error from k groups of m_j replicates is

$$\text{Pure Error SS} = \frac{\sum\sum(y_{i,j} - \bar{y}_i)^2}{\sum m_j - k}$$

The lack of fit of the model to the data is then

$$\text{Lack of Fit} = \sum(y_i - \bar{y})^2 - \text{Pure Error SS}$$

The confidence interval for the regression line is

$$\hat{y}^0 \pm t_{n-2,\alpha} s \sqrt{\frac{(x^0 - \bar{x})^2}{\sum(x_i - \bar{x})^2}}$$

The confidence interval for the new y is

$$\hat{y}^0 \pm t_{n-2,\alpha}\, s\left(1 + \sqrt{\frac{(x^0 - \bar{x})^2}{\sum(x_i - \bar{x})^2}}\right)$$

11.10.2 Matrix equations for linear regression

The response values are a column vector $\mathbf{Y}$ and the predictor values are a matrix $\mathbf{X}$. In $\mathbf{X}$, each column represents one predictor and the rows are the values which give the corresponding values in $\mathbf{Y}$. Each element in column 1 of $\mathbf{X}$ is set to 1.000 if an intercept is needed in the model. If $\boldsymbol{\beta}$ is a column vector of the model parameters then the model equation is

$$\mathbf{Y} = \mathbf{X} \cdot \boldsymbol{\beta}$$

The least squares solution of the matrix equation for $\boldsymbol{\beta}$ is,

$$(\mathbf{X}^T\mathbf{X})^{-1}\mathbf{X}^T\mathbf{Y} = \boldsymbol{\beta}$$

where ^{T} indicates the transpose of the matrix and $^{-1}$ represents the inverse of a matrix.

The mean sum of squares about the regression, s^2, is the estimate of variance of the responses about the regression line,

$$s^2 = \mathbf{Y}^T\mathbf{Y} - \mathbf{b}^T\mathbf{X}^T\mathbf{Y}$$

The standard errors in the b values are given by the diagonal elements from

$$se(b) = s\sqrt{(\mathbf{X}^T\mathbf{X})^{-1}}$$

The confidence interval for the regression is

$$y^0 \pm t_{n-p,\alpha}\, s\sqrt{\mathbf{X}^0(\mathbf{X}^T\mathbf{X})^{-1}\mathbf{X}^{0T}}$$

The confidence limits for a new y is

$$y^0 \pm ts\sqrt{1 + \mathbf{X}^0(\mathbf{X}^T\mathbf{X})^{-1}\mathbf{X}^{0T}}$$

Bibliography

Bell, R.P. and Tranter, R.L. (1974) Rates and hydrogen isotope effects in the ionisation of 1,1-dinitroethane. *Proceedings of the Royal Society London A*, **337** 517-527.

Draper, N.R. and Smith, H. (1981) *Applied Regression Analysis*, Wiley, Chichester, ISBN 0 471 02995 5.

Gemperline, P.J., Long, J.R. and Gregoriou, V.G. (1990) Spectroscopic calibration and quantitation using artificial neural networks. *Analytical Chemistry*, **62** 1791-1797.

Gemperline, P.J., Long, J.R. and Gregoriou, V.G. (1991) Non-linear multivariate calibration using principal component regression and artificial neural networks. *Analytical Chemistry*, **63** 2313-2323.

Ryan, T.P. (1997) *Modern Regression Methods*, Wiley, Chichester, ISBN 0 471 52912 5.

Sillen, L.G. and Ingri, N. (1962) High speed computers as a supplement to graphical methods. Part I. *Acta Chimica Scandinavica*, **16** 172-191; (1964) Part II. *Arkiv fur Kemi*, **23** 97-121.

Tranter, R.L. and Davies, B. (1993) Spectroscopic multicomponent analysis, in *UV Spectroscopy Techniques, Instrumentation and Data Handling*, (eds. B.J. Clark, T. Frost and M.A. Russell), Chapman and Hall, London.

Wasserman, P.D. (1989) *Neural Computing Theory and Practice*, Van Nostrand Reinhold, New York, ISBN 0 442 20743 3.

Wentworth, W.E. (1965) Rigorous least squares adjustment: application to some non-linear equations. Part I. *Journal of Chemical Education*, **42** 96-103, 1965; Part II, **42** 162-167.

12 Latent variable regression methods
O.M. Kvalheim

12.1 Introduction

The most commonly used method for generating regression models, multiple linear regression (MLR), is usually applied in situations with a single response and, perhaps, a small number of predictors. It also includes a number of assumptions about the data that must be adhered to (more or less strictly) if reliable models are to be formed.

Most real experiments, particularly where spectroscopic measurements are made or where pilot plant or full-scale production is involved, can generate large quantities of data for many predictors and several responses. In addition, the data may have non-Normal distributions, may not have zero error in the predictors and may have colinearity between predictors. It would be a shame not to be able to generate models from the data because of the limitations of MLR.

This chapter concentrates on latent variable regression (LVR) methods as a means of handling these problems in an economic and practical way. The goal is not to cover all possible situations, but rather to give some hints and explanation of when and why these methods are useful.

12.2 Where to go

Information needed	Go to Section(s)
What sort of data can the methods handle?	12.3, 12.10, 12.11, 12.12
Why shouldn't I use MLR?	12.4, 12.10.5, 12.11
What is PCR?	12.6, 12.7
What is PLS?	12.7, 12.10.7
Can I compare a PCR or PLS model with one I already have from MLR?	12.8
How can I get a feel for what my data is like, e.g. possible outliers or other features?	12.9, 12.10.4, 12.10.8, 12.11.2, 12.11.4
Which is better to use, PCR or PLS?	12.10, 12.11

12.3 Aims of regression analysis

Data produced in laboratories and factories are commonly multivariate. The reason for this is the ever-increasing availability and use of sensors for measuring processes and products on hundreds or even thousands of variables. The measurements are, for example, utilised to construct models for predicting quality characteristics from raw material characteristics and process conditions. Figure 12.1 illustrates a common data-analytical situation.

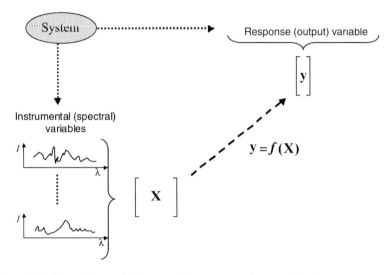

Figure 12.1 Regression modelling to predict a response from multivariate measurements.

The aim of regression analysis is to establish a predictive model between one or several response variables (also called output variables) that we can label $\{y_1, y_2, \ldots, y_L\}$ and one or several input variables (also called predictive variables) that we can label $\{x_1, x_2, \ldots, x_M\}$. Each y is related to the x:

$$\begin{aligned}
y_1 &= f(x_1, x_2, \ldots, x_M) \\
y_2 &= f(x_1, x_2, \ldots, x_M) \\
&\cdots \\
y_L &= f(x_1, x_2, \ldots, x_M)
\end{aligned} \quad (12.1)$$

For example, y_1 maybe the yield of reaction, y_2 the purity of product, y_3 the average crystal size of the product. Each y generates a vector comprising the measurements on the product from several experiments. Similarly for each x. x_1 could be reaction temperature, x_2 the volume of solvent, x_3 the pH of the quenching solvent, x_4 the temperature at which crystallisation starts. Each x generates a vector of values corresponding to the experiment conditions that generated the y values.

We can collect the input variables in a vector **x**. This leads to the following relation for the response variables:

$$y_i = f(\mathbf{x}), \quad i = 1, 2, \ldots, L \quad (12.2)$$

The relation expressed by equation (12.2) is called a response model. In order to determine the relation between the responses and the input variables (the response models), the responses and the input variables are measured for a training set. This can be expressed as

$$\mathbf{Y} = f(\mathbf{X}) \quad (12.3)$$

With N measurements for the responses, the dimensions of the matrices **Y** and **X** are $N \times L$ and $N \times M$, respectively.

Under the constraint of linear models, we can write equation (12.3) as

$$\mathbf{Y} = \mathbf{XB} + \mathbf{E} \quad (12.4a)$$

The matrix **B** contains the regression coefficients, i.e. the parameters needed for estimating the output variables from the input variables. For one output variable, $\mathbf{y}_i$, the model reduces to

$$\mathbf{y}_i = \mathbf{Xb}_i + \mathbf{e}_i \quad (12.4b)$$

In this chapter, we will focus on the latter situation since it is always possible to model each response independently and the principles are generally the same whether we look at one or several responses. Thus, we drop the subscript i and write the regression relation in equation (12.4b) as

$$\mathbf{y} = \mathbf{Xb} + \mathbf{e} \quad (12.5)$$

The reason for establishing regression models may be reduced 'response' time in the information process, e.g. by moving from off-line to at-line or on-line quality measurements, or to substitute expensive and/or subjective methods with cheap and/or more objective ones. Although the predictive aspect is stressed here, we shall not forget that predictive modelling is also used for exploratory purposes, i.e. for uncovering correlation structures in the x-variables related to one or several y-variables.

The performance of the various regression methods differs depending on the specific problem at hand. It must always be kept in mind that a method that is excellent for certain purposes, under certain conditions and premises, may be useless when the aims of the analysis, the premises and/or the experimental conditions and the data structure change. We should also note that regression modelling is a stepwise and iterative approach, and that partly different strategies and methods may be preferred by different data analysts. Having said that, different tastes in model validation, outlier detection and variable selection criteria for a particular regression problem should not lead to highly different results in terms of the final prediction model.

12.4 Multiple linear regression

The multiple linear regression (MLR) solution of equation (12.5) is obtained by minimising the sum of squared response residuals, i.e. by minimising $\mathbf{e}^T\mathbf{e}$ (see also Chapter 11). This leads to

$$\mathbf{b}_{MLR} = (\mathbf{X}^T\mathbf{X})^{-1}\mathbf{X}^T\mathbf{y} \qquad (12.6)$$

By means of the regression coefficients from equation (12.6), we can predict the response variable

$$\hat{y}_{MLR} = \mathbf{x}^T\mathbf{b}_{MLR} \qquad (12.7)$$

The 'hat' (ˆ) on y symbolises a predicted value.

Multiple linear regression is of limited usefulness for analysing and modelling data in such situations as shown in Figure 12.1. One reason for this is that it may not be possible to calculate the inverse $(\mathbf{X}^T\mathbf{X})^{-1}$. With an increasing number of variables, the variables can be expected to be partially correlated. This may lead to linear dependences between x-variables. This is easy to understand if we think of an infrared spectrum measured on a process stream. Such a spectrum may consist of absorbance recorded on thousands of wavelengths, but the inherent complexity of the sample may be reflected in only a few ingredients being blended in the process. Thus, the variation in absorbance can be described by a few underlying factors (sometimes slightly more than the number of blended ingredients owing to interactions and other influencing factors).

The colinearity problem leads to a breakdown of the basic assumption for MLR modelling that the inverse $(\mathbf{X}^T\mathbf{X})^{-1}$ can be calculated. Colinearity also arises when the number of training set samples N is smaller or equal to the number of input variables M. Other methods are needed to solve the regression problem under such circumstances.

12.5 Use of generalised inverse to circumvent the colinearity problem

The solution to equation (12.5) in case of colinearity problems is achieved by introducing the so-called generalised inverse, $\mathbf{X}^+$. The generalised inverse, also called the Moore–Penrose inverse, is defined as

$$\mathbf{X}^+ = (\mathbf{X}^T\mathbf{X})^{-1}\mathbf{X}^T \tag{12.8}$$

The formal solution to equation (12.5) thus becomes

$$\mathbf{b}_{GI} = \mathbf{X}^+\mathbf{y} \tag{12.9}$$

The subscript GI is an abbreviation for generalised inverse.

The predicted response values are obtained as

$$\hat{y}_{GI} = \mathbf{x}^T\mathbf{b}_{GI} \tag{12.10}$$

12.6 Principal component regression

The general inverse represents a formal solution to the problem of colinearity. Principal component regression (PCR) (see for example Jolliffe, 1986) is another method for coping with the colinearity problem. With this method, the dimensionality of the input data $\mathbf{X}$ is first reduced to produce A orthogonal (uncorrelated) principal components:

$$\mathbf{X} = \mathbf{T}_{PC}\mathbf{P}_{PC}^T = \sum_{a=1}^{A}\mathbf{t}_{PC,a}\mathbf{p}_{PC,a}^T + \mathbf{F}_{PC} \tag{12.11}$$

The matrices, $\mathbf{T}_{PC}$ and $\mathbf{P}_{PC}$ consist of score $\{\mathbf{t}_{PC,a}\}$ and loading $\{\mathbf{p}_{PC,a}\}$ vectors, respectively, for each principal component. The principal components of the matrix $\mathbf{X}$ are calculated by minimising the residual variation in $\mathbf{X}$, $\mathbf{F}_{PC}$, under the constraint of orthonormal loadings and orthogonal scores vectors. The x-variables are assumed to be mean-centred (see Chapter 10).

With the transformation of data into principal components, the true complexity of the data is revealed and MLR can be carried out on these reduced data. Thus, the scores $\{\mathbf{t}_{PC,a}\}$ for the a principal components, are used for establishing a regression model:

$$\mathbf{y} = f(\mathbf{t}) = f(\mathbf{t}_1, \mathbf{t}_2, \ldots, \mathbf{t}_A) \qquad (12.12a)$$

$$\mathbf{y} = \mathbf{T}_{PC}\mathbf{b}_{PCR} + \mathbf{e}_{PCR} \qquad (12.12b)$$

Equations (12.12a) and (12.12b) show that the scores contain the quantitative information in $\mathbf{X}$ about the (mean-centred) response variable $\mathbf{y}$. From equation (12.12), the regression coefficients are calculated as

$$\mathbf{b}_{PCR} = (\mathbf{T}_{PC}^T \mathbf{T}_{PC})^{-1} \mathbf{T}^T \mathbf{y} \qquad (12.13a)$$

Owing to the orthogonality between score vectors, the matrix $\mathbf{T}_{PC}^T \mathbf{T}_{PC}$ is diagonal and the regression coefficient on each principal component, $b_{PCR,a}$ can be calculated independently of the others as

$$b_{PCR,a} = \mathbf{t}_{PC,a}^T \mathbf{y} / (\mathbf{t}_{PC,a}^T \mathbf{t}_{PC,a}), \quad a = 1, 2, \ldots, A \qquad (12.13b)$$

Thus, by decomposing $\mathbf{X}$ into principal components, an algorithm similar to Yates's algorithm for calculating regression coefficients in orthogonal designs can be used (see, e.g., Box et al., 1978). This property of principal components (and other latent-variable regression methods as well, see below) is utilised when we build and validate models by means of the so-called cross-validation technique (see, e.g., Martens and Næs, 1989).

The elements of the diagonal matrix $\mathbf{T}_{PC}^T \mathbf{T}_{PC}$ are the eigenvalues of the matrix $\mathbf{X}^T \mathbf{X}$. These eigenvalues are proportional to the variance explained by each principal component and play a role in assessing the dimension (number of significant principal components) of the model.

In Figures 12.2a and 12.2b the PCR procedure is illustrated for a one-analyte system measured at two wavelengths. We measure spectra for a set of samples with known concentrations y_i of the analyte. In the absence of interferents, the scores of the first principal component are proportional to the concentration (Figure 12.2a). The spectral residuals correspond to instrumental noise.

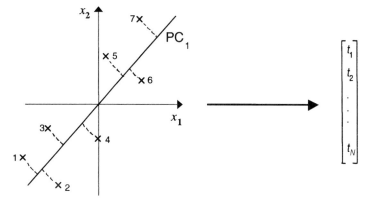

(a)

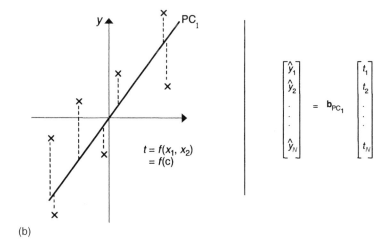

(b)

Figure 12.2 (a) Principal component model of the input variables. (b) Use of PC scores as input variables to the regression model.

Note the way that the principal component and the regression models are calculated. For the principal component model, it is the sum of squared residuals perpendicular to the principal component that is minimised (Figure 12.2a), while for the regression model, it is the sum of squared residuals in the response residuals that is minimised (Figure 12.2b). When decomposing into principal components, all variables are treated equally with regard to minimising the difference between the model and the measured data. For the regression model, we want to minimise the difference between the measured and predicted response. The two steps in the PCR modelling process are thus defined as

$$\text{Minimise} \sum_{i,j}(x_{i,j} - \hat{x}_{i,j})^2 \qquad (12.14\text{a})$$

$$\text{Minimise} \sum_{i}(y_i - \hat{y}_i)^2 \qquad (12.14\text{b})$$

The use of principal components scores instead of the original variables reduces the noise in the x-variables. Scores are also important for identifying outliers, grouping of data and bad design of training sets.

If interferents are present, we need more than one principal component to model the response variable. However, if the spectral signals are additive for the quantified analyte and the interferents, we can still model the concentration of the analyte in terms of scores describing the desired analyte and the interferents.

The loadings are important for interpretative purposes and variable reduction, i.e. for identifying the most important input variables for variable reduction and for interpretation of the regression models. In this sense, the loadings together with the regression coefficients play an important role for a 'knowledge-based' validation of the model in addition to the statistical one, usually, cross-validation and explained variance in input and response variables (Figure 12.3).

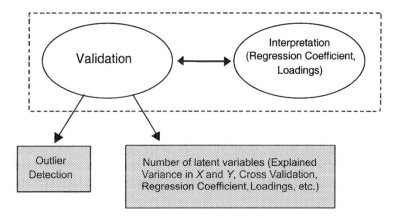

Figure 12.3 Pictorial representation of the LVR modelling process.

Different criteria are available for selecting the principal components to be used in the regression:

- Statistical criteria such as variance explained in **X**, variance explained in **y**, and predictive ability of the principal components.
- Knowledge of the variables, e.g. correspondence between spectral and structural features, and shape, i.e. smoothness of loadings and regression coefficients for spectra x-variables.

Usually, these criteria are used simultaneously in order to make the best possible decision.

12.7 Latent-variable regression methods

We shall now show that PCR is only one of a class of regression methods based on decomposing **X** into so-called latent variables. The term latent variable signifies a linear combination of the measured variables. Actually, the number of different ways of decomposing data into latent factors and then constructing a regression model from these latent factors is unlimited, as we shall see in this section. The key point is to select an appropriate and robust regression method for the problem at hand. We shall illustrate this by some examples.

Decomposition into principal components represents a least squares solution that minimises the residuals $\mathbf{F}_{PC}$ in equation (12.11). Other criteria are available, e.g. using the covariance ('overlap') between the input variables and one or more response variables as in partial least squares (PLS) decomposition (see, for example Martens and Næs, 1989) to obtain components more relevant to the responses, or using so-called markers (variables or objects, see Kvalheim, 1987). An algorithm for a general decomposition can be obtained through a slight modification of the so-called NIPALS algorithm, i.e. an algorithm for the decomposition of **X** by successive projections, each projection providing a latent variable (Kvalheim, 1987):

$$\text{Define } \mathbf{X}_1 = \mathbf{X}.$$

Repeat for $a = 1, 2, \ldots, A$ (A = number of latent variables extracted).

1. Select a latent variable, i.e., a weighting vector $\mathbf{w}_a$; $\|\mathbf{w}_a\| = 1$ and its dimension is $M \times 1$.
2. Project the row vectors in **X** on $\mathbf{w}_a$, i.e. calculate the column vector of scores;

$$\mathbf{t}_a = \mathbf{X}_a \mathbf{w}_a.$$

3. Calculate the row vector of variable loadings, $\mathbf{p}_a$, expressing the covariances between the latent variable a and the original variables, i.e.

$$\mathbf{p}_a^T = \mathbf{t}_a^T \mathbf{X}_a / (\mathbf{t}_a^T \mathbf{t}_a)$$

4. Remove the dimension of $\mathbf{X}_a$ associated with $\mathbf{w}_a$: $\mathbf{X}_{a+1} = \mathbf{X}_a - \mathbf{t}_a \mathbf{p}_a^T$.

Different decompositions of $\mathbf{X}$ can be defined through different choices of $\{\mathbf{w}_a\}$. Some methods for decomposition are:

- Principal component decomposition,

$$\mathbf{w}_a = \mathbf{p}_a / \|\mathbf{p}_a\|$$

where $\mathbf{p}_a$ is the (converged) loading vector for the latent variable a.
- Partial least squares decomposition,

$$\mathbf{w}_a = \mathbf{u}_a^T \mathbf{X}_a / \|\mathbf{u}_a^T \mathbf{X}_a\|$$

where $\mathbf{u}_a$ are the (converged) latent variable column vectors for the response variable block (equal to $\mathbf{y}$ with one predicted variable only).
- Marker variable projection (MVP),

$$\mathbf{w}_a = \mathbf{g}_j$$

where $\mathbf{g}_j$ is the vector of zeros except for position j, which is 1 when variable j is projected on (one can also use linear combination of selected variables).
- Marker object projection (MOP),

$$\mathbf{w}_a = \mathbf{x}_i / \|\mathbf{x}_i\|,$$

when object i is projected on (one can also use linear combinations of selected variables).

As long as the orthogonalisation (step 4) in the algorithm above is included, equations (12.12a), (12.12b), (12.13a) and (12.13b) are valid for any latent variable decomposition.

12.8 Relationship between regression coefficients in LVR and MLR

If $\mathbf{X}$ is not colinear, i.e. of full rank, it follows that $\mathbf{y} = \mathbf{X}\mathbf{b}_{MLR} = \mathbf{T}\mathbf{P}^T\mathbf{b}_{MLR}$ and $\mathbf{y} = \mathbf{T}\mathbf{b}_{LVR}$. By combining these two expressions, it follows that

$$\mathbf{b}_{LVR} = \mathbf{P}^T\mathbf{b}_{MLR} \qquad (12.15)$$

Equation (12.15) states that the regression coefficients for the latent variables can be expressed as weighted sums of the regression coefficients for the original variables calculated by MLR. The loadings represent the weight factors. Large loading on a variable implies a large contribution to the latent-variable regression coefficient if the variable in question simultaneously has a reasonably large regression coefficient in MLR. Since a large coefficient in MLR means a large contribution to the prediction of the response $\mathbf{y}$, it is clear that variables that simultaneously explain well the variance in $\mathbf{X}$ and the prediction of $\mathbf{y}$ can be selected by LVR. Thus, LVR represents a way of variable selection. Furthermore, since we do not need to include all the latent variables from the decomposition of $\mathbf{X}$, we can select only those latent variables with a high explanatory power both in $\mathbf{X}$ and $\mathbf{y}$. In this way, noise reduction is achieved. On the other hand, if a large part of the variance relevant for $\mathbf{y}$ is obtained in the minor PCs, the prediction ability is reduced, i.e. there is a trade-off between including relevant information and noise.

It is useful to express the regression coefficients in LVR in terms of the original variables. By a straightforward calculation (Kvalheim and Karstang, 1989) it follows that

$$\mathbf{b}_{LVR,x} = \mathbf{W}_{LV}(\mathbf{P}_{LV}^T\mathbf{W}_{LV})^{-1}\Lambda^{-1}\mathbf{T}_{LV}^T\mathbf{y} \qquad (12.16)$$

The subscript LVR,x on the regression coefficients indicates that the regression vector is expressed in the original input variables. The matrices $\mathbf{W}_{LV}$, $\mathbf{P}_{LV}$ and $\mathbf{T}_{LV}$ are the weightings, loadings and scores obtained from the decomposition of $\mathbf{X}$ (see Section 12.5). The matrix Λ is diagonal with elements being the eigenvalues of the matrix $\mathbf{X}$. Comparison with equation (12.9) shows that the general inverse for a latent variable regression model can be written as

$$\mathbf{X}^+ = \mathbf{W}_{LV}(\mathbf{P}_{LV}^T\mathbf{W}_{LV})^{-1}\Lambda^{-1}\mathbf{T}_{LV}^T \qquad (12.17)$$

12.9 Residual standard deviation and leverage for outlier detection

The residual standard deviation (RSD) for a sample is defined through

$$\text{RSD}_i^2 = \mathbf{f}_i^T \mathbf{f}_i / (M - A) \tag{12.18}$$

The vector $\mathbf{f}_i$ is the residual vector of sample i after being fitted to the latent variable model of the x-variables. A large RSD for a sample means that the sample is dissimilar to the modelled samples and maybe rejected as an outlier (see Chapter 2 for a discussion of outliers and their rejection). However, for a sample included in the calculation of the regression model, a small RSD may be indicative of an outlier. This is the case also if the sample simultaneously has a small RSD and a large leverage. By substituting equation (12.13a) into equation (12.12b), we obtain

$$\mathbf{y} = \mathbf{T}_{\text{LV}}(\mathbf{T}_{\text{LV}}^T \mathbf{T}_{\text{LV}})^{-1} \mathbf{T}_{\text{LV}} \mathbf{y} + \mathbf{e}_{\text{LV}} \tag{12.19a}$$

which can also be written as

$$\hat{\mathbf{y}} = \mathbf{T}_{\text{LV}}(\mathbf{T}_{\text{LV}}^T \mathbf{T}_{\text{LV}})^{-1} \mathbf{T}_{\text{LV}}^T \mathbf{y} \tag{12.19b}$$

From this we can define the so-called 'hat' matrix,

$$\mathbf{H} = \mathbf{T}_{\text{LV}}(\mathbf{T}_{\text{LV}}^T \mathbf{T}_{\text{LV}})^{-1} \mathbf{T}_{\text{LV}}^T \tag{12.20}$$

The term 'hat' matrix stems from the result of applying $\mathbf{H}$ to the response vector $\mathbf{y}$: we obtain $\mathbf{y}$ with a hat, that is the predicted response vector.

From equations (12.19b) and (12.20), we observe that

$$\hat{y}_i = \mathbf{h}^T \mathbf{y} = h_{i,1} y_1 + h_{i,2} y_2 + \cdots + h_{i,N} y_N \tag{12.21}$$

Equation (12.21) shows that the predicted value of the response for a sample is a weighted sum of the responses of all samples in the training set. The weighting factor is the leverage of the sample in the x-variables. If a sample has a large leverage, it means that it influences the regression model greatly by 'grabbing' a lot of the explained variance in the model. Strong outliers will have large leverage and, at the same time, a low RSD. A plot of RSD against leverage is, therefore, a useful tool for detecting strong outliers.

12.10 Analysis of historic data from an industrial process

12.10.1 The problem

Our first example is from an industrial process in which it was wanted to relate variations in the amount of product produced by the process to variations along the process line. The process variables are masked, but this does not constrain our ability to demonstrate various approaches to data evaluation and modelling.

12.10.2 The data

The data were collected during half a year and daily averages for the 27 weeks are given in Table 12.1.

It should be mentioned that this a slow process with many steps and that it takes several days from start to the end of process line. Therefore, daily averages for each week are appropriate in this application. Furthermore, Table 12.1 provides the variables that had been selected

Table 12.1 Data for industrial process

Week	Amount	Water	Pore-1	Cryst	Pore-2	Stop
18	325	9.2	109	140	33.1	0.29
19	304	9.4	105	130	34.1	0.29
20	306	9.1	108	154	32.7	0.43
21	313	9.8	87	242	32.5	0.29
22	284	9.4	79	290	32.7	0.43
23	267	8.2	74	283	31.2	0.14
24	236	8.3	73	253	30.9	0.57
25	215	9.6	81	159	33.1	0.14
26	225	10.1	73	195	34.4	1.00
27	232	9.8	79	194	33.6	0.14
28	240	9.6	87	152	34.7	0.57
29	261	10.3	101	131	35.3	0.86
30	297	10.8	108	120	35.0	0.29
31	272	11.0	103	132	33.2	0.29
32	274	10.0	109	118	33.7	0.43
33	268	10.7	89	164	34.1	0.71
34	284	10.6	92	153	33.3	0.14
35	309	12.2	94	139	34.2	0.57
36	307	11.8	92	154	33.5	0.43
37	338	13.1	90	152	34.9	0.29
38	330	14.0	86	163	34.3	0.43
39	350	12.4	90	150	34.9	0.29
40	367	12.4	90	150	36.0	0.14
41	347	13.7	87	142	37.5	0.43
42	360	12.3	85	198	34.7	0.57
43	349	12.4	81	225	36.5	0.43
44	344	11.8	80	218	35.5	0.29

using engineers' knowledge as important for variation in amount produced by the process.

12.10.3 Weekly production, the response

A plot of the daily average production for weeks 18–44 (Figure 12.4), shows a large variation. Integrating the production curve from a base line of 367 ton (representing the maximum daily production during a week) and dividing by the number of weeks, estimates the average loss to 70 ton per day for the recorded period.

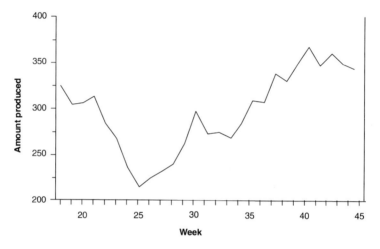

Figure 12.4 Average daily production during weeks 18–44.

12.10.4 Exploratory analysis by PCA

The data were first analysed by principal component analysis to look for outliers and variable correlations. Figure 12.5 shows a biplot of the data standardised to unit variance and projected onto the first two PCs. Evidently, no strong outliers are observed. We observe, however, a strong positive correlation between water content (%Water), porosity (Pore-2), and, amount produced (Amount), showing that high water content and high pore number provide high production. Weeks with high daily average production are located in the same region of the biplot as these three variables.

The variables crystallinity (Cryst) and another porosity measure (Pore-1) make an almost 90° angle with Amount and are therefore of minor importance for the variation in production. The variable Stop

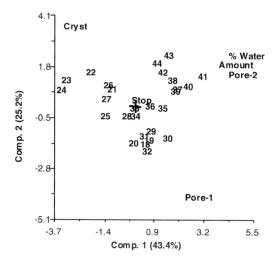

Figure 12.5 Biplot of industrial data.

measures the time when a special part of the process is stopped and feed is taken from silos. The biplot shows that *Stop* is located close to origin, indicating that this variable does not influence the overall amount produced by the process. However, a plot of the contributions of each variable to each PC (Figure 12.6), shows that *Stop* is not at all explained by the two-component PC model. Therefore, we cannot make any statements about this variable from the two-component PC model.

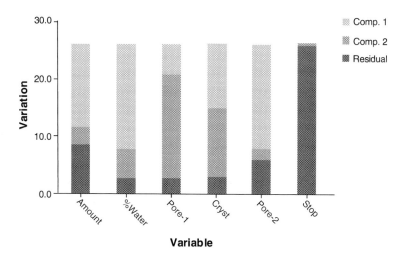

Figure 12.6 Distribution of variation in the PC model.

12.10.5 *Regression analysis, MLR*

Figure 12.7 shows a bar diagram of the regression coefficients for the MLR model with *Amount* as response and the other five variables as input variables.

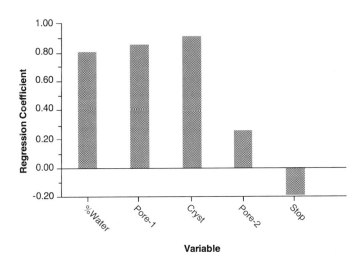

Figure 12.7   Regression coefficients from MLR.

Tall bars indicate high importance, while low bars represent variables of small importance. A bar on the positive side of the baseline indicates that the input variable is positively correlated to the response, while a bar on the negative side indicates a negative relationship. The regression coefficients for MLR thus indicate that *Cryst*, *Pore-1* and *%Water*, in descending order, are the major contributors to the variations in amount produced. In passing, we note that this result contradicts the results from the exploratory PCA. Predicted versus measured production shows a reasonable model with a correlation of 0.9. A Normal probability plot of the deviations between measured and predicted *Amount* shows a tendency for grouping that might be taken as a warning that the MLR model is not reliable.

12.10.6 *Regression analysis, PCR*

Table 12.2 summarises the result of regressing the principal components on produced *Amount*. The explained variance in *Amount* by each PC implies that the first, third, and fifth PCs are most significant for

Table 12.2 Summaries of variance explained by principal components

% Variance explained		Improve prediction?
in x	in y	
44.1	25.3	Yes
28.4	6.5	Yes
19.7	17.0	Yes
4.6	1.1	No
3.3	31.8	Yes

predicting the average daily production. These three PCs also show good predictive ability for the response. However, the low variance in input variables for PC5 indicates that, despite the high correlation with the response, this PC may not be significant after all. Figure 12.8 shows the variables' contributions to each PC.

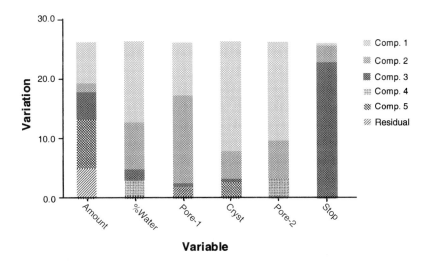

Figure 12.8 Distribution of variation in the five-component PC model.

We observe that PC5 has only a very small contribution from *Pore-1* and *Cryst*. The strong correlation with *Amount* is probably a result of a 'lurking variable', i.e. a variable not included in the data but strongly influencing the response and to some degree correlated with minor parts of the variance patterns in *Cryst* and *Pore-1*.

A plot of scores on PC5 in the time direction (Figure 12.9) shows a pattern of high positive values in the first weeks and then a change to high negative scores around week 25. The pattern seems to repeated from week 29, but with much smaller deviations from the zero mean.

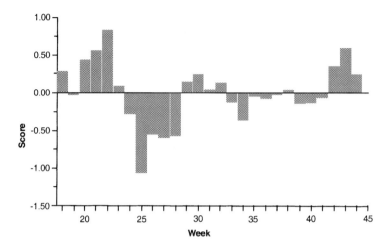

Figure 12.9 Scores on PC5 showing a systematic pattern with time.

The regression model with the three PCs explaining most in terms of the input variables is shown in Figure 12.10. This is in agreement with the

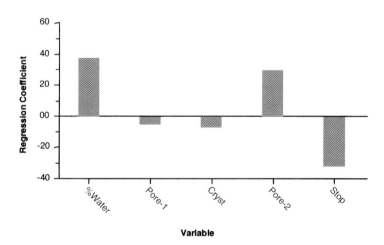

Figure 12.10 Regression coefficients for three-component PCR model.

PCA biplot from the exploratory analysis (Figure 12.5), with *%Water* content and *Pore-2* as the most important variables. However, the variable *Stop* is now also revealed as important. Furthermore, *Stop* is negatively correlated to amount produced, providing us with crucial process information. The regression coefficients for this three-component PCR model are completely different from those of MLR (Figure 12.7). The MLR approach is overfitting by including the small variation in PC5, which happens to correlate to a large extent with the response. Thus, MLR may point to input variables with little or no real explanatory significance when a lurking variable imprints its correlation structure on some of the input variables! On the other hand, PCR provides the capability for noise reduction because principal components with small contributions from the input variables and/or small contributions to the response can be removed from the model.

A plot of predicted and measured produced *Amount* (Figure 12.11) shows that the PCR model is not good. In the first production weeks (18–22) the model predicts too low a produced *Amount*, then from week 25 to week 28 the model predicts values that are too high. From week 29, the correspondence between predicted and measured production is excellent, except for week 42. Note that the deviation pattern between measured and predicted is almost identical to what we observed in the score plot of the fifth principal component (Figure 12.9), both plots pointing towards a lurking variable.

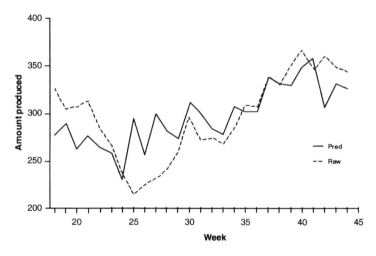

Figure 12.11 Measured and predicted *Amount* from three-component PCR model.

12.10.7 Regression analysis, PLS

The modelling is next done with PLS. However, in order to avoid too much repetition, we will now do another type of model validation. Instead of using all weeks for building the regression model, we select every second week in Table 12.1 and use the other weeks as an independent prediction set. Cross-validation of the 14 weeks thus selected showed that only the first PLS component was significant, explaining 58.5% of the variation in produced *Amount* and 42.4% of the input variables standardised to unit variance. This is further confirmed by inspecting the average prediction error (Figure 12.12). The average prediction error or, more precisely, the standard error of validation (SEV) is defined through

$$\text{SEV}_a^2 = \frac{\sum_i (y_i - \hat{y}_i)^2}{N - a - 1}, \quad a = 1, 2, \ldots, A \quad (12.22)$$

Cross-validation is assumed in calculating SEV_a, excluding samples successively and predicting them from the model of the others. A major

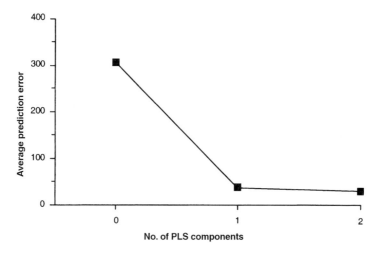

Figure 12.12 Average prediction error plotted as a function of number of PLS components.

improvement is observed on moving from zero to one PLS component, but only minor improvement is achieved by adding the second PLS component.

The regression coefficients (Figure 12.13) show that %*Water* content, *Stop* and *Pore-2* (in descending order) are the most critical variables for predicting produced *Amount*. This is consistent with PCR (Figure 12.10), but PLS needs only one component compared to three for PCR. This example shows the superiority of PLS not only compared to MLR but

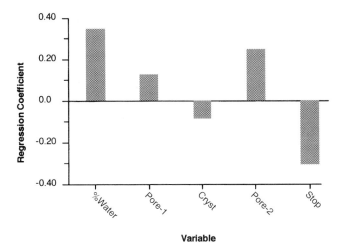

Figure 12.13 Regression coefficients for one-component PLS model.

also with respect to PCR in creating a much more parsimonious latent variable regression model.

Predicted versus measured *Amount* for this PLS model gives the picture shown in Figure 12.14. Following the points in the time direction, we observe that at first the PLS model predicts lower *Amount* produced than measured. This pattern changes at week 26. From this week and onwards to week 34, the predicted values are all significantly higher than the measured values.

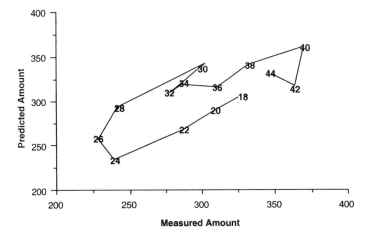

Figure 12.14 Predicted versus measured amounts for the one-component PLS model.

In Figure 12.15, there is a pattern of too low predictions (positive residuals) at the beginning and end of the period, and too high predictions (negative residuals) in the middle. Residual standard deviation (RSD) plotted versus leverage for the samples shows only a weak outlier (week 24) which cannot be responsible for this effect. Thus, the systematic deviation between predicted and measured values shows that a lurking variable is in action. Prediction of the objects not included in the building of the PLS model confirms this conclusion. Furthermore, the plot of RSD shows that the effect cannot be explained by objects being outliers in the input variables.

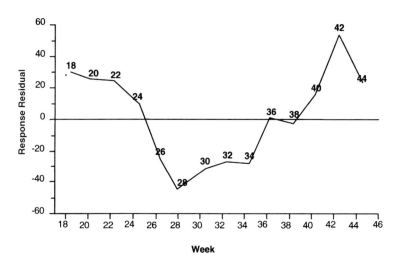

Figure 12.15 Residuals between predicted and measured amounts plotted in the time direction.

12.10.8 *Revealing the lurking variable*

The large negative effect of the variable *Stop* implies that the lurking variable may be related to the intermediate product transformed in this step of the process. Another reason for this suspicion was that, owing to difficulties in measuring at this process step, there was no direct measurement of the quality of the intermediate product. Thus, we needed to identify an indirect variable telling the same story as the desired one. An environmental variable, SO_2, turned out be a possible candidate. Increased use of sulphur dioxide was a clear sign of low quality on the intermediate product. A plot of sulphur dioxide versus produced *Amount* shows a negative correlation (Figure 12.16).

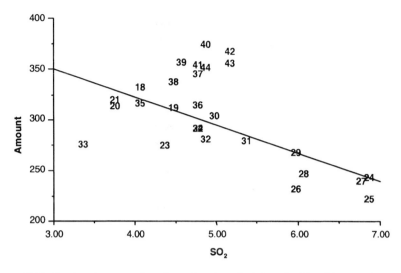

Figure 12.16 Produced amount plotted as a function of sulphur dioxide. PLS regression model with lurking variable included. Slope = −27.612, Intercept = 432.153, $R = -0.577$, $R^2 = 0.333$.

PLS modelling of the data with the lurking variable included gave two components explaining 77.6% of the variation in produced *Amount*. Both the plot of average prediction error and cross-validation gave unambiguously two significant PLS components.

Regression coefficients and predicted versus measured *Amount*s are shown in Figures 12.17 and 12.18, respectively. From Figure 12.17 we observe that the lurking variables together with the %*Water* content comprise the dominating input variables. The plot of predicted versus measured *Amount* further shows that the systematic over- and under prediction has disappeared. Plotting RSD versus leverage identifies week 24 as a weak outlier. However, removal of this object makes only a slight change to the model.

12.11 Mixture design, spectroscopy and PLS regression for predicting concentrations

The need for fast methods for on-line quality control has led to the use of various spectroscopic techniques to map chemical composition or properties related to chemical structure as spectral profiles. These profiles are next linked to concentrations or quality characteristics by regression modelling. This approach has several advantages compared with traditional approaches. For one, no selective wavelengths are needed for

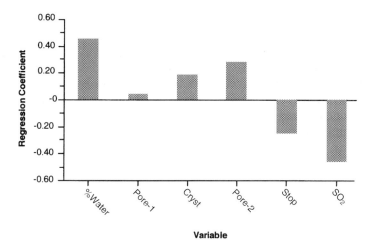

Figure 12.17 Regression coefficients for two-component PLS model.

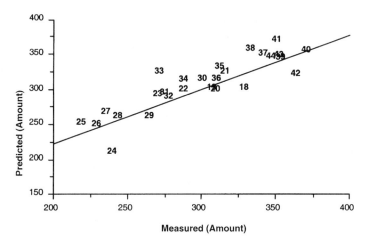

Figure 12.18 Predicted vs measured production. Slope = 0.775, Intercept = 66.578, $R = 0.881$, $R^2 = 0.775$, Adj. Corr. = 0.757.

prediction of a response. The only requirement is that there is a signal part in the profile for the response uncorrelated to other responses mapped by the profile and with an S/N ratio that is good enough for quantitative prediction. Furthermore, since the spectral profiles change when the system is changing, spectral profiling and modelling can provide quantitative information about deviation from normal spectral patterns and thus provide tests for outlier detection.

Our next example is such a case: multivariate calibration is used to determine by PCR the concentration of one analyte in the presence of two interfering analytes. First we illustrate the robustness of latent variable regression methods towards interferents included in the model. Next we illustrate the breakdown of the model when unmodelled interferents enter the scene. Finally, we demonstrate how selection of spectral windows for calibration can improve the results when unmodelled interferences appear. PLS gives almost exactly the same results as PCR and is not shown here. MLR cannot be used owing to the strong colinearities in the spectral variables.

12.11.1 The data

Forty-five samples were prepared as mixtures of three analytes: methylcyclohexane, ethylbenzene and dibutyl ether. Table 12.3 shows the fraction of each constituent. The samples were analysed by mid-infrared spectroscopy in the C–H stretch region (from 2990 to 2780 cm^{-1}). Spectra were recorded with a data resolution of $1\,cm^{-1}$, thus providing spectral profiles at 211 wavelengths (Toft *et al.*, 1993). The spectra of the three pure analytes are shown in Figure 12.19. We observe that there are no selective wavelength regions for any of the analytes.

12.11.2 Exploratory analysis of spectra, PCA

Principal component analysis of the raw spectra after variable centring gave the score plot shown in Figure 12.20. The triangular shape of the score plot reflects the mixture design used for preparing the samples. The samples at the corners of the triangle represent spectra of pure analytes. Mixtures of two analytes are located on the sides of the triangle connecting the corners representing the pure spectra of the analytes. Except for noise, all the variation in spectral profiles is mapped in the score plot. This is a result of the centring of the data. Samples labelled by the same number, e.g. 8a and 8c, are replicates, that is they are mixed using the same fractions of the analytes. We note from the score plot that samples 8a, 8c and 10a, 10c, are separated in the score plot, while all other replicates plot jointly as they should. This indicates that one or both replicates for mixtures 8 and 10 are outliers.

On calculating and drawing a dendrogram (Figure 12.21) from the scores on PC1 and PC2, we observe that all replicates link together except those for 8 and 10. This confirms that some of these runs are outliers.

The RSD of all samples (Figure 12.22) show that the samples 8a and 10a have a significantly larger RSD than the other samples. It turned out

Table 12.3 Mixing ratio of 45 samples

Sample label	Methylcyclo hexane	Ethylbenzene	Dibutyl ether
Me_Cy_He	1.00	0.00	0.00
EthylBen	0.00	1.00	0.00
DiBu_Eth	0.00	0.00	1.00
4a	0.50	0.50	0.00
4b	0.50	0.50	0.00
5a	0.50	0.00	0.50
5b	0.50	0.00	0.50
6a	0.00	0.50	0.50
6b	0.00	0.50	0.50
7a	0.33	0.33	0.33
7b	0.33	0.33	0.33
1b	1.00	0.00	0.00
2b	0.00	1.00	0.00
3b	0.00	0.00	1.00
2c	0.00	1.00	0.00
8a	0.75	0.00	0.25
9a	0.25	0.00	0.75
10a	0.75	0.25	0.00
11a	0.51	0.25	0.25
12a	0.25	0.25	0.50
13a	0.00	0.25	0.75
14a	0.27	0.48	0.25
15a	0.25	0.75	0.00
16a	0.00	0.75	0.25
8c	0.75	0.00	0.25
10c	0.75	0.25	0.00
15c	0.25	0.75	0.00
17a	0.87	0.00	0.13
18a	0.62	0.00	0.38
19a	0.37	0.00	0.63
20a	0.13	0.00	0.87
21a	0.87	0.13	0.00
22a	0.64	0.12	0.24
23a	0.37	0.13	0.50
24a	0.25	0.13	0.62
25a	0.00	0.12	0.88
26a	0.62	0.38	0.00
27a	0.50	0.37	0.12
28a	0.12	0.38	0.50
29a	0.00	0.36	0.64
30a	0.38	0.62	0.00
31a	0.17	0.64	0.19
32a	0.00	0.63	0.37
33a	0.12	0.88	0.00
34a	0.00	0.87	0.13

that both of these samples had been left for some time in the laboratory before the IR spectrum was acquired. The reason for the outlying behaviour may thus be evaporation, slightly changing the proportions between the analytes.

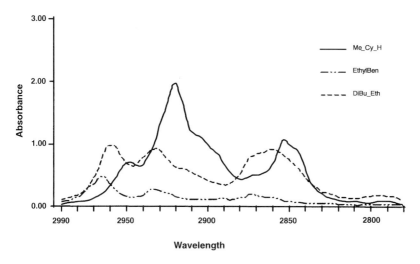

Figure 12.19 Spectra of the three pure analytes.

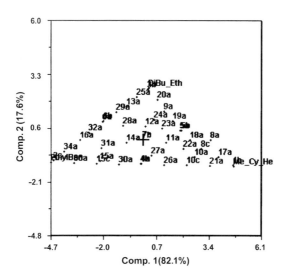

Figure 12.20 Score plot of the 45 samples on the two major principal components.

12.11.3 *Regression analysis, PCR*

The first 27 samples in Table 12.3, excluding 8a and 10a, were modelled by PCR. Two components explained almost 100% of both the spectral variation and the variation in concentration of methylcyclohexane.

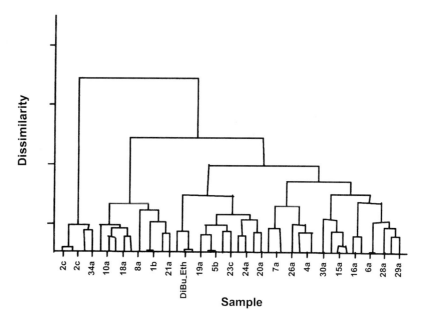

Figure 12.21 Dendrogram of samples calculated from the scores on the two major principal components.

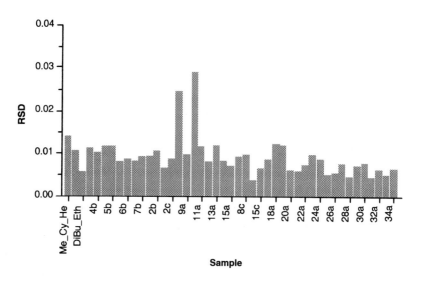

Figure 12.22 Residual standard deviation (RSD) for each sample in the two-component model.

A plot of the average prediction error confirms that two PCs are optimal. Predicted versus measured concentration values showed perfect positive correlation (0.999). Thus, although the two interferents are not explicitly modelled, their presence in the calibration samples ensures good prediction of methylcyclohexane.

Comparison of the regression coefficients (Figure 12.23) with the spectrum of methylcyclohexane reveals a positive relationship between the part of the spectrum where methylcyclohexane is dominant, while the negative parts corresponds to parts of spectra where the two interferents dominate.

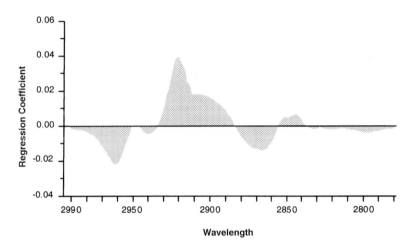

Figure 12.23 (Spectral) regression coefficients for the two-component PLS model.

One can question whether the prediction is just as good for samples not used in the calibration set. For samples outside the calibration set, the prediction is still perfect with a correlation of 0.999. The prediction is also good for 8a and 10a, although the RSD indicates that these samples are outliers. There are two reasons for this. First, the spectral profiles in the region of the largest signal and largest regression coefficients dominate the prediction. For our case, this is the region where methylcyclohexane has the largest signal. This makes the prediction more robust than expected for this analyte. Secondly, the limit of RSD for rejecting samples as outliers is too narrow. This is a result of (i) the replication of samples in the calibration set, and (ii) the very strong colinearities in spectral data. Both effects exaggerate the number of degrees of freedom, resulting in too small a critical value for the F-test used to assess samples as outliers (Grung and Kvalheim, 1994). The sample replication effect is also observed if samples are similar because of the measurement technique or colinearities inherent in the system from which the samples are collected.

12.11.4 Influence of unmodelled interferents

The calibration was repeated including only two-component mixtures containing methylcyclohexane and ethylbenzene. This gave a calibration set with 10 samples selected among the first 27 samples in Table 12.3 (sample 10a was excluded as outlier). Only one PC was necessary to account for almost 100% of the variation in both spectra and concentration of methylcyclohexane. The prediction of the 10 calibration samples was also perfect.

However, if we now predict all the other samples, the result is as shown in Figure 12.24. The predictions are bad, and, as expected, especially bad for the low concentrations of methylcyclohexane. Four samples (21a, 26a, 30a and 33a) show excellent agreement between measured and predicted

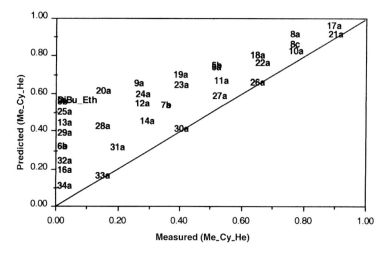

Figure 12.24 Prediction of methylcyclohexane for samples not included in the calibration set.

concentration of methylcyclohexane. Not surprisingly, these samples have no contribution from the unmodelled interferent (see Table 12.3).

If we look at the RSD for the prediction samples, we find that almost all samples are rejected as outliers. Only those with zero concentration of dibutyl ether fall inside the rejection limit. The reason for this is the spectral signal from the unmodelled interferent. We can confirm this by looking at the residual spectrum of a sample with 100% of the interferent: Figure 12.25 shows such a residual spectrum and we observe that this residual spectrum is similar to the spectrum of pure dibutyl ether (Figure 12.19) outside the region where methylcyclohexane is dominating.

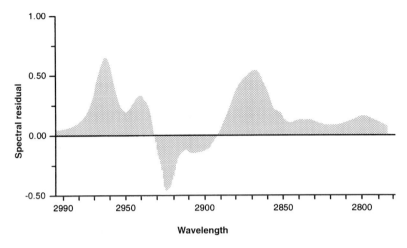

Figure 12.25 Residual spectrum of sample after fitting to the one-component PCR model. The fitted sample contained only the unmodelled interferent.

12.11.5 Reducing the influence of unmodelled interferences by wavelength selection

We can use the regression coefficients (Figure 12.26) to select the spectral window where methylcyclohexane dominates (2930–2890 cm^{-1}), to obtain a PCR model with better predictive abilities in the presence of the unmodelled interference.

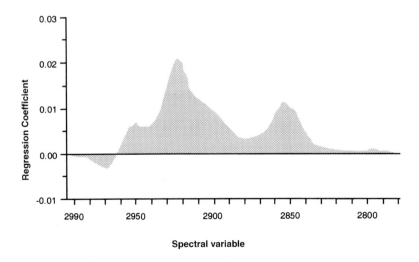

Figure 12.26 Regression coefficient for one-component PCR model.

Figure 12.27 shows the prediction for the samples not included in the calibration model. The results are much better for the 'window' calibration, than for the full spectrum calibration, although not good enough for quantitative use.

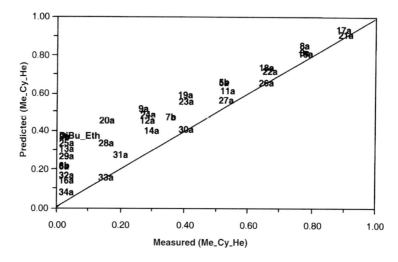

Figure 12.27 Prediction of methylcyclohexane based on 'window' calibration for samples not included in the calibration set.

12.11.6 Correcting for unmodelled interferents

We have seen that to achieve good prediction from a spectral calibration model, the training set must span all the possible interferents in the future samples in addition to the responses we want to predict. Furthermore, the possible range of responses and interferents in the prediction samples must be varied correspondingly in the training set. The problem with degradation of prediction ability caused by unmodelled interferents can, however, sometimes be solved by means of differentiation (see Chapter 2). Thus, if the spectral profile of an interferent has maxima (or minima) at wavelengths where the modelled analytes have signal, the first (and third) derivative of the interferent will have zero contribution at these wavelengths.

This observation can be utilised to produce good prediction even if unmodelled interferents turn up. Unmodelled interferents will show up in the residual spectrum after fitting of a prediction sample to the model. Correction can then be performed by differentiation. The key point is to determine the spectral maxima for the unmodelled interferent. One approach, called local curve fitting (LCF), has been demonstrated by Karstang and Kvalheim (1991). The detection of the maxima in the interferents is crucial for the method. This step can be aided by so-called sequential rank analysis (SERA) (Liang *et al.*, 1994) on the training set augmented by the prediction sample with the unmodelled interferent. The SERA techniques detect the presence of maxima as a drop in local rank of the matrix after differentiating in the spectral direction. Comparison of the result from SERA with the result of local rank analysis on the training set data reveals the maxima for the unmodelled interferent.

12.12 Additional hints for spectral profiles as input variables in industrial applications

There are several additional important aspects in industrial applications of spectral profiling and regression modelling. A good example of these is an application of PLS regression to model the mixing ratios of three constituents for a polymerisation process (Toft *et al.*, 1992). The challenge was to make a model that could predict the mixing ratios of three constituents within a time interval of 2 minutes from sampling. This prediction was then used to adjust the mixing ratios towards the target value (further details of the application are not needed for the purposes of this section). Complication arises when the range of the response is small. This was the case here and is almost a rule in industrial applications. In these cases, it is wise to use a training set spanning a slightly larger range of the response if possible. A good idea is to use a design around the target region for the response. It is also necessary to have a sensitive measurement technique that can provide precise and accurate instrumental data. Replication and repeated measurements are a must when the range of the response is small in order to shrink the imprecision and inaccuracy in measurements and thus in models.

Spectral data are usually influenced by drifting baselines and other instrumental artefacts. These effects have to be taken care of by pretreatment of data. Drifting baselines can be subtracted. Curved baselines can be removed by differentiation. For the spectra from the blending process, second-order differentiation transformed a curved baseline into a linear intensity shift. Normalisation was subsequently performed to remove 'size' effects in the spectra. Many other forms of pretreatment are available and are used. The key point is that the user needs experience in order to carry out the right pretreatment.

Acknowledgement

Reidar Arneberg, Pattern Recognition Systems Ltd, High-Tech Centre, Thormøhlensgt. 55, N-5008 Bergen, Norway, is thanked for assistance in generating the Sirius graphics used in this chapter.

References

Box, G.E.P., Hunter, W.G. and Hunter, J.S. (1978) *Statistics for Experimenters: An Introduction to Design, Data Analysis and Model Building*, Wiley, New York.

Grung, B. and Kvalheim, O.M. (1994) Rank determination of spectroscopic proifles by means of cross validation. *Chemometrics and Intelligent Laboratory Systems*, **22** 115-125.

Karstang, T.V. and Kvalheim, O.M. (1991) Multivariate prediction and background correction using local modelling and derivative spectroscopy. *Analytical Chemistry*, **63** 767-772.

Kvalheim, O.M. (1987) Latent-structure decomposition (projections) of multivariate data. *Chemometrics and Intelligent Laboratory Systems*, **2** 283-290.

Kvalheim, O.M. and Karstang, T.V. (1989) Interpretation of latent-variable regression models. *Chemometrics and Intelligent Laboratory Systems*, **7** 39-51.

Liang, Y.-Z., Manne, R. and Kvalheim, O.M. (1994) Resolution of embedded chromatographic peaks by sequential rank analysis of the first-order differentiated elution profile in the domain. *Chemometrics and Intelligent Laboratory Systems*, **22** 229-240.

Martens, H. and Næs, T. (1989) *Multivariate Calibration*, Wiley, Chichester.

Jolliffe, I.T., *Principal Component Analysis*, Springer-Verlag, Berlin.

Toft, J. and Kvalheim, O.M. (1993) Eigenstructure tracking analysis for demonstration of heteroscedastic noise in instrumental profiles—application to transmittance and aborbance IR. *Chemometrics and Intelligent Laboratory Systems*, **19** 65-73.

Toft, J., Kvalheim, O.M., Christy, A.A., Karstang, T.V., Kleveland, K. and Henriksen, A. (1992) Analysis of nontransparent polymers: mixture design, 2nd derivative attenuated total internal reflectance FT-IR and multivariate calibration. *Applied Spectroscopy*, **46** 1002-1008.

13 Data reconstruction methods for data processing
A.G. Ferrige and M.R. Alecio

13.1 Introduction

Every user of a scientific instrument wishes to obtain the best possible results that the instrument is capable of providing, and generally some form of post-processing is applied to the acquired data in an attempt to extract the last residue of information. However, the user is normally restricted to the software provided by the instrument manufacturer and the facilities are often rather limited. Whether this arises because the manufacturer believes that his product is adequate or because he is unaware of alternative algorithms available is debatable. The net result is that the user is frequently frustrated in being unable to solve his or her problem even with the best available data. It is under these circumstances that some more powerful form of data processing is required as a means of extracting information that may be contained in the data but is not immediately apparent.

In this chapter, we deal with advanced data reconstruction methods that efficiently uncover information by fitting a theoretical model to the experimental data. Using this approach, measurement noise and occasional errors can be rejected, and data processing artefacts largely eliminated. These new methods break convention in that, instead of applying inverse filters like many traditional data enhancement methods, they generate a series of mathematically plausible results that fit the actual data within the noise level. These methods are therefore not filters and they provide both the underlying sharp information contained in the data and a reconstruction of the data themselves.

The ability to fit a theoretical model to experimental data makes data reconstruction methods remarkably versatile. Most applications fall into one of two classes. In the first, the experimental data need processing to enhance the signal-to-noise ratio and/or the resolution in order to facilitate interpretation. In the second class of application, the required result has to be calculated from the experimental measurements using a theoretical understanding of the experiment. For example, a distribution of decay components can be calculated from relaxation measurements over a period of time in fluorescence lifetime studies or NMR experiments.

Data reconstruction methods also allow researchers to develop new methods specific to their own problems, although a reasonable level of programming competence is required. In analytical chemistry, a most important application is the deconvolution of spectroscopic and chromatographic data, which forms the main focus of this chapter.

We do not cover the potential imaging applications of data reconstruction techniques. Because these methods work equally well in more than one dimension, they may be applied to many techniques including high-resolution X-ray diffraction data, multidimensional NMR measurements (Laue *et al.*, 1986), high-resolution scanning electron microscopy (Ferrige *et al.*, 1991) and astronomical imaging (Gull and Daniell, 1978) as well as more general imaging problems. However, we shall not cover these applications in this chapter as they rely on the concepts developed for the essentially one-dimension applications we do cover.

At present, three defined data reconstruction methods are commercially available. These are the maximum entropy-based MaxEnt products and the fully Bayesian Massive Inference^{TM} (MassInf) products offered by MaxEnt Solutions Limited of Cambs, UK and the ReSpect^{TM} products (an abbreviation of REconstructed SPECTra) offered by Positive Probability Limited of Isleham, Cambs., UK. Although the methods achieve similar goals, there are significant differences in the ways these are attained. We shall highlight important differences within the discussions, but for reasons of consistency we shall use ReSpect^{TM} for all examples. When you come to apply the methods yourself, however, you should consider carefully the type of problem you have and the results you want to achieve before deciding on the specific method to use.

13.2 Where to go

Information about	Go to Section(s)
What does data reconstruction mean?	13.4.2, 13.5
What about using Fourier Transforms and other filters?	13.4.1, 13.6.1
What sort of data can I reconstruct?	13.7 case studies
Are there advantages in smoothing data first?	13.3, 13.4.1
What sort of peak shapes can be handled?	13.7
There is a lot of background in my signals.	13.5.2, 13.7.3, 13.7.5
Am I restricted to simple peak shapes?	13.7.8
Does the data collection rate affect things?	13.7.4
The peak widths change during the data collection.	13.7.6, 13.7.9
Can I get peak areas and positions?	13.5.1, 13.7.1

13.3 Information content of data

In order to appreciate the difference between the traditional methods and the new data reconstruction methods, it is useful to understand what data contain.

It is a pointless exercise to attempt to extract information that is simply not present in the data to begin with. For example, if the instrument noise level is such that 2 femtomoles of compound are completely swamped by instrument noise, then the signals of interest are indistinguishable from the noise and there is insufficient information for them to be revealed. One must therefore consider what makes up the 'wiggly line' that requires interpretation.

Generally, all data contain noise from the experiment and the measurement system, including the electronics, whether this be Gaussian distributed or counting noise (Poisson). In fact, it is very rare for noise to conform to theoretical expectation for a number of reasons. With respect to Poisson noise, there is normally some element of conventional, random electronic noise present so that it contains a Gaussian component. Where Gaussian noise is expected, it will only be truly Gaussian if the bandwidth of the electronics is sufficient that the highest noise frequencies are accurately recorded and no internal filtering has been applied. However, it is quite common for manufacturers to filter the acquired data before presenting them to the user so that the presented results are 'cleaner' than they would otherwise be. This is particularly true for many optical instruments (IR, UV and fluorescence) where automatic filtering is governed by the scan speed and the slit width. Therefore, the user does not often see 'raw' data: they are already damaged beyond recovery by filtering techniques.

So what makes up the data as seen by the user? There are two components:

- *noise* which may or may not be representative of what is actually generated;
- *signals* of interest.

However, the signals also rarely conform to what the user wishes to see. Ideally, the user wants to see sharp signals with no overlaps, so that interpretation and quantification are straightforward. More often signals are blurred or damaged by the instrument and the experimental conditions. For example, in chromatographic techniques, diffusion and the finite sample volume both contribute to producing peaks of a finite width. In NMR spectroscopy, the sample temperature, solvent viscosity and any magnetic field inhomogeneity all contribute to generating peaks that are much broader than their natural width. In optical spectroscopies, the true peak width is convolved with the slit width to provide broader peaks. In mass spectrometry, peaks are broadened by the energy distribution of ions with the same mass-to-charge ratio, by imperfections in magnetic or quadrupole electric fields and by gating times in time-of-flight instruments.

Therefore, *signals* comprise two parts—their natural width as determined by the nature of the sample and a second, often much broader, component that is determined by the experimental conditions and the limitations of the instrument. It follows that, irrespective of the type of spectroscopy, spectrometry or chromatography, there will be a range of factors that broaden peaks, and this frequently leads to peak overlaps that inhibit interpretation and quantification of the data.

13.4 The basics of data reconstruction methods

The emerging data reconstruction methods involve iterative, nonlinear techniques to fit combinations of theoretical and/or empirical peak shapes to the peaks in the data. We look at the basics of the techniques in this section. So that we can appreciate better the processes involved and the benefits to be gained, we shall first consider some of the features of the traditional approaches.

13.4.1 The traditional approach to data processing

Traditional methods that involve smoothing or resolution-enhancement functions include Savitsky–Golay and other linear smoothing functions, Gaussian sharpening, Fourier smoothing and deconvolutions, convolution difference, wavelets, etc. Such methods, attempt to undo the damage introduced by noise or peak broadening by applying some form of inverse filter. They are mathematically simple and the processes involved are easily understood. Because they are linear methods, they are generally fast to compute (say, up to a few seconds). Such methods are therefore commonplace in the software provided by manufacturers.

If the data are too noisy to reveal the required information, some form of smoothing is applied. Where peaks overlap, some form of sharpening or deconvolution is attempted to resolve them. However, there is always a trade between signal-to-noise ratio (S/N) and resolution. It is universally true that any filtering process aimed at improving S/N will broaden peaks and increase the risk of overlap with nearby signals. Conversely, peak-sharpening routines only sharpen peaks at the expense of an increased noise level in the sharpened result. Naturally, some resolution-enhancement filters are more efficient than others, but it is inescapable that the noise will increase. In practice, it is only under exceptional circumstances that the width of peaks can be halved without the noise exploding to unacceptable levels.

Among the most successful filters for both S/N and resolution enhancement are Fourier filters (Lindon and Ferrige, 1980). Exponential and Gaussian Fourier smoothing efficiently reduce noise but at the expense of broadening peaks and introducing respectively a Lorentzian or Gaussian component into peak shape in the smoothed result. Lorentzian to Gaussian Fourier deconvolutions and other similar methods are very effective at enhancing resolution but suffer the disadvantage that the starting S/N must be high and the peaks in the data must have a strong Lorentzian component. Although Gaussian to Gaussian Fourier transforms may be applied successfully to data where the peaks are essentially Gaussian to begin with, the possible improvements in resolution are much smaller because the applied deconvolution function must be much more severe to overcome the faster Gaussian decay rate in the Fourier domain. As a consequence, the noise in the result increases rapidly for even modest gains in resolution.

Unfortunately, filtering methods designed to enhance resolution are unsatisfactory when the peaks in the data are asymmetric, as encountered in most chromatographic techniques and also in poorly set-up NMR and MS instruments. This is because the applied filter is designed to enhance a specific frequency range and the frequency components for the left and right sides of asymmetric peaks may be quite different.

Another major failing of filters is that they are generally subjective. The user tries a range of input parameters and mentally iterates towards an acceptable result. In addition, filtered data are normally difficult to quantify. In the case of smoothed results, it is frequently difficult to determine the limits over which to measure the peak area and it is also not possible to assess the error introduced by noise. For resolution-enhanced data, the peak limits are generally straightforward to define but the increased noise level can make assessments of intensity errors meaningless.

What is required is a data processing method that does not broaden peaks when the S/N is enhanced and does not introduce noise when the resolution is enhanced. In addition, it should be possible to quantify the results and provide realistic assessments of position and intensity errors. Peak-fitting techniques potentially have these advantages in that parametrised peak shapes, positions and intensities can be determined for small-scale problems when the number of peaks is known at the outset. However, most problems of interest to the analytical chemist involve large data sets containing an unknown number of peaks and a different approach is required.

Because the analytical chemist is primarily interested in peak positions and intensities and is normally able to determine peak shapes adequately directly from the data, it is only necessary to devise a fitting method that provides peak positions and intensities. Data reconstruction methods are well-suited for large-scale problems of this type and are particularly powerful data processing tools.

13.4.2 Data reconstruction methods

Information is lost from experimental measurements through noise, peak broadening caused by the experiment and the instrument, and any filtering beyond the control of the user. Filtering would further damage the data and reduce the chance of recovering the desired information. The modern generation of data reconstruction algorithms represent the state of the art in quantitative data processing and they are designed to mathematically reconstruct the data to within the noise level. These techniques are not magic, however, and can recover only the information that remains in the data to be processed.

In none of the three commercially available algorithms is there any attempt to undo directly the damage in data. Instead, these methods always work forwards, reconstructing the information in the data. Apart from a few standard instructions, the only input to these deconvolution programs is a peak profile—the *Model*—which must adequately match the profile of the peaks in the data. The programs are designed to converge or terminate when the mathematical reconstruction of the data fits the actual data within the noise level.

Three results are then available to the user:

1. The *Deconvolution*—the deconvolved result, including fully quantified estimates of the peak position and intensity errors. This result represents a resolution enhancement of the data and shows the underlying detail contained in them.
2. The mathematical reconstruction of the data—the *Reconstruction*. This is obtained by convolving the *Deconvolution* with the *Model*, and represents a S/N enhancement of the data.
3. The *Misfit*, which is the difference between the data and the Reconstruction.

In all of these methods the data are simply used for purposes of comparison with the mathematical reconstruction. As the iterations progress, the *Deconvolution* is refined so that the *Reconstruction* more closely fits the actual data. The calculation continues until the best possible fit within the noise level is obtained. Therefore, the data themselves are *never* processed and the results obtained are synthetic reconstructions. The data are used only as a reference with which to assess the quality of the results at any point in the iteration cycle.

It might seem that this relatively new approach to data processing is the panacea to all problems. However, there is a penalty—time. As with any iterative procedure, the computations can be lengthy. In addition, the computation time is very data dependent. This is because noise provides freedom for calculations of this type to wander around without converging. This is best understood from the following example.

Consider two simple data files, each containing an identical single peak but with different S/N values. Assume that one file has a low S/N and the other a very high S/N and that the *Model* used is a little too wide. For the low S/N data, the *Model* will adequately fit the data within the noise level because the noise level is high. However, it is a different story for the high S/N data. Here, the same *Model* does not fit the data peak profile adequately and the *Misfit* over the peak may be greater than the noise level. Therefore, as far as the computation is concerned, it can never be possible for the *Reconstruction* to fit the data within the noise level. It follows that, unless provisions are made to deal with this specific situation, the computation will continue indefinitely. In fact, all three commercially available programs are able to deal with this, but they do so in different ways and such differences are discussed in Section 13.6. Therefore, depending on the S/N and the quality of the *Model*, these computations can be quite fast or interminably slow.

It is clear that the quality of the *Model* is the most important input to the process and is vital if high-quality results are to be obtained. It is therefore essential that suitable *Model* design tools are available to the user. Providing there is a single peak somewhere in the data, it is a simple process to zoom in to the peak and calculate the best four parameters that represent its width and shape—i.e. the left and right width and the left and right shape. Typically, the width parameters are measured at half-height and this presents no problem. However, determining the best shape parameters is dependent on the standard shapes available in a shape library and the way in which they are mixed. Obvious shapes for the library would be square wave, triangle, Gaussian, Lorentzian and Super-Lorentzian, the Super-Lorentzian having more intense wings than a Lorentzian. Unfortunately, there will be many experimental shapes that cannot be modelled from these standard shapes. In particular, it is common for some chromatographic techniques to produce peaks that cannot be matched by mixing Gaussian and Lorentzian shapes, the trailing edge reducing more rapidly than a mixture of these will allow. In some situations, a better match can be achieved with an exponentially modified Gaussian. Therefore, the shape library should contain profiles that suit the technique involved.

Figure 13.1 illustrates an ideal case. Here, the peak in question is a 50:50 mixture of Gaussian and Lorentzian. A small amount of noise has been added for realism. Pure Gaussian and Lorentzian profiles are overlaid for comparison. Even for these extreme shapes, it is clear that the top half of the peak is fitted adequately when the width is correct. However, the match in the wings for either shape is poor and it is obvious that the shape must also be adequately fitted if good deconvolutions are to be obtained.

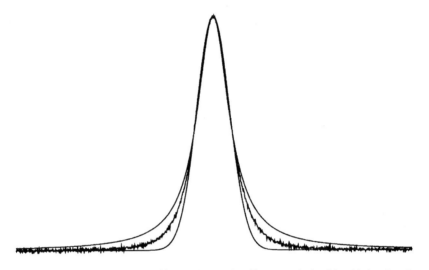

Figure 13.1 A synthetic peak (50% Gaussian and 50% Lorentzian) with added noise. Pure Gaussian and Lorentzian profiles of identical width are overlaid for comparison.

Of course, more often than not there is no isolated peak that can be parametrised and it is necessary to resort to other methods of *Model* design. Where peaks overlap, it is often possible to model the left-hand and right-hand profiles from different peaks in the data. One method of achieving this is to overlay a parametric curve first with one side of one peak and then the other side of another and interactively change the parameters until an adequate match is obtained.

Where peak overlap is severe, it is simply not possible to design an adequate *Model* directly. In these situations there are two alternatives. In the first, the *Model* is estimated, the deconvolution program is run and the standard deviation of the *Misfit* is noted. One or more of the parameters are changed, the program run again and the *Misfit* is noted. There will be little difference in *Misfit* standard deviations for *Models* that are narrower and have smaller wings than the correct profile because data reconstruction programs will always be able to include extra detail in the *Deconvolution* so that its convolution with the *Model* provides a

Reconstruction that fits the data. However, when the width or the peak wings are increased beyond those for the actual profile of the peaks in the data, it is no longer possible for the *Model* to fit the data. As a consequence, the standard deviation of the *Misfit* rises sharply. Therefore, a series of carefully designed trial experiments allows suitable *Model* parameters to be determined.

This procedure can be very time consuming and it would never be used unless the solution to the problem was particularly important and could not be achieved through other means. The second alternative makes use of the fact that it is possible to program at least one of the data reconstruction methods to optimise the *Model* iteratively. In this method it is only necessary to start with a *Model* that is somewhere near the truth. The program then makes its first estimate of the result, which is compared with the data so that a new, better *Model* is produced for use in a second iteration. Each iteration cycle yields a better *Model* and the process terminates when there is no significant improvement in the standard deviation of the *Misfit*.

To summarise, these programs are provided with the *Model* and an estimate of the noise level. Both are generally obtained directly from the data. The former requires user intervention, but the noise may be determined automatically. The program then makes its first estimate (a guess) at what the *Reconstruction* should look like by convolving its first estimate of the *Deconvolution* with the *Model* and comparing its result with the data. This represents one iteration cycle. The iterations continue until the *Reconstruction* fits the data within the estimated noise level or until the quality of the fit cannot be improved. This process is normally computationally intensive and relatively time consuming to perform. Depending on the data, the application and the algorithm used, the computation times may be anything from a few seconds to over an hour using a laboratory PC. It is therefore unlikely that these methods will be treated as 'on-line' in the near future. Even so, the additional information that may be recovered using these techniques can be dramatic and can often outweigh the time penalty. The general scheme is shown in Figure 13.2.

We now look more closely at the mathematical processes involved, as some knowledge of the details of the methods is useful in practice. It is also appropriate to compare and contrast the commercially available methods.

13.5 Theory of data reconstruction techniques

A helpful way to visualise the data reconstruction process is provided by Figure 13.3. This is a sketch of the way that the *Misfit* depends on the

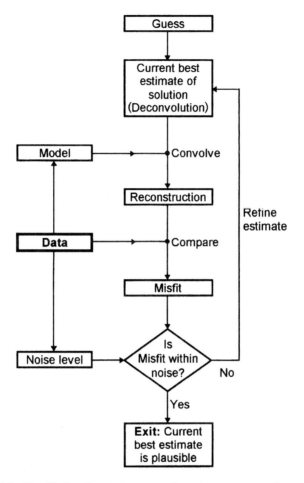

Figure 13.2 Simplified outline of data processing using a reconstruction technique.

possible deconvolutions. The axes of the figure represent the intensity of any two points in the deconvolution. (Naturally, the deconvolution generally involves a large number of points, and Figure 13.3 is actually a partial view of a multidimensional graph). Every possible deconvolved result is represented by a point in Figure 13.3. The contours illustrate the misfit to the data conveniently summarised by the single number χ^2 (the square of the misfit at each data point, summed over all the data after taking account of measurement errors).

The closest possible fit to the data within the constraints of the model has the lowest value of χ^2 and is represented by point A. However, this is unlikely to be the optimum solution, since some of the noise in the data is

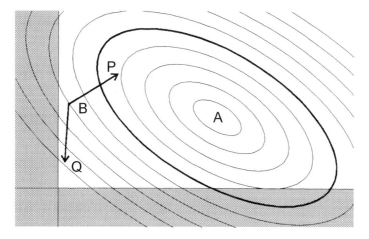

Figure 13.3 The dependence of data misfit on the possible deconvolutions. The axes represent the intensities of two points in the deconvolution. Shaded regions indicate unrealistic solutions. The contours show data misfit, measured by χ^2. The bold contour gives the most likely fit to the data. A: The closest possible fit to the data. B: The estimated most likely deconvolution. BP: Direction of steepest reduction of χ^2. BQ: Direction of steepest reduction of deconvolution complexity. The shaded areas indicate deconvolutions that contain negative features. In many applications, such deconvolutions are unrealistic solutions since the signal intensities are generally proportional to analyte concentrations and are thus necessarily nonnegative. Acceptable deconvolutions are thus restricted to the unshaded regions of the figure.

misinterpreted as signal and gives rise to spurious detail in the *Deconvolution*. At the desired solution, each datum point is fitted only within its measurement error, so that the most likely value of χ^2 is not necessarily the lowest. By estimating the errors from the noise in the data or from the known characteristics of the measurement technique, the most likely value of χ^2 is readily found and is illustrated by the bold contour in the figure. Every point in the unshaded region near this contour is consistent with the data and therefore could represent the desired solution.

To progress further, the data reconstruction techniques now look more closely at the possible deconvolutions and impose additional constraints to deduce the most probable results. These constraints differ in the three commercially available techniques, but all embody the idea of favouring the simplest deconvolutions, containing the fewest features, over more elaborate alternatives. In this way, we reconstruct only those features that are essential for agreement with the data, and avoid overinterpreting the data or introducing artefacts. Combining this general principle with the requirement for consistency with the data, we conclude that the likely solutions are distributed near the point representing the simplest solution on the target contour of Figure 13.3.

Both MaxEnt and ReSpect™ set out to identify the single most likely solution as the basis for further analysis. Starting with a first guess at the solution, these algorithms find the vectors **BP** and **BQ**. **BP** is the direction in which the estimated solution needs to be moved in order to improve the agreement between the corresponding reconstruction and the actual data. On the other hand, movement along **BQ** would result in a simpler deconvolution. Using this (and other) information, the algorithms are able to improve the estimated solution and thus to iterate towards the most likely deconvolution.

Rather than identifying the single most likely solution, MassInf attempts to characterise the entire statistical distribution of all possible solutions. This is achieved by generating a large number of representative examples. Each example is assigned a likelihood of being the actual solution on the basis of the quality of its fit to the data. Additionally, each is given a Bayesian prior probability reflecting its complexity, so that more sparse deconvolutions are considered more likely. The most likely solution is then obtained by taking a complex weighted average of the example solutions (Sibisi and Skilling, 1997).

13.5.1 Quantified results

At the end of its calculation, MassInf uses the many possible solutions it has found to calculate peak areas and positions. Since a range of such solutions is considered, a distribution of values for the intensity and position of each peak is obtained, from which expected values and reasonable error estimates for these quantities can be deduced. For example, features that are well defined and highly probable exhibit relatively little variation between samples, and can be quantified with small position and intensity errors. This powerful approach to the problem can in principle deliver reliable error estimates even in difficult circumstances (e.g. when the distribution is highly asymmetric). The main problem is probably the practical difficulty of adequately characterising a complex multidimensional distribution of possible solutions using a manageable number of examples and in a conveniently short computation time.

MaxEnt also attempts to capture the statistical distribution of possible deconvolution results. After identifying the single most likely solution, MaxEnt finds a simplified approximation to the required distribution that matches it at the most likely solution. MaxEnt then explores this approximation to generate a number of possible solutions, from which expected values and error estimates of peak areas and positions are deduced as before (Skilling, 1990, 1991). In practice, this procedure generally works well, though unrealistic error estimates can arise in certain circumstances through a breakdown of the approximations involved.

In the ReSpect™ algorithm, peak positions and intensities are determined directly from the *Deconvolution*. However, their errors are determined using both the deconvolved result and the experimental data. The *Misfit* over any feature provides a reliable estimate of the total noise over that feature. The position and intensity of nearby peaks are obtained from the *Deconvolution* and this information, along with the actual misfit, provides all that is necessary to compute robust errors. This approach has the advantage that the task of identifying the single most likely solution is considerably simplified and can be completed very rapidly, resulting in much shorter computation times.

13.5.2 Separating signals from noise

Because these methods reconstruct the data within the noise level, it follows that the *Deconvolution* is essentially free of noise. Consequently, its corresponding *Reconstruction* contains almost no noise, the noise residing in the *Misfit*. Therefore, as far as the user is concerned, the data have been separated into two channels that separate the signals from the noise. The *Reconstruction* is effectively the signal channel and the *Misfit* is the noise channel.

13.5.3 The significance of enforcing positivity

Although all data reconstruction methods are able to deal with data containing both positive and negative signals, it is more common to express the deconvolved result using only positive peaks. Negative signals in the data are then fitted only if the *Model* contains negative features, otherwise they are treated as errors or noise. Thus, enforcing positivity is, in many cases, a more realistic treatment of the data. Furthermore, positivity prevents the formation of negative-going side-lobes that often result from attempts to enhance resolution, and it therefore allows reliable and readily quantifiable high-resolution deconvolutions to be obtained.

13.6 Practical issues

13.6.1 Unrealistic noise assessments

Data reconstruction methods aim to fit the data within their noise levels. However, this is often not possible because the data may have been filtered so that the noise may no longer conform to theoretical expectations. This would not present a problem if one knew exactly what the noise had been before the data had passed through any hardware or software filter. However, without knowing how the data were filtered, it is not possible to compute the original, true noise level. Thus, the data reconstruction algorithm is required to fit the data unrealistically closely.

Under these circumstances, the algorithm may still converge because of detail in the noise that allows the *Reconstruction* to fit the data within the given noise level. However, where data are more severely filtered, the noise level determined from the data will be substantially lower than the true level and there will be no possibility of achieving a fit within such an unrealistic assessment. The consequences of this are that

- the data are characteristically overfitted, i.e. additional irrelevant features are detected in the noise that do not contribute to the quality of the result;
- single peaks may be resolved into more than one component, particularly where the noise unfavourably distorts the profiles of the peak in question.

The different data reconstruction methods solve the problem of an unrealistically low noise estimate in different ways. MaxEnt combats this situation by automatically increasing the assessment of the noise level until convergence can be attained. MassInf expands the boundary of the space containing plausible results, and hence the magnitude of the *Misfit* that is acceptable, until it is able to fit the data.

The ReSpect^{TM} algorithm is able to deal with filtered data because the initially determined noise level is used simply as a starting guide for the algorithm to achieve. The program monitors its progress towards its goal and an on-going analysis of its internal diagnostic information provides it with a continuously updated estimate of the noise level it must achieve, including any mismatch between the *Model* and the actual peak profile (see Section 13.6.2). Therefore, the computation does not go too far and converges reliably on the correct *Misfit* contour.

13.6.2 Model *mismatch*

When the *Model* is an inadequate match to the profile of the data peaks, one of two situations arises, each with its problems. If the *Model* is narrow compared with the peaks in the data, then the *Model* underfits all features of the actual profile. In these circumstances, single peaks are very likely to be resolved into more than one signal. However, the convolution with the *Model* still provides a *Reconstruction* that is faithful to the data.

On the other hand, if the *Model* is wider than the true profile, it follows that the *Misfit* will always exceed the assessed noise level even when the noise conforms to the anticipated distribution. Therefore, even when the noise estimate is correct, it is too low for the *Model*. Indeed this type of *Model* mismatch behaves like an additional form of noise that must be taken into account if the computation is to stop at the correct point. By splitting the problem into two phases and analysing the program's progress in the first phase, ReSpect^{TM} is able to determine the total noise level (i.e. instrument noise plus any filtering effects plus model mismatch 'noise') that must be achieved by the computation in order to successfully reach the correct *Misfit* contour. Of course, the wider model eliminates the possibility of single peaks being split, but close, severely overlapped peaks may then be reported as being single peaks.

13.6.3 Underfitted peaks

All data reconstruction methods tend to underfit very weak peaks. This is a consequence of favouring the simplest deconvolutions containing the minimum information consistent with the data. All of the signals tend to be underfitted by the standard deviation of the noise and it follows that only part of the true intensity is recovered for signals comparable to the noise level. The underfitting problem arises because it is easier to reliably fit intense peaks and these tend to be overfitted at the expense of weak peaks so as to favour the simplest deconvolutions. Even so, some of the methods recover the intensity of weak signals more reliably than others because the three different strategies lead to somewhat different solutions. Indeed, the strategy for some methods may be readily modified and the kernel designed to meet specific requirements.

13.7 Data reconstruction applications

In this section we show what can be achieved using the data reconstruction strategy as opposed to traditional methods. A range of examples is discussed covering different spectroscopic and chromatographic techniques. The developing sequence of examples shows an increasing complexity of problem, and details are given on how these are dealt with. For consistency, the ReSpect^{TM} algorithm has been used in all of the following examples.

It should be noted that any positive background is interpreted as signal and will give rise to additional peaks in the *Deconvolution* so that, when they are convolved with the *Model*, the background will be reproduced. Such peaks confuse the genuine signals and any background is therefore subtracted before processing. Except where specifically stated, the following examples have been background-corrected using a spline.

The *Model* parameters may generally be determined quickly and easily directly from the data by inspection. Although this procedure rarely produces the ideal *Model*, it is unlikely that it will be in sufficient error to give erroneous results. The quantified errors will naturally be larger when the *Model* is not ideal and it is only where the smallest possible errors are required that special care must be taken. In these circumstances, peak-fitting routines or a self-optimisation process may be used.

Centroid plots are presented in some of the following examples. The centroid of a peak is the position at the centre of its area such that the peak areas to the left and right of the centroid are equal. Centroids have two main benefits. The first is that a peak position may be readily determined for asymmetric peaks. Secondly, the total intensity of a peak is collapsed into a single spike and its height is therefore directly proportional to its intensity.

Mass spectrometrists have used centroids for decades, but the algorithms have evolved and have become extremely complex in an attempt to deal with peaks that are noisy, asymmetric or overlapped. In addition, a threshold is necessary to determine the peak limits. A major benefit of data reconstruction methods is that, because the *Deconvolutions* are sharp and peaks are well-defined, the simplest centroiding algorithm suffices and no threshold is required.

13.7.1 Synthetic data

This example illustrates the benefit of data reconstruction techniques over the traditional methods. It also shows how the uncertainty in the quantified results is dependent on S/N and with peak overlap. The results are also qualitatively compared with alternative treatments involving triangular smoothing to enhance S/N and with Fourier deconvolution to enhance resolution.

The data comprise four identical peaks with a width at half-height of eight sampling intervals and a shape of 50% Gaussian + 50% Lorentzian. The peak positions, their RMS S/N (the RMS S/N ignoring outliers is 20% of the peak-to-peak values) and their absolute intensities are given in Table 13.1. The peak-to-peak S/N for the weakest peak is therefore only 1.25. Peaks are labelled from left to right.

Table 13.1 Theoretical positions and intensities

Peak	Position	S/N	Intensity
1	100	25.00	244.07
2	150	25.00	244.07
3	160	12.50	122.03
4	170	6.25	61.02

S/N enhancement
Figure 13.4 compares conventional triangular smoothing with the ReSpect™ *Reconstruction*. The top trace shows the starting data. Triangular smoothing is particularly effective at removing high-frequency noise and the result of a nine-point triangular smoothing is shown below the data. Although high-frequency noise has been reduced dramatically, the peaks are broadened to the point where the signal resolution is being reduced.

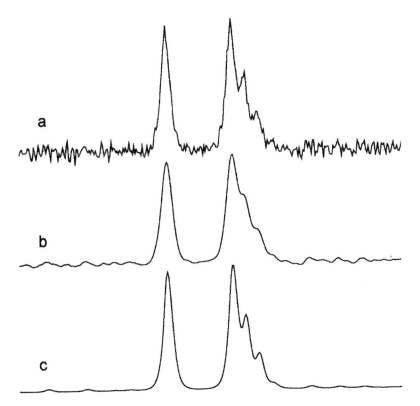

Figure 13.4 Comparison of data *Reconstruction* and filter methods for S/N enhancement. (a) Synthetic data: 350 data points containing four peaks. (b) Result of applying a nine-point triangular smoothing filter. (c) *Reconstruction*.

The bottom trace shows the *Reconstruction* using a slightly narrower *Model* than the ideal. For enhancing S/N, the width reduction is automatic and the computation terminates at the end of the first phase. The reason for using a slightly narrow *Model* is to ensure that the quality of the *Reconstruction* is not compromised by unfavourable noise characteristics, which may arise from its random nature, in the regions where detail is present. Because the algorithm correctly identifies the *Misfit* contour of interest and does not significantly overfit peaks in the data, only those significant low-frequency features in the noise that conform to the *Model* are recovered. Therefore, there is much less noise in the *Reconstruction*. Also, because the method is not a filter, the peaks in the *Reconstruction* have the same width as those in the data. Detail is consequently retained.

Resolution enhancement and quantified results
The top trace of Figure 13.5 shows the data (the same as for Figure 13.4). The centre trace shows the corresponding Lorentzian to Gaussian Fourier deconvolution. Although the peaks are now well resolved, two

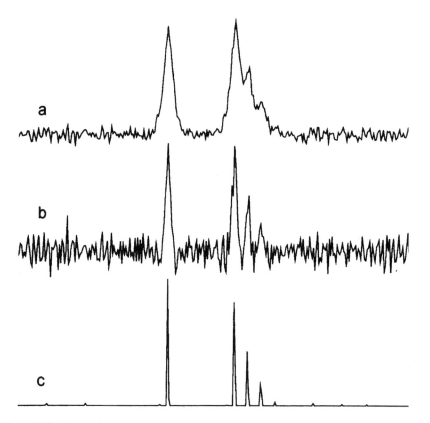

Figure 13.5 Comparison of data *Reconstruction* and filter methods for resolution enhancement. (a) Synthetic data: 350 data points containing four peaks. (b) Result of Lorentzian to Gaussian Fourier enhancement. (c) *Deconvolution*.

features are apparent. First, the noise level, as expected, has increased substantially. Secondly, because the data peaks have a 50% Gaussian component, the method introduces negative side-lobes on both sides of every peak. A compromise by trial and error is therefore required to minimise these while, at the same time, obtaining an adequate enhancement without introducing more noise than necessary. Although useful results may be obtained, the method becomes rather subjective. Furthermore, it is impossible to quantify results of this type because the limits of each peak cannot be adequately defined and the effect of side-lobes between peaks is unknown.

The bottom trace shows the *Deconvolution* using the correct *Model*. Here, all peaks are deconvolved to baseline and the noise is minimal, allowing straightforward quantification. The fully quantified results are shown in Table 13. 2. Peaks are labelled from left to right and the errors are for one standard deviation. For these particular data, the found positions are correct within statistical expectation even though the position for peak 1 is outside the standard error. There are four measurements and one measurement is therefore expected to lie between 1 and 2 standard deviations. It so happens that all the absolute intensities are within the standard error, but this is not surprising bearing in mind that there are only four measurements.

Table 13.2 Found positions and intensities with their errors

Peak	Position	Intensity
1	99.84 ± 0.14	239.69 ± 5.60
2	150.01 ± 0.19	244.65 ± 5.93
3	159.98 ± 0.29	125.52 ± 10.56
4	170.13 ± 0.40	56.43 ± 9.22

The assigned errors have significance and require further explanation. It is important to appreciate that errors are directly related to the evidence for the feature in question. Since peak 1 is intense and isolated from the rest, there is considerable certainty about both its position and intensity. It therefore carries the smallest error. Although peak 2 is intense, it is overlapped by the weaker peak 3 and its computed errors are therefore greater. Peak 3 is both weak and overlapped by two peaks, and there is therefore less certainty about its position and intensity. Its errors are correspondingly even higher. Although peak 4 is weaker still, it is only overlapped by one peak. Its errors happen to be of the same order of magnitude as those for peak 3, and this is certainly reasonable. While it would be unwise to read too much into a single analysis, these particular results do conform to Normal expectations. Another synthetic data set with different Gaussian noise could give unexpected results, but this would be a result of unfortunate noise characteristics over the individual peaks.

13.7.2 Standard 1-D deconvolutions and reconstructions

This example shows the quality that may be obtained from a data reconstruction method on experimental data. The top trace of Figure 13.6 shows the background-corrected data that are part of one of the four tracks (the G track) from gel electrophoresis DNA sequencing. The data were provided by the Wellcome Research Laboratories. Apart from relatively weak 'cross-talk' between the fluorescence filters, the other tracks (A, C and T) do not contain peaks coincident with those present in the data shown here.

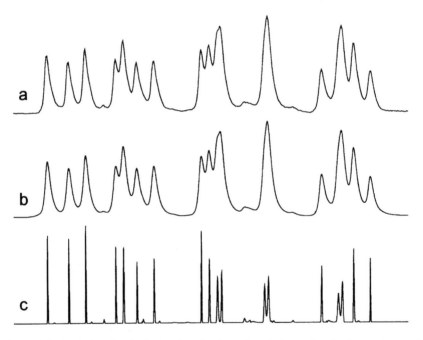

Figure 13.6 Analysis of gel electrophoresis data (supplied by the Wellcome Research Laboratories). (a) Data: part of fluorescence trace of a single track of DNA sequencing experiment. (b) *Reconstruction*. (c) *Deconvolution*.

The *Model* was estimated directly from the data, using primarily the peaks near the left-hand end of the trace. For these data, the variation in peak width across the data is small and the *Model* is an adequate match to the data peaks. The centre and bottom traces respectively show the

Reconstruction and the *Deconvolution*. The individual bases are clearly resolved and the *Reconstruction* is virtually noise-free. The different peak widths in the *Deconvolution* are a combination of the slightly variable peak width in the data and the different certainties resulting from peak overlaps. Where peak overlap is particularly severe, the increased uncertainty of position and intensity is reflected in the peaks being broadened in the *Deconvolution*. The more isolated peaks are sharper because there is an increased certainty about their position and intensity.

13.7.3 Peak detection

One of the major benefits of data reconstruction methods is their ability to recover weak features that are present in data. Therefore, these methods potentially offer a means of reducing detection limits and providing fully quantified results. Naturally, it is important that the constraints be designed so that weak features are not compromised by overfitting of strong ones.

In this example, provided by PerSeptive Biosystems, the data are the time-of-flight mass spectrum of a supposedly 22 kilodalton polymer. The polymer is a mixture of species differing in mass by the repeating unit that forms the fundamental polymer unit. The distribution of masses is required in order to define the polymer's composition. Figure 13.7 shows the acquired data, which are particularly noisy.

Figure 13.8 shows a horizontal expansion to the right of the envelope maximum and covers the region where the signals are beginning to be swamped by noise. In fact, the S/N of the peaks towards the right limit is very much less than unity. Below this is the *Reconstruction* obtained from a straightforward 1-D deconvolution in which the genuine signals are cleanly recovered. The bottom trace shows the *Deconvolution* and here it is apparent that the found peaks have different widths. This is because the noise introduces uncertainty in both position and intensity and the precise noise characteristics over each peak therefore determine how well it may be defined. As the uncertainty increases, so the width of the deconvolved peak also increases.

The top trace of Figure 13.9 shows the result of an 11-point triangular smoothing and it is quite clear that the noise destroys any hope of recovering information at the right-hand limit of the displayed data using traditional data processing methods. The lower trace shows a different type of presentation for the *Deconvolution*. Here, the found peaks have been centroided (the peak width is collapsed to a single line whose position is at the weighted centre of the mass range covered by the peak) so that they become delta functions. Because the height of the centroids is directly proportional to the found intensities, this presentation does not

528 DESIGN AND ANALYSIS IN CHEMICAL RESEARCH

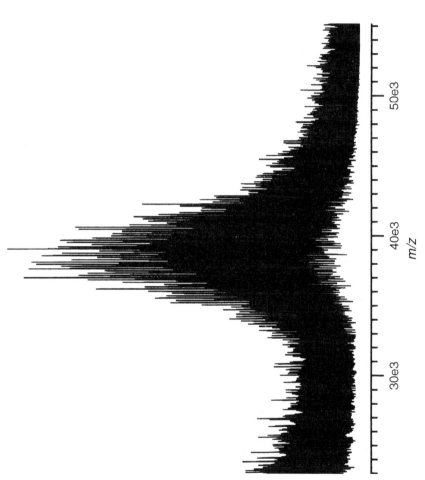

Figure 13.7 Time-of-flight mass spectrum of a 22 kDa polymer (data supplied by PerSeptive Biosystems).

DATA RECONSTRUCTION METHODS

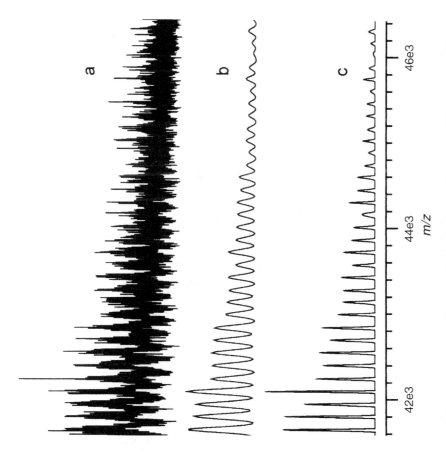

Figure 13.8 Analysis of time-of-flight mass spectrum (data supplied by PerSeptive Biosystems). (a) Data: part of the time-of-flight mass spectrum of a 22 kDa polymer. (b) *Reconstruction*. (c) *Deconvolution*.

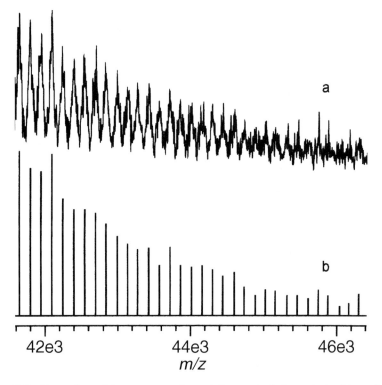

Figure 13.9 Comparison of data reconstruction and filter methods for analysis of time-of-flight mass spectrum (data supplied by PerSeptive Biosystems). (a) Result of applying a 11-point triangular smoothing filter to data shown in Figure 13.8. (b) Centroided *Deconvolution* of the data shown in Figure 13.8.

suffer from width variations in the *Deconvolution*. In this particular example, the centroid plot is a useful means of visually assessing the uncertainty of both peak intensities and positions. The molecular weight distribution of the polymer is expected to change smoothly and the increasing uncertainty introduced by noise as the S/N degrades is increasingly apparent for both intensities and m/z values.

13.7.4 Undersampled data

In principle, data reconstruction methods operate in the continuum limit (the *Model* is composed of continuous functions that have a value at any x-value). Therefore, the result may be output using more points than there are in the data. Because these methods are not filters and do not involve interpolation of the original data to generate the *Models*, using additional output points may reveal more detail, providing the information is in the data to begin with. For example, if an adequate *Model* can be designed, high S/N data may contain the necessary information for peaks to be resolved even when the acquisition interval is too coarse for this to be apparent in the data. In practice, additional information may be retrieved under these circumstances by increasing the output data resolution up to four times. Increasing the number of points beyond this does not provide more information.

The example discussed here uses part of the proton NMR spectrum of 5-nitroisoquinoline to illustrate this. The data were provided by the Wellcome Research Laboratories. The top trace of Figure 13.10 shows the H8 proton acquired with a sufficiently fine sampling rate to provide a *Deconvolution* in which the peaks are resolved to the baseline (lower trace). The intensity of the central peaks of each multiplet is actually twice that of the other peaks. However, their peak heights are reduced because they comprise two very severely overlapped peaks and there is insufficient information in the data for them to be resolved. Therefore,

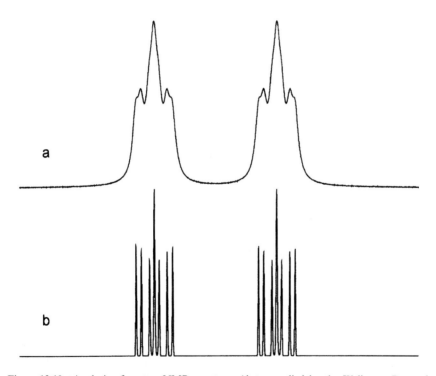

Figure 13.10 Analysis of proton NMR spectrum (data supplied by the Wellcome Research Laboratories). (a) Data: the H8 proton of 5-nitroisoquinoline. (b) *Deconvolution*.

these peaks are slightly broader than the others in the *Deconvolution*. The pattern, thus, involves four different proton coupling constants.

If the sampling rate is reduced by 16 times, the data are inadequately digitised and there are then insufficient points to reveal the information contained in them unless more points are used to compute the results. Because fast Fourier transforms are used in the iteration cycle in calculating convolutions, the number of additional points in the deconvolution may readily be increased in factors of 2. The top trace of Figure 13.11 shows the inadequately sampled data and the centre trace shows its *Deconvolution*. Only some of the peaks are now resolved. However, by computing the *Deconvolution* over 4 times the number of points, most of the information is recovered (bottom trace). Now there is simply insufficient information in the inadequately sampled data to show any hint that the central peaks of each multiplet are close doublets; their heights are therefore close to double that of the other peaks.

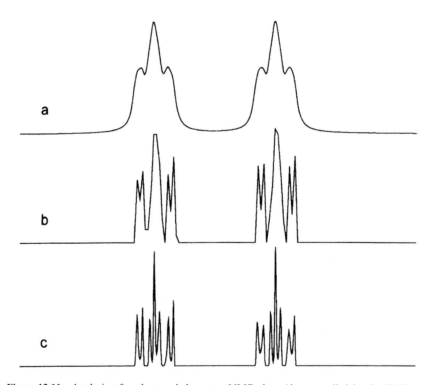

Figure 13.11 Analysis of undersampled proton NMR data (data supplied by the Wellcome Research Laboratories). (a) Data: the H8 proton of 5-nitroisoquinoline, using 16 times fewer points than Figure 13.10. (b) Standard *Deconvolution* of the data. (c) *Deconvolution* computed using 4 times more points than in the data.

13.7.5 Correcting backgrounds (low frequencies)

Separating signals from backgrounds is a particularly difficult problem and no method is universally applicable. In fact, a number of methods are supplied by instrument manufacturers using Fourier transforms, derivatives or linear prediction. Depending on the data, they can work very well or fail miserably. However, data reconstruction methods offer an alternative approach that may prove to be more reliable.

Where the frequencies that make up a background are significantly lower than those that represent the signals (i.e. the background is considerably broader than the signals), it is generally possible to compute a background of sufficient quality to meet most needs. In principle, it should be possible to separate a background from signals even when the background features are only a little broader than the signals. Unfortunately, this cannot be achieved in practice because there is always a certain degree of cross-talk between similar frequencies. At present two basic methods are under development.

The first requires estimates of the width of the peaks and also the width of the narrowest components that comprise the background. Because data reconstruction programs are not restricted to working with only one channel, they can be modified to generate simultaneously a reconstruction that best fits the background in one channel and, in another channel, a reconstruction that best fits the signals, with the sum of the two being the best fit to the actual data. Two potential disadvantages arise. The first is that such computations are much more intensive and there is a significant time penalty. Much more serious is the fact that the number of degrees of freedom is doubled, which provides additional freedom in the calculation. The consequence is that convergence to a reliable end-point is more difficult to achieve.

An alternative method involves splitting the problem into two parts so that the number of degrees of freedom is not excessively large. Here, a noise correlation method is used first to identify where signals occur in the data. A benefit of the noise-correlated spectrum or chromatogram is that the average peak width for the data peaks can be determined automatically when the noise is not excessively high.

A standard one-dimensional deconvolution can then be run on the noise-correlated data so that only the signals are recovered. This emerging method is not perfect and the computed background is somewhat spiky. However, it is of sufficiently high quality that a simple curve-fitting routine successfully removes any residual high-frequency undulations. Filtering of the computed background is not adopted since this broadens its undulations slightly and reduces the quality of its fit to the data. Figure 13.12 shows the results of this approach using a development program.

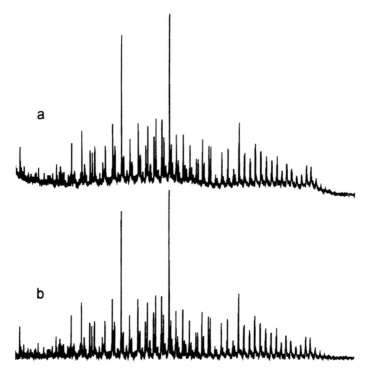

Figure 13.12 Automatic background correction (data supplied by PerSeptive Biosystems). (a) Time-of-flight mass spectrum of DNA sequencing experiment. (b) Result of automatic background correction.

The top trace shows the data that were provided by PerSeptive Biosystems and shows a DNA sequencing experiment run on one of their time-of-flight mass spectrometers. The bottom trace is the result of an automatic background correction using a modified ReSpect^{TM} program.

13.7.6 Variable peak width

When the peak width in the data varies, a problem for any linear deconvolution method (inverse filters, Gaussian sharpening and Fourier transforms) is posed because each signal comprises a different frequency distribution and the methods generally available are designed to operate over a very limited and defined set of frequencies. A major benefit of data reconstruction methods is that they are nonlinear and they are able to deal with situations where the peak width varies in a defined or rational manner. This is the case for many chromatographic measurements and, providing the peak width variation can be determined, data reconstruction

DATA RECONSTRUCTION METHODS 535

programs can be modified to accommodate a *Model* that varies continuously in step with the changes that occur in the data.

The data presented in this example (top trace of Figure 13.13) were provided by the Central Science Laboratory of the MAFF and represent the gradient liquid chromatogram of a standard mixture of 15 compounds. The experiment run time was reduced by 10 times so that the peaks were severely overlapped. The peak width increases across the data, the width increasing by about 50% from left to right. For these data, the width variation was assumed to be linear and its rate of change was determined from the first and last peaks. The shape parameters for the *Model* were

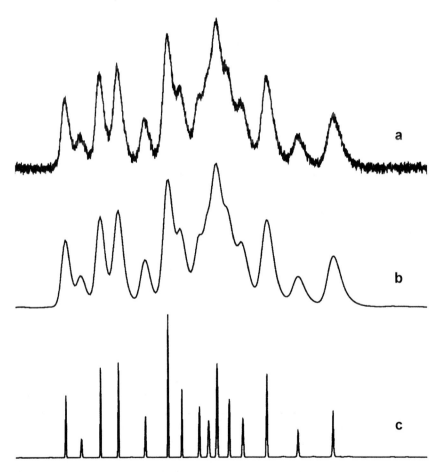

Figure 13.13 Analysis of gradient liquid chromatography data (supplied by the Central Science Laboratory of the MAFF). (a) Chromatography of a standard 15-component mixture, exhibiting systematic variation in peak width. (b) *Reconstruction* of the data. (c) *Deconvolution*.

determined from the last peak. Using a nonlinear *Model* and running the one-dimensional deconvolution program provides both the *Reconstruction* (centre trace) and the *Deconvolution* (bottom trace).

As can be seen, all peaks are resolved in the *Deconvolution*, including the five severely overlapped components near the centre of the data. *Deconvolutions* of this type therefore provide a useful means of quantifying severely overlapped peaks in chromatographic data, allowing chromatography run times to be substantially reduced without loss of information.

13.7.7 Pattern recognition

There is no limit to the complexity of the model that may be used for a deconvolution and complex models may often be used to simplify data so that interpretation is more straightforward. Depending on the application, the pattern may indeed be extremely complex. One obvious application could be the search and quantification of a specific component in a mixture, and this would apply to most spectroscopic techniques. The *Model* in this case would be the complete spectrum of the component in question. Another application would be to eliminate recurring experimental artefacts. This particularly applies to MALDI (Matrix Assisted Laser Desorption Ionisation) mass spectrometry where solvent and matrix adducts accompany the signals of interest and confuse the interpretation of mixtures. Here, the *Model* would be designed to represent the pattern of adducts so that its intensity was deconvolved into the parent signals; the following example illustrates this application.

The data were provided by PerSeptive Biosystems and are the MALDI mass spectrum of a purchased sample of bovine ribonuclease B, which is a glycoprotein, using a sinapinic acid matrix. The data are shown in Figure 13.14 and the requirement is to determine the distribution of mannose units in the sample.

It is clear that the peaks are asymmetric and comprise more than one component. This is a frequent problem with MALDI experiments and is due to the nature of the ionisation process. Therefore, for these data, designing a suitable *Model* to match the peaks is slightly more complicated. Initially the intensities of the MALDI adducts with respect to the primary signals are unknown and these must be established and incorporated into the *Model*. Another complication is the presence of a strong background, but the peak overlaps make it impossible to decide how this should be corrected manually. The data were therefore prepared and processed in the following way.

The background was first corrected using the automatic method described previously, (see Example 13.7.5). The *Model* was assessed

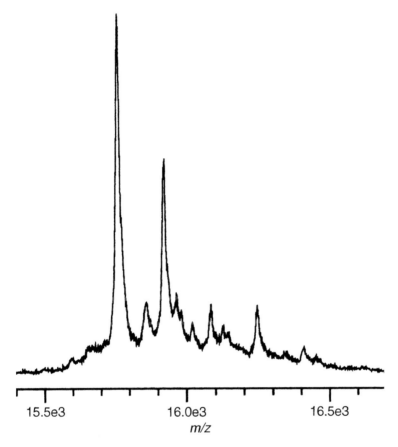

Figure 13.14 MALDI mass spectrum of glycosylated bovine ribonuclease B (data supplied by PerSeptive Biosystems).

directly from the data and designed to match the overall profile of the most intense signals and a centroided *Deconvolution* was obtained. Because the mass difference of the adducts from the main peaks is predetermined by the matrix used, the adducts are easily identified and their relative intensities determined. A simple convolution of the determined values with the initial *Model* provides a new *Model* that incorporates the variations introduced by the MALDI process and the formation of adducts. The computed *Model* is shown in Figure 13.15. The data reconstruction program was then run on the background-corrected data using this *Model*.

The top trace of Figure 13.16 shows the background-corrected data and the lower trace shows its centroided *Deconvolution*, in which the

538   DESIGN AND ANALYSIS IN CHEMICAL RESEARCH

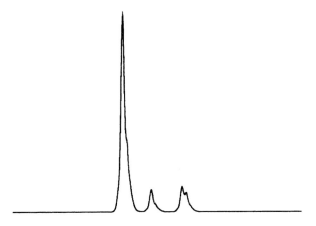

Figure 13.15 Designed *Model* showing pattern of adducts produced by MALDI mass spectrum experiment (data supplied by PerSeptive Biosystems).

centroid heights represent the relative intensities of the differently glycosylated components. This type of pattern-recognition application provides a clean, noise-free result that is simple to interpret and quantify. In this particular case, however, the quantified errors would not be robust, because it is not known that the adduct intensities are truly identical for all components.

13.7.8 Model *optimisation*

There are generally two situations where it is not possible to determine the *Model* to adequate precision directly from the data. The first is where there is significant peak overlap throughout the data and the second is where the shape cannot be modelled by mixing standard functions, such as Gaussian and Lorentzian peak profiles. However, provided that a reasonable starting estimate of the *Model* can be made, the *Model* can be optimised. A major benefit of this approach is that a *Model* may be generated that cannot be produced by mixing standard profiles. Also, the process works equally well for asymmetric peaks.

In the example presented here, the data were provided by the Wellcome Research Laboratories and are the result of size exclusion chromatography of a mixture of nine high-molecular-weight peptides. Over the region covered, the peak widths and shapes are expected to be very similar and it was therefore not necessary to use a variable *Model*. The data are shown as the top trace of Figure 13.17. The initial *Model* for optimisation was estimated from the best-resolved peak near the centre of the data. A *Deconvolution* containing 11 major peaks is obtained with this

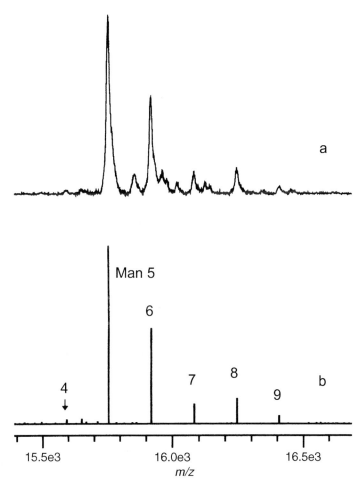

Figure 13.16 Analysis of MALDI mass spectrum. (a) MALDI mass spectrum of ribonuclease B after automatic baseline correction. (b) The centroided *Deconvolution* using designed model (Figure 13.15).

Model and this is shown below the data. The fact that more peaks are found than expected is an indication that the *Model* is too narrow and/or that the profile has insufficient wings.

Optimisation of the *Model* required five iterations of the method and the final *Deconvolution* is shown as the lower trace of Figure 13.17. Here, only the expected nine components were found, apart from a very minor peak that could be either an impurity or a residual mismatch of the *Model*. The overlaid integral trace shows the found intensity ratios to be close to 2:1:4:2:4:2:2:4:2, in agreement with the known proportions of the

540   DESIGN AND ANALYSIS IN CHEMICAL RESEARCH

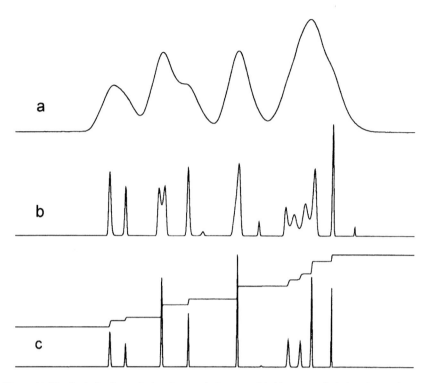

Figure 13.17 Optimisation of the deconvolution *Model* (data supplied by the Wellcome Research Laboratories). (a) Size exclusion chromatogram of a mixture of nine peptides. (b) *Deconvolution* using the initial *Model*. (c) Deconvolution using self-optimised *Model*.

components in the mixture. A more detailed quantification is inappropriate because the response factors of the individual peptides were not determined.

13.7.9 *Electrospray charge deconvolutions*

A major advantage of reconstruction methods is that complicated questions may be asked and a highly plausible solution found. In the case of electrospray mass spectrometry, the requirement is to produce a zero-charge mass spectrum from the data that contains a range of discrete charges for each species. At the same time, it would be beneficial to enhance the resolution and S/N. Problems of this type are mathematically challenging, but solving them in a single, complex calculation is not necessarily the most suitable approach. Such calculations, although possible, fall foul of certain difficulties. The most serious of these are:

- No single *Model* can adequately represent the profile of all the peaks in the data because the peak width varies appreciably with charge state.
- To simplify the calculation and improve the computation time, it is usually necessary to specify the mass range in which the result is expected to lie and the number of points required in the output. However, the data to be processed may contain extraneous peaks due to low-molecular-weight components such as solvents. Accounting for these peaks properly in the specified mass range may be impossible and the computed result consequently contains artefacts to accommodate their intensity. Furthermore, such artefacts reduce the intensities of legitimate weak peaks in the solution.

An alternative approach is therefore required.

For the method described here, the calculation uses a step-wise strategy resulting in a transparent procedure that allows consistency with the fundamental rules of electrospray and allows the result to be checked at every stage. The data are first deconvolved using a nonlinear *Model* that takes into account the way the peak width changes with charge state. This provides a fully quantified peak table that has not been compromised by the *Model*. The table is then used to determine every plausible zero-charge mass from each peak whether it be a genuine signal or noise, and then to examine the *Deconvolution* to check the evidence for each possible mass. The vast majority of possible results are rejected because they do not conform to the fundamental rule that charge states in an electrospray mass spectrum must be consecutive.

The algorithm is then used in a different context to compute the most plausible intensities in the zero-charge spectrum that best fit the *relevant* peaks in the data. This is performed iteratively as irrelevant correlations are progressively rejected. The major advantages of this approach are:

- The *Model* does not compromise the result.
- The settings of output mass range and number of points are unnecessary.
- Extraneous signals in the data have no effect on the *Deconvolution* so that only those signals that contribute to the zero-charge spectrum are reconstructed.

In the example discussed here, the data were provided by the Genzyme Corporation and are the electrospray mass spectrum of a sample of horse heart myoglobin of molecular weight 16 951.5. The mass/charge data are shown in Figure 13.18. The different charge states for each species present

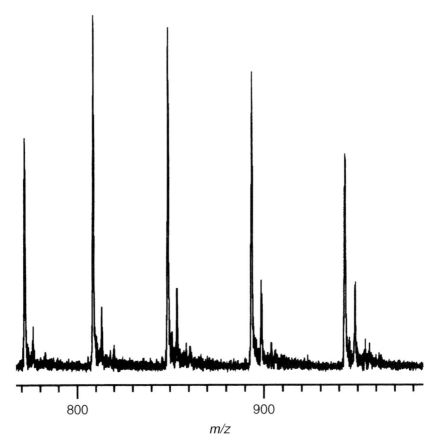

Figure 13.18 Electrospray mass spectrum of horse heart myoglobin (data supplied by the Genzyme Corporation).

are clear and the major series corresponds to the horse heart myoglobin. The weaker signals to the right of each major signal are not high-molecular-weight contaminants. They are adducts that arise from the experimental conditions used and low-molecular-weight contaminants in the sample as purchased. However, it is not possible to assign these adducts without first generating a zero-charge *Deconvolution* in which the resolution and the S/N are both improved.

Figure 13.19 shows the zero-charge result using the strategy described above. Its quality is independent of the output mass window and, because there is no requirement to fit the total data intensity, artefacts are reduced and weak peaks reach their full intensity within the constraints. Also, because the model is not a compromise, any model mismatch is minor and mismatch artefacts are not observed. As can be seen, there are the expected dramatic gains in both resolution and S/N.

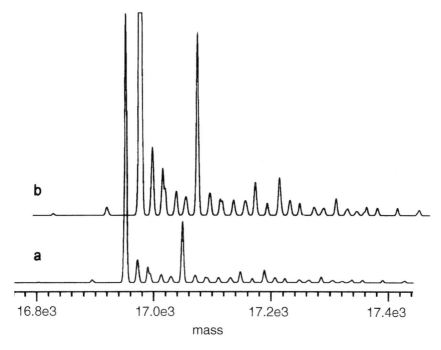

Figure 13.19 The zero-charge mass spectrum of horse heart myoglobin. (a) The zero-charge result obtained from the data shown in Figure 13.18. (b) Vertical expansion, ×3, offset for clarity.

Figure 13.20 shows part of the centroided zero-charge result along with the mass differences. The full peak assignments are shown in Table 13.3. (For further details also see Alecio *et al.*, 1998).

13.8 Concluding remarks

It has only been possible to provide the reader with a somewhat brief and limited insight into the theory of data reconstruction methods and what they are able to achieve. There are, of course, many other applications where these methods provide dramatic improvements over traditional methods. The data reconstruction method concentrated upon in this chapter is now being used to detect and quantify peaks for powder X-ray diffraction data, as well as to provide *Deconvolutions* and *Reconstructions*. It has been used to analyse centroided mass spectrometric peptide mapping data in order to determine the mass of the fragments present in complex mixtures (Zhang *et al.*, 1998). A more general application of current interest is the potential to remove backgrounds automatically in a wide range of spectroscopic and chromatographic data not discussed in this chapter.

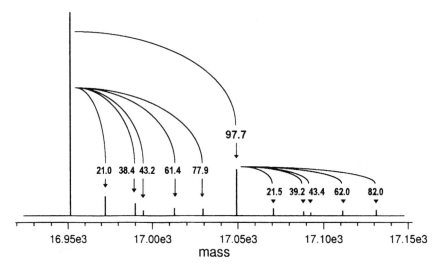

Figure 13.20 The centroided zero-charge mass spectrum of horse heart myoglobin. Peaks are labelled with their mass differences from the base peak.

Table 13.3 Adduct peak assignments. The adduct mass differences are from the main peaks (indicated in italic type)

Mass	Mass difference	Assignment	Expected
16 951.5	–	–	–
16 972.5	21.0	Na	22.0
16 989.9	38.4	K	38.0
16 994.7	43.2	2Na or HNCO	44.0 or 43.0
17 012.9	61.4	Na + K or urea	60.0
17 029.4	77.9	2K or 2Na + K	78.0 or 82.0
17 049.2	–	–	–
17 070.7	21.5	Na	22.0
17 088.4	39.2	K	38.0
17 092.6	43.4	2Na or HNCO	44.0 or 43.0
17 111.2	62.0	Na + K or urea	60.0
17 131.2	82.0	2K or 2Na + K	78.0 or 82.0
16 951.5	–	–	–
17 049.2	97.7	H_3PO_4	98.0

Of interest to the analytical chemist and process control engineer is the potential to reduce the sampling time necessary for many quality control measurements. Depending on the particular application, a *Reconstruction* of satisfactory quality may be obtained on data acquired in a small fraction of the normal acquisition time. Furthermore, where chromatographic methods are employed, it may no longer be necessary to repeat a time-consuming experiment several times in order to estimate the errors. A single quantified *Deconvolution* has the potential to do a better job and, as an added bonus, can reduce the measurement time of a single run as well.

Finally, many spectroscopic and chromatographic measurements are grossly oversampled and data reconstruction methods are equally effective on correctly sampled data. These methods may therefore be used in the design of time-efficient experiments. Also, the potential gains in both resolution and S/N allow these methods to be applied to problems that would not otherwise be attempted.

References

Alecio, M.R., Zhang, X.K. and Ferrige, A.G. (1998) ReSpect™—A new algorithm for generating artefact-free zero-charge spectra from ESIMS macromolecule data. *Proceedings of the 46th ASMS Conference on Mass Spectrometry and Allied Topics*, p. 144.

Ferrige, A.G., Seddon, M.J. and Pennycook, S.J. (1991) The application of MaxEnt to electron microscopy, in *Maximum Entropy and Bayesian Methods* (eds. C.R. Smith *et al.*), Kluwer Academic Press, Seattle.

Gull, S.F. and Daniell, G.J. (1978) Image reconstruction from incomplete and noisy data. *Nature*, **272** 686-690.

Laue, E.D., Mayger, M.R., Skilling, J. and Staunton, J. (1986) Reconstruction of phase sensitive two-dimensional NMR spectra by maximum entropy. *Journal of Magnetic Resonance*, **68** 14-29.

Lindon, J.C. and Ferrige, A.G. (1980) Digitisation and data processing in Fourier transform NMR, in *Progress in NMR* (eds. J.Emsley *et al.*), Pergamon Press, Oxford, pp. 22-66.

Sibisi, S. and Skilling, J. (1997) Prior distributions on measure space. *Journal of the Royal Statistical Society B*, **59** 217-235.

Skilling, J. (1990) Quantified maximum entropy, in *Maximum Entropy and Bayesian Methods* (ed. P.F. Fougere), Kluwer Academic Press, Seattle.

Skilling, J. (1991) On parameter estimation and quantified MaxEnt, in *Maximum Entropy and Bayesian Methods* (eds. W.T. Grandy, Jr and L.H. Schick), Kluwer Academic Press, Seattle.

Zhang, X.K., Ferrige, A.G. and Alecio, M.R. (1998) ReSpect™—a novel approach to the automated analysis of peptide mapping data. *Proceedings of the 46th ASMS Conference on Mass Spectrometry and Allied Topics*, p. 206.

Index

alternative hypothesis 16-19, 137, 143, 170
analogue to digital 43, 74
analysis of variance 54, 58, 107, 109, 142, 175, 201, 279, 285, 288, 292, 293, 305, 441, 456, 458
    between group sum of squares 57, 143, 299, 300
    components 10, 142-143, 300, 312
    confounding 307, 352
    co-variates 25, 230, 298, 308
    cross classification 177-183, 302-303
    error sum of squares 287, 289, 292, 293, 295, 299
    fixed effects 296-99, 305
    F-test 143, 257, 442
    hierarchical 301, 304, 311
    interactions 303, 311
    Kruskall Wallis test 175
    lack of fit 217, 231, 256, 257, 260, 295, 443, 446
    linear regression 279, 285, 288, 293, 305, 441, 456, 458
    mean residual sum of squares 299
    mean square error 226
    mean treatment sum of squares 300
    median polish 180, 181, 182
    mixed effects 297
    multiple comparisons 175, 176, 179, 180
    nested 46, 109, 301, 304-305, 311-15
    one way 57, 99, 175-77, 279, 296, 301, 310
    pure error 266, 443
    random effects 296, 299-301, 305, 312
    randomised blocks 301-302, 309-310
    residuals sum of squares 142, 287, 292, 295, 297, 299, 300, 303
    special causes 286, 287
    sum of squares about the regression 442, 443
    table of results 142, 202, 205, 211, 292, 293, 299, 302, 304, 309, 311, 312, 441-443, 445, 456, 459, 460
    total sum of squares 284, 286, 287, 292, 293
    two-way 99, 177, 182, 279, 301-305, 308-313
    within group 57, 299
autocorrelation 290, 295
autoscaling 75, 77, 409, 410-411
average 15, 100, 117, 118, 149, 205, 284, 292, 308, 320, 325, 383, 424
    arithmetic 147, 149, 284
    confidence interval 54, 100, 122, 124, 157, 158
    geometric mean 149
    robust/resiliant estimators 149-153, 156, 159, 174, 182

Bayes' Theorem 14, 145
Bayesian statistics 99
Bayesian Massive Inference 508, 518
Bayesian prior probability 14, 518
bias 8, 26, 37, 40, 86, 91, 98, 105, 171, 172, 193, 229, 265, 281, 282, 285, 290, 295, 302, 401, 436, 451, 466
    unbiasedness 121, 184, 283
blocks 25, 135, 177, 194, 203-205, 229, 230, 250, 301, 302, 305, 309, 341
    experiment design 8, 25, 193, 209, 270, 290, 301
    multi-block PLS 341
    orthogonal 270
Box and Whisker plot 53, 60, 163

calibration
    inverse regression 448
    linear 286, 447
    multivariate 495
    zero intercept 249, 450-452, 457
cause and effect diagram 194, 228, 233
causes
    assignable 193, 317, 318, 319, 321, 339
    common 7, 279, 282, 286, 291, 296, 299, 300, 305
    special 7, 279, 282, 285, 291, 305, 319, 329, 333, 337, 340
Central Limit Theorem 61, 144, 145, 289
centroid 292, 355, 388, 390, 392, 402, 415,

INDEX 547

416, 522, 527, 538
centroid plots 522
Chi square 285, 328, 334, 516, 517
Chi square control charts 328
Chi square distribution 125, 131, 143, 158, 178, 285, 327
classification 367, 371, 385, 391, 393, 412
   class membership 393, 394, 398
   neural networks 368, 376-378, 401, 403
   supervised 367, 368
   unsupervised 367, 368
   validation 405-407
cluster analysis 52, 79, 365, 367, 368, 372, 374, 381, 385, 389, 412
   agglomerative 381, 386
   average linkage method 367, 383-385
   centroid 389, 390, 415, 416, 538
   city-block distance 414
   classification 367, 371, 385, 412
   complete linkage method 367, 383, 385
   dendrogram 383, 385, 386, 388, 390, 497
   error sum of squares 386
   Euclidean distance 396, 414-416
   Forgy's method 388-391
   hierarchical 367, 381, 386-388, 390
   K method 389
   Minlowski distance 413, 416
   neural networks 367, 368, 376-378, 399-405
   non-hierarchical 367, 381, 388-391
   nonlinear mapping 373-376
   similarity 367, 382, 383, 386, 413-417
   single linking method 367, 383-85
   supervised methods 367
   Ward's method 386-388
clusters 58, 341, 342, 365, 368, 372, 374, 389, 392, 398
collinearity 273, 497, 501
comparison
   Kolmorgorov-Smirnov 174
   Mann-Whitney U test 173
   of data sets 119-132, 173-174
   paired 140-142, 169-172, 310
   permutation tests 99, 174
   sign test 290
   Wilcoxon Mann Whitney test 174
   Wilcoxon signed rank test
confidence interval 16, 23, 54, 99, 100, 101, 106-130, 132, 147, 151, 259, 291, 293, 299, 440, 443, 445, 446, 447, 450, 471
   bootstrap method 159
   Hotelling 328, 330, 334, 335

medians 158, 176
   robust for location 157-158
   robust for spread 158-160
   single-sided 131, 132
   two-sided 131, 157
   variance 101, 125-130, 158
confidence level 17, 123, 132, 157, 158, 257, 262
confidence limits 16, 23, 471
confounding 203, 204, 208, 209, 211, 213, 221, 231, 307, 352, 455
consumer risk 103
control chart 49, 50, 51, 93, 109, 168, 317, 319-323
   average (xbar) 320, 322
   latent variable 333-338
   multivariate 319, 328-334, 343-345
   range 51, 320
   standard deviation 51, 320
correlation 78, 336, 423, 428-432, 455
correlation coefficient 416, 430, 439, 470
   accidental relationship 428
   cause and effect 194, 424-428, 432
   functional relationship 182, 426, 433
   of noise 533
   R^2 217, 226, 256, 259, 428-432, 439, 456, 470
   R^2 adjusted 428-432, 439, 470
   significance 431
   structures 476
covariance 78, 229-328, 329, 331, 334, 335, 336, 337, 428-432, 481
covariates 15, 25, 230
cross validation 336, 407, 454, 478, 480, 492, 495
cusum 54-58, 71, 106, 323-325, 327
   plot 54, 169, 325, 330
   breakpoints 56, 57, 58
   historic data 55, 56, 342
   multivariate 330-332
   of Hotelling T 331
   V-mask 56, 324

data
   bivariate 292
   centroid 292, 355, 389, 390, 392, 416, 522, 527, 538
   checking 22, 34, 48-66, 161-166, 226, 341, 369-381
   historical 6, 24, 55, 56, 61, 63, 289, 321, 325, 327, 337-343, 344, 350-360, 485
   multiblock 340

548   INDEX

multivariate 75-81, 327, 368, 378, 380, 392, 396, 416
multiway 341
prediction set 492
rational subgroup 321, 330
recording 42, 44-48, 114, 254
reference 337, 342, 368
reporting 44-48
test set 396, 402, 406-407
training set 405-407, 475, 477, 480, 484, 504, 505
variability 7, 21, 283, 421, 431
visualisation 369-381
data cycle 35-36
data errors 22, 49, 342
data pre-processing 34, 66-82, 405, 412, 505
data reconstruction 265, 507-545
    applications 521-545
    basics 510-515
    theory 515-519, 543
data resolution 42, 216, 497, 508, 510, 523
data transformation 75, 82, 166-168, 175, 176, 183, 249, 262, 367, 369, 404, 405, 406, 436, 462, 478
data transformation, logarithmic 217, 226, 368, 408, 410, 416, 463
decision making 1, 2, 10, 15, 18, 31, 113, 116, 130, 136, 279
deconvolution 507, 510, 512, 514
defining statistics 1, 2, 3, 6, 7, 12, 29, 281
degrees of freedom 64, 100, 101, 123, 124, 125, 127, 130, 133, 157, 175, 284, 293, 299, 302, 303, 312, 319, 330, 398, 440, 443, 533
differentiation 44, 70-75, 463, 504, 505
    digitisation levels 74
    Savitsky-Golay filter 75
    simple differences 74
digitisation 42-44, 365, 532
    frequency 43, 44
    level 43, 44, 74
    Nyquist frequency 44
discriminant analysis 368
    canonical variates 369
    centroid 392, 415, 416
    class membership 393, 394, 398
    classification 367, 368, 393
    discriminant line 392, 393, 394, 397, 398, 402
    linear 367, 368, 391-396, 399, 401, 405, 407
    local model 398

    neural networks 367, 368, 376-78, 399-405
    principal component methods 368, 381, 396, 398
    significance 398
    SIMCA 367, 368, 393, 396-399, 407
    similarity 367, 396, 413-417
    supervised methods 367
    test set 396, 406
    training set 405-407
    validation 405-407
    weight vector 376, 377, 392
distance measures
    city-block 414
    error function 374
    Euclidean 396, 414, 415, 416
    Mahalanobis 328, 396, 415, 416
    Manhatten 414
    Minkowski 413, 416
    multivariate 328, 396, 416
    weighted 329
distribution 9, 14, 38, 101, 148-150, 174, 327, 367, 380, 507, 518
    bimodal 59
    binomial 14, 15, 59, 170
    Boltzman 359
    Boxcar, see uniform
    Cauchy 360
    Chi square 125, 131, 143, 158, 285, 329
    F 127-128, 131, 312, 330
    frequency 58, 521, 534
    Gaussian 15, 509, 513, 522
    kurtosis 148, 164
    lognormal 102, 148, 164
    Lorentzian 513, 522
    mode 148
    Normal 57, 59, 60, 61, 64, 99, 101, 109, 117, 118, 145, 255, 269, 285, 288, 322, 410, 435, 463
    Poisson 15, 59, 509
    population 16, 99, 282
    residuals 9
    sampling 94
    skewness 54, 59, 148, 164, 165, 260
    Student's t 122, 123, 131, 180
    symmetry 59
    uniform 59, 144
dot plot 54, 97

effect plot 212, 225, 227
eigenvalue 78, 369, 371, 478, 483
    variance explained 77, 78, 79, 478, 480

eigenvectors 335
equivalence 138-140
  test for 139
error 3, 7, 61, 62, 281, 401, 407, 434, 475
  assignable 193, 317-319, 321, 339
  assumptions 117
  common cause 7, 279, 282, 286, 291, 296, 299, 300, 305
  estimates 454, 455, 518
  experimental 193
  noise 8, 38, 52, 57, 66, 74, 75, 188, 189, 193, 227, 232, 251, 253, 269, 279, 282, 286, 288, 302, 317, 356, 371, 421, 431, 454, 479, 509, 511, 519
  pure 266, 443
  random 8, 9, 11, 39, 324, 335, 436
  special cause 7, 279, 282, 285, 291, 305, 318, 329, 333, 337, 340
  systematic 8, 11, 38, 39, 193, 282, 436
  transcription 49
  Type I 17, 22, 103, 104, 133, 169, 259, 307, 310, 325
  Type II 17, 22, 103, 104, 134, 307, 325
estimates 3, 9, 12, 23, 98, 101, 106, 123, 127, 134, 142, 201, 283, 423, 486, 512
evolutionary operation 353-355
experiment design 8, 19, 22, 24, 27, 29, 34, 36, 38, 63, 87, 91, 106, 107, 117, 121, 132, 136, 170, 188, 192, 208, 260, 279, 282, 285, 296, 308, 316, 317, 342, 345, 347, 359, 394, 445, 455, 486, 505, 545
  algorithmic designs 271
  alias 217, 307
  all combinations 458
  blocking 8, 22, 25, 135, 177, 193, 194, 203-205, 209, 229, 250, 269, 270, 271, 290, 301, 302, 305, 309, 341
  Box Behnken designs 252, 253, 269, 271, 350, 353
  Central Composite design 235, 249-251, 252, 253, 269, 270, 350, 353
  centre points 192, 216, 217, 227, 251, 253, 256, 260, 266, 270, 271, 351
  choice of factors 228
  choosing a design 227-230, 253
  confirmation experiments 259
  confounding 203, 204, 208, 209-13, 221, 231, 307, 351, 455
  constrained regions 272
  design matrix 255, 306
  design space 197, 219, 220, 229, 231, 275, 351
  diagnostics 226
  D-optimal designs 215, 271, 350, 351, 353
  duplication 193
  extrapolation 105, 244, 259
  factorial 22, 99, 107, 176, 188, 351, 353
  fold over 208, 213, 217, 227, 230, 251
  fractional factorial 107, 206-209, 211, 214, 221, 227, 232, 240, 244, 350, 351
  full factorial 197-203, 207, 240, 249, 303, 311
  G-optimal designs 271
  interaction effect 200, 241, 244, 247
  interaction of factors 197, 206, 215, 220
  irregular fraction designs 213
  main effect 193, 198, 200, 204, 206, 208, 215, 220, 268, 351
  missing runs 230, 231, 270
  mixed level 225
  mixture designs 273, 277, 495, 497
  nested 96
  optimal 271, 272, 275
  optimality 347
  orthogonal blocking 270
  Plackett-Burman 107, 135, 213, 214, 240, 249, 350, 351
  prediction error 265, 266, 492, 495, 500
  principal properties 275
  problems 231-232
  random effects 312
  randomisation 8, 11, 25, 26, 38, 39-41, 87, 193, 194, 230, 253, 302
  randomised block 302, 309
  replicated axial designs 251, 270
  replication 8, 11, 17, 18, 22, 25, 26, 28, 38, 39, 133, 134, 195, 229, 289, 300, 301, 312, 501
  resolution 209-214, 221, 224, 229, 248, 249
  response surface modelling 238, 240, 277, 353
  robust designs 216, 276
  rotatability of designs 268, 270
  run standard order 198, 254
  sampling schemes 87
  screening 188, 196, 206, 213, 227, 232, 244, 248, 251, 258, 270, 350, 351
  sequential experiments 227, 232-255, 256, 353
  small composite designs 272
  space filling designs 257, 272, 273

star points 227, 249, 253, 261, 267, 270, 353, 464
Taguchi 135, 214, 276
uniform precision 251, 266, 268
validation of model
experiment run order 8, 11, 25, 26, 38, 39-41, 87, 153, 193, 194, 198, 226, 230, 302
exploratory data analysis 148, 161, 490, 497

F distribution 127-128, 178, 131, 312, 330
factors 4, 7, 16, 24, 25, 188, 190, 197, 240, 258, 281, 282, 285, 291, 331, 333, 372, 385, 421, 422, 431, 434, 436, 462, 473, 477, 510, 513
   assignable 193, 317, 318, 319, 321, 339
   choosing 194-196, 228
   continuous 190, 227, 281
   controlled 117, 170, 214
   discrete 190, 227, 281
   effect 119, 191, 194, 208
   external 37
   independent 428
   interaction 191, 199, 203, 208, 217, 222, 241, 244, 247, 253, 268, 303, 305, 306, 311, 340, 341, 428, 454, 477
   levels 92, 191, 196, 197, 231, 296, 455
   noise factors 214
   nuisance 25, 231, 282, 390
   predictor 291, 333, 422, 431, 436, 471
   qualitative 21, 24, 147, 191, 274
   quantitative 21, 24, 147, 190, 274
   rational subgroup 321, 330
   response 10, 24, 118, 191, 214, 226, 240, 279, 291, 343, 352, 431, 471, 473, 540
   treatments 24, 178, 190, 296, 302, 308
   uncontrolled 5, 10, 37, 121
fault diagnosis 342
filters 66, 67, 68, 507, 509, 519, 524
   inverse 507, 510, 534
   Savitsky-Golay 42
   window 47, 68
fishbone diagram 194, 228, 233
fixed effects 296, 297, 305
F-test 127-130, 143, 398, 442, 443, 460, 501
   confidence interval 127-130, 131, 312
functions 427
   desirability 350
   empirical 240, 421
   objective 346, 347
   series approximation 463
   Taylor series approximation 241, 464
general linear models 279, 289, 305, 306, 308

genetic algorithms 350, 357, 461, 468
   fitness function 359

help
   expert assistance 280, 289, 305
   self help 29-30
   when to seek expert 15, 16, 29-32, 64, 124, 194, 228, 280, 403, 436
   where to go sections 1, 35, 85, 113, 146, 190, 237, 280, 315, 364, 422, 473, 508
histogram 54, 58, 60, 62, 117, 118, 120, 161, 172
   cumulative 174
   standard deviations 118
Hotelling T^2 statistic 328, 330, 334, 335
hypothesis testing 16-19, 103, 136-38, 257, 293, 331, 440
   Alternative Hypothesis 16-19, 103, 137, 139, 170, 257
   Null Hypothesis 16-19, 103, 137, 139, 171, 172, 180, 257, 331
   Type I error 17, 133, 169, 259, 310
   Type II error 17, 134

integration 70-75
interaction 191, 197, 199, 206, 208, 217, 220, 222, 241, 244, 247, 253, 268, 303, 305, 306, 311, 341, 428, 454
interaction plots 192, 201, 227
intercept 433, 434, 439, 451, 453, 466, 471
   standard error 439
   zero intercept 450-452
interpretation of statistical tests 125, 230, 431, 455-59, 480, 507, 509, 536

kurtosis 148

lack of fit 217, 231, 256, 257, 260, 295, 358, 443, 446, 455, 459, 470
lack of fit and pure error 266, 443
latent variable control charts 333-337
latent variables 319, 333, 338, 369, 370, 481, 482
least median square regression 185, 434
least squares 175, 255, 284, 433, 434, 453, 469
   influential points 105, 289, 294, 307, 351, 436, 449, 484, 502
   leverage 105, 289, 294
   linear 227, 279, 284, 285, 287, 290, 291, 292, 294, 305, 344, 423, 434, 445, 447, 453, 475

maximum likelihood criterion 284, 286
minimum variance criterion 283, 286
outliers 184, 289, 290, 295, 434, 448, 484
stepwise 183, 454, 459-461
weighted 448, 463
L-estimators 150, 151
letter values analysis 153-156, 164-166, 167
limit of detection 104
limit of quantitation 104
limits
    control 50, 319, 321, 325, 326, 327, 331, 332, 342
    out of control signal 330, 331, 335, 339
linear discriminant analysis 367, 368, 391-396
linear programming 347, 354
linear regression 227, 279, 285, 287, 290, 291, 292, 305, 345, 421, 422, 423, 434, 445, 469, 475
    accidental relationship 428
    analysis of variance 279, 285, 288, 293, 305, 441, 456, 458
    assumptions 258, 286, 294, 307, 435, 448, 449
    best fit criterion 435
    bias 290, 295, 436, 451
    calibration 286, 447
    cause and effect 424, 428, 432
    co-linearity 253, 273, 454, 477
    confidence intervals 260, 262, 291, 293, 299, 440, 443, 445-46, 447, 456, 470, 471
    confounding of predictors 307, 351, 455
    cross validation 454
    degree of association 256, 259, 423, 428-32, 439, 455, 456
    empirical functions 240, 421, 468
    empirical models 314, 318, 347, 350-353, 359
    errors in x-values 449-450
    extrapolation 105, 244
    fit 217, 226, 256, 259, 428-432, 439, 442, 443, 456
    functional models 182, 318, 347, 421, 426, 433, 466, 468
    general linear models 279, 305, 306
    genetic algorithms 358, 461, 468
    influential points 105, 289, 294, 307, 352, 436, 449, 484, 502
    interactions 301, 305, 340, 351, 353, 428, 477
    intercept 60, 105, 183, 205, 291, 433, 434, 439, 451, 453, 466, 471
    inverse 448
    lack of fit 217, 231, 236, 257, 259, 295, 357, 443, 446, 455, 459, 470
    leverage 105, 185, 289, 294
    linearisation of response 464
    local model 240, 398, 452, 464, 504
    model fit 431
    multiple 225, 341, 350, 353, 361, 452-61, 473, 476, 488
    neural networks 466-468
    nonparametric 449
    normal equations 297
    Normal probability 258, 295, 437
    over fitting 257, 260, 403, 454, 491, 520
    overall error 439
    parameters 433, 441, 443, 445, 456, 461, 471
    parsimonious model 459
    polynomial model 241, 244, 259, 454, 456
    prediction error 265, 266, 492, 495, 500
    prediction interval 259, 260, 262, 445
    principal components 453, 461, 477-81, 488, 499
    pure error 266, 443, 470
    R^2 217, 226, 256, 259, 428-432, 439, 456, 470
    R^2adjusted 428-432, 439, 470
    replication 443, 446
    residuals plot 183, 437
    robust 449
    Root Mean Square Error 256, 257, 260
    significance 441, 442
    slope 60, 105, 165, 183, 291, 293, 432, 435, 434, 439
    standard errors 439-440, 456, 459, 470, 471
    stepwise 454, 459-461
    sum of squares about the mean 442
    sum of squares about the regression 442, 443, 471
    sum of squares due to the regression 442, 458
    underfit 520, 521
    weighted 448
    zero intercept 249, 450-452, 457
loadings 79, 335, 337, 370, 372, 411, 412, 478, 480, 482, 483
loadings plot 372, 411, 412
lurking variable 231, 489, 494, 495

Mahalanobis distance 328, 396, 415, 416
Mann-Whitney U test 173
mathematical statistics 289
matrix algebra generalised inverse 477
maximum entropy 508, 518
maximum likelihood 283, 284, 286
mean absolute deviation 153
measurement strategy 107
measurement uncertainty 100, 260, 522, 527
measurement
   design 36, 232
   design space 197, 231, 275, 351
   error 193, 517
   noise 38, 42, 370, 442, 507, 509
   system 4, 19, 21, 23, 24, 25, 36-38, 42, 90, 127, 128, 436, 509
   variability 7, 18, 21, 116, 282, 289, 305, 421
measures, statistic 3, 16, 121, 170, 173, 175, 214, 259, 282, 284, 293, 321, 323, 325, 332, 338
median 53, 54, 67, 106, 147, 150, 153, 163
   confidence intervals 158, 176
   median polish 180-182
   of absolute deviation 152, 153
M-estimator 151
methods
   nonparametric 174
   statistical 4, 7, 12, 15, 19, 26, 28, 29, 32, 279, 281, 291, 361
   variability 18, 21, 289, 305
misfit 512, 515, 519
missing values 182, 230, 231, 270
mixed effects 297
model 1, 4-7, 21, 37, 117, 160, 168, 180, 215, 239, 255, 258, 285, 287, 333, 338, 347, 407, 423, 514, 536
   artefacts 561, 504, 507, 520, 541
   best fit 286
   best fit criterion 435
   classification 393, 396
   degree of association 217, 226, 256, 259, 423, 438-432, 439, 455
   empirical functions 314, 318, 343, 347-348, 351-354, 360, 421, 468
   extrapolation 105, 244, 259
   fit 217, 225, 226, 256, 259, 338, 442, 443, 456, 464, 469, 488, 514, 520
   fundamental 314, 318, 346, 347
   general linear 308
   inductive 4-7
   inferential 342

latent variable 484
linear 225, 227, 285, 287, 290, 291, 292, 294, 305, 343, 353, 433, 434, 445, 475
local 241, 504
location 284
mathematical 6, 319, 347, 354
misfit 515, 519, 542
mixed effects 296
multiple linear 488
nonlinear 434, 436, 536, 541
Normal probability 23, 258, 295, 434, 488
parameters 433, 441, 443, 456, 461, 471, 522
parsimonious 459, 493
polynomial 241, 244, 259, 268, 353, 454, 456
prediction uncertainty 367
predictive 421
quality 256
R^2 217, 226, 256, 259, 428-432, 439, 456, 470
R^2adjusted 428-432, 439, 470
random effects 296
regression 476, 478, 490
residuals plot 183
response 475
sequential 307
series approximation 463
Taylor series approximation 241, 464
theoretical 244, 421, 432, 447, 507
validation 6, 22, 257, 279, 288, 396, 405-407, 476, 478, 480, 491
validation check samples 344
modelling
   empirical 350-353
   influential parameters 188, 436
   influential points 105, 146, 185, 194, 289, 294, 307, 351, 484, 502
   leverage 105, 185, 289, 294, 484, 494
   linear least squares 279, 284
multiblock data 340
multiple regression 225
   stepwise 183
multivariate data 75-81, 327
multiway data 340

neural networks 341, 359, 360-366, 368, 393, 396, 464-468
   back propagation 402, 466
   bias level 401, 466
   classification 400, 403
   hidden layer 400, 401, 403, 466

INDEX 553

inputs 400, 466
Kohonen 368, 376-378
multi-layer feed-forward 399-405
neurons 376, 400, 401, 466
nonlinear 400
output layer 400, 401, 466
test data 396, 406-407
threshold level 401, 402
training 402, 405-407
transfer function 400, 401
validation 405-407
weight vector 376, 377, 403
NIPALS method 79, 336, 481
noise 8, 37, 52, 57, 66, 74, 75, 188, 189, 193, 227, 232, 251, 254, 269, 279, 282, 286, 288, 302, 317, 353, 371, 421, 431, 454, 509, 519
   correlation 533
   Gaussian 509, 526
   instrument 42, 479, 509
   level 57, 251, 254, 275, 421, 507, 511, 519
   Poisson 509
   reduction 483, 491
   signal to noise ratio 21, 215, 272, 496, 507, 510, 512, 522, 540
nonlinear mapping 368, 373-376
nonlinear model 541
nonlinear regression 183, 421, 461-469
   fit 464
   genetic algorithms 357, 461, 468
   neural networks 464-468
   pit mapping method 464, 468
   residuals mapping method 464, 468
   series approximation 463
   steepest descent method 466, 468
   Taylor series approximation 241, 464
   weighted 463
nonparametric methods 146, 169, 173, 174, 175, 182, 290, 449
nonparametric statistics 174
Normal distribution 51, 60, 61, 64, 99, 101, 109, 117, 118, 145, 259, 269, 285, 288, 322, 410, 435, 463
normality testing 60
Null Hypothesis 16-19, 103, 137, 139, 171, 172, 180, 331
Nyquist frequency 44

observational studies 91, 98, 103, 170
optimisation 248, 276, 314, 316, 318, 345-347, 360, 374, 405, 538
   criterion 346
   desirability function 350
   empirical models 347, 350-353, 360
   Evolutionary Operation 353-355
   fitness function 359
   genetic algorithms 350, 357
   global optimum 346, 347, 352, 357, 358, 516
   interactions of factors 135, 203, 248, 340, 348, 351, 353
   linear programming 347, 354
   local optimum 346, 352
   multicriteria 276, 349, 350
   multicriteria decisions 349, 350
   one variable at a time 27, 348
   overlay contour plots 349
   Pareto optimal 350
   process 314
   real time 353
   sequential methods 350, 353
   simplex method 347, 352, 353, 354
   simulated annealing method 357-358
   steepest ascent method 350, 352, 356
outlier 11, 51, 61-66, 79, 99, 146, 148, 150, 151, 153, 161, 182, 184, 185, 289, 290, 294, 295, 341, 342, 343, 353, 364, 434, 436, 448, 480, 484, 497
   detection 63-66, 341, 350, 476, 480, 484, 486, 496
   discordant value 11, 22, 61-65, 66, 289, 436
   multivariate 185
   multivariate detection 161-166
   outside values 161, 163

paired comparisons 140-142, 169-172, 182, 310
paired comparisons sign test 169, 172
parallel co-ordinates 368, 378-381
   correlation of parameters 379, 380
   similarity of parameters 380
parameters 16, 188, 190, 277, 281, 282, 285, 372, 385, 421, 423, 431, 434, 443, 475, 513
   location 148, 150-153
   nuisance 282, 290
   spread 148, 152-156
   statistic 3, 16, 103, 121, 170, 173, 175, 259, 282, 284, 293, 321, 323, 325, 332, 338
Pareto chart 200, 212, 217, 221, 227

554   INDEX

partial least squares 336, 337, 342, 350, 353, 360, 361, 434, 453, 461, 481, 491
   calibration 497
   colinearity 477
   contribution plots 339, 341
   cross validation 478, 492, 495
   leverage 484
   local model 504
   lurking variable 494, 495
   model fit 338
   multi-block data 342
   nonlinear 340, 343, 361
   number of components 336
   over fitting 403, 454, 520
   prediction error 338, 339, 344, 492, 495
   underfit 520, 521
   variables selection 476, 483, 502
   weight vector 336
pattern recognition 536
planning 19, 34, 36, 85, 92, 114-17, 139, 194-96, 230, 232, 305, 351
plots
   biplot 486, 490
   Box and Whisker 53, 60, 163
   boxplot 161, 163, 176
   centroid 522
   contour 238, 239, 259, 277
   contour overlay 349
   contribution (Partial Least Squares) 339, 341, 487
   control 318
   control, Western Electric Rules 323
   Cusum 54, 71, 106, 319, 324, 327, 331
   Cusum, V-mask 52, 324
   dendrogram 383, 385, 386, 388, 390, 497
   dot 54, 97
   effect 212, 225, 227
   histogram 54, 58, 60, 62, 117, 118, 120, 172
   interaction 192, 201, 227
   loading 372, 411, 412
   multivariate charts 318, 327, 328-334, 341, 343-345
   multivariate Cusum 330
   Normal probability 23, 60, 71, 164, 258, 295, 437, 488
   parallel co-ordinates 378-381
   Pareto 200, 212, 217, 221, 227
   range chart 51, 320
   residuals 183, 226, 258, 294, 337, 436, 437, 438, 457, 463, 484, 502
   response surface 239, 241-247, 262
   scatter 23, 52, 53, 62, 75, 77, 79, 183, 429, 432, 437
   score 372, 396, 489, 497
   scree 371, 398
   Shewhart 51, 55, 106, 319-323, 327
   standard deviation chart 51, 320
   star 88, 260
   stem and leaf 161
   suspended rootogram 161, 164
   trend 49, 50
   univariate charts 319, 327, 344
   xbar (average) chart 320, 322,
population distribution 16, 99, 282
population variance 283, 284
power 17, 18, 22, 25, 26, 38, 94, 99, 103, 104, 106, 134, 172, 483
   number of experiments 251, 260, 275
predictors 7, 24, 190, 279, 291, 333, 422, 431, 434, 436, 462, 471, 473
   errors in 288
principal components analysis 75-81, 275, 276, 334, 337, 342, 368, 369-373, 381, 396, 398, 410, 411, 486, 497
   colinearity 477, 501
   eigenvalue 78, 369, 371
   eigenvectors 336
   graphical representation 77, 370, 479
   latent variables 319, 333, 338, 369, 370, 481
   linear combinations of variables 334, 461, 482
   loadings 79, 335, 337, 370, 372, 411, 412, 478, 480, 482, 483
   NIPALS method 79, 335, 481
   regression 453, 461, 477-481, 488, 489, 499
   scores 79, 336, 337, 369, 372, 396, 478, 480, 482
   scree plot 371, 398
   singular value decomposition 369
   variance explained 77, 78, 79, 478
principal components regression 361, 453, 461, 477-481, 488, 499
   calibration 495
   co-linearity 501
   cross validation 478, 480, 492, 495
   leverage 484
   local models 504
   lurking variable 489, 494, 495
   over fitting 403, 454, 491, 520
   prediction error 492, 495, 500
   underfit 520, 521

INDEX 555

variables selection 476, 480, 483, 502
x-value correlation 477
probability 12-19, 28, 50, 60, 90, 103, 125, 137, 145, 161, 258, 358, 443
   examples of use 14
   rules 12-14
process control 94, 189, 295, 316, 544
   automatic 317, 318
process investigation 214, 276
process monitoring 24, 93, 334, 338, 340
producer risk 103
projection to latent structures, see Partial Least Squares
p-value 12, 23, 137, 171, 217, 257, 309, 310

quartiles 154, 164

random effects 296, 299, 300, 305, 312
random error 8, 9, 11, 39, 324, 335, 436
random numbers
   generation 144
   Normally distributed 117, 144
   uniform 144
randomisation 8, 11, 25, 26, 38, 39-41, 87, 193, 194, 230, 253, 302
   run order 198, 226
   schemes 11, 40
   standard design 198, 253
randomness 168-169
range 15, 53, 126, 147, 148, 320, 342, 346, 350, 352, 409, 505
   squared ranks test 174
rank 148, 154, 170, 175, 177, 483
rank regression 183
recording numbers 44, 45
regression 105, 183, 255, 285, 432, 474
   accidental relationship 424-428
   cause and effect 422, 423, 432
   coefficient 475, 476, 478, 488, 492, 501, 502
   diagnostics 226
   fitting 225
   functional relationship 182, 426, 433, 466, 468
   lack of fit 217, 231, 256, 257, 260, 295, 359, 443, 446, 455, 470
   latent variable methods 481
   least median squares 185, 423, 434
   model 476, 478, 490
   multiple linear 225, 452-461
   nonlinear 183, 241, 275, 340, 343, 360, 421, 436, 461-469

nonparametric 182-186
Partial Least Squares 334, 336, 337, 342, 350, 353, 434, 453, 461, 481, 491
pictorial representation 433, 434, 441, 447, 449
prediction error 338, 339, 344
Principal Components 453, 461, 477-81, 488, 499
rank 183
residuals plots 183
resistant 182-186
robust 105, 182-186, 290, 481
stepwise 183
x onto y 433, 448
y onto x 453
replication 8, 10, 11, 17, 18, 22, 25, 26, 28, 38, 39, 40, 121, 133, 192, 193, 201, 227, 229, 257, 289, 296, 300, 301, 304, 312, 432, 449, 453, 470, 497, 501
   centre points 192, 216, 217, 251, 253, 256, 260, 266, 270, 352
   measurements 38, 55, 86, 123, 443, 446
reporting numbers 44, 46
residuals 22, 23, 141, 166, 180, 183, 231, 259, 285, 287, 297, 306, 308, 434, 435, 457, 476, 478, 539
   nonlinear regression method 464
   plots 183, 226, 259, 337, 436, 437, 438, 457, 463, 484, 502
   Studentised 259, 261
resistant methods 146, 150, 151, 166-168, 177
resolution enhancement 72, 74, 510, 512, 519, 524, 534, 540
response 10, 24, 118, 191, 214, 216, 229, 240, 238, 258, 279, 291, 344, 346, 352, 421, 423, 431, 434, 471, 475
response surface modelling 238, 277, 350, 352
   algorithmic designs 271
   blocking 25, 270, 290
   Box Behnken designs 252, 253, 269, 271, 350, 353
   central composite design 235, 249-51, 253, 270, 350, 353
   centre points 192, 216, 217, 251, 253, 256, 260, 266, 270, 271, 351
   confirmation experiments 260
   constrained regions 272
   design matrix 255, 306
   designs 240, 249, 252, 265, 358

D-optimal designs 215, 271, 350, 351, 353
G-optimal designs 271
interactions 214, 244, 247
mixture designs 273, 277
optimal 271, 272, 275
optimality 346
optimisation 259, 276, 314, 317, 345, 346, 351, 353, 358, 539
prediction error 265, 266, 492, 495, 500
prediction intervals 260, 262
principal properties 275
replicated axial designs 270, 271
resolution 249
robustness 276
rotatability of designs 268, 270
small composite designs 272
space filling designs 272, 273, 275
star 227, 249, 253, 261, 267, 270, 353, 464
uniform precision 251, 266, 268
response surface plots 239, 241-247, 262
response multivariate 368, 372, 377, 378, 380, 385, 403, 416
responses 191, 195
responses multivariate 75, 79, 290, 314, 327, 329, 330, 337, 344, 474
risk 6, 12, 17, 89, 103, 113, 131, 249
robust designs 276
robust methods 145, 166-168, 177, 182
robust regression 105, 182, 183, 290, 481, 497
Root Mean Square Error 256, 257, 260
rounding 10, 42, 45-49, 61, 124, 135, 260, 386, 437
runs
   average length 94, 106, 322, 326, 330, 331, 332
   Markov chain 169
   number 251, 260, 275, 351
   standard order 198, 253

sample size 18, 22, 25, 98, 99, 101, 103, 106, 107, 109, 132-35, 322, 323
sampling 4, 9, 10, 24, 62, 86, 106, 107, 159, 299, 305, 310, 505, 544
   acceptance 93
   distribution 94
   equipment 100
   frequency 93, 321, 322
   invasive 92
   physical 86, 95-98, 100

process control 93-95, 109
random 100, 102, 159, 160
rate 531
sample size 18, 22, 25, 98, 99, 101, 106, 107, 109, 133, 134, 322, 323
sequential 102
statistical 86, 87
strategies 9, 87-88, 90, 94, 98, 100, 107, 121
stratified 100, 107
system 21, 23, 25, 37
systematic 102, 107
variability 116, 131
scaling 55, 60, 244, 334, 368, 378, 407
   autoscaling 75, 77, 409, 410, 411, 486
   logarithmic 408, 410
   Mahalanobis 328, 396, 415
   mean centred 334, 343, 409, 478, 497
   range 378, 404, 409, 410, 411
scatter plot 23, 52, 53, 62, 75, 77, 79, 183, 429, 432, 437
score plot 372
scores 79, 336, 337, 369, 372, 376, 478, 480, 482, 489
scree plot 371, 398
sequence value 49, 57
sequential experiments 227, 232-235, 250, 353
Shewhart chart 55, 106, 319-324, 326, 327
sign test 106, 290
signal 8, 38, 480, 509, 519
signed rank 106, 158, 172
significance critical value 16, 17, 64, 65, 66, 171, 172, 501
   level 17, 103, 136, 157, 180, 310, 399, 443
   practical 16-19, 22, 23, 28, 58, 310
   p-value 137, 171, 217, 309, 310
   regression 441, 442
   statistical 16-19, 28, 58, 137, 258, 398, 399
   testing 16, 296
significant figures 45, 47
   relation with standard deviation 47
similarity 367, 386, 396, 413-417
   correlation coefficient 382, 416
   Euclidean distance 396, 414, 415, 416
   matrix 382, 383
simulation 117, 241, 316, 347
skewness 54, 59, 148, 164, 165, 261
slope 433, 434, 439
slope standard error 439

INDEX 557

smoothing
    exponential 511
    exponentially weighted moving
        average 66, 69, 319, 325, 327
    Fourier 511
    Gaussian 511
    linear 510
    moving average 57, 66, 67
    moving median 67
    moving range 320
    moving standard deviation 320
    multivariate EWMA 332
    noise reduction 483, 491
    Savitsky Golay 66, 68, 510
    triangular 523, 527
    window filter 67, 68
software 22, 26, 27, 29, 32, 46, 47, 53, 64, 75,
        99, 123, 144, 185, 196, 258, 260, 270,
        275, 276, 290, 296, 305, 507, 510
    errors 32
    statistics 30, 32, 60, 75, 78, 106, 142, 145,
        156, 160, 164, 174, 298, 308, 309,
        445, 457, 460
special cause 318, 329, 333, 337, 341
specifications 224, 318, 344, 349, 353
specifications multivariate 344, 349
spread, see range, standard deviation 174
standard deviation 9, 15, 37, 42, 47, 50, 55,
        64, 78, 100, 101, 107, 117, 118, 121,
        130, 147. 148, 149, 153, 155, 165,
        265, 283, 320, 392, 409, 430, 436,
        439, 484, 514, 521
    bootstrap 159
    combined 119
    confidence interval 101, 125-130, 158
    jack-knife 159
    residual 484, 494
    robust/resistant estimators 153-157, 174
    sample 157
    successive difference 55
standard error 121, 139, 151, 174, 265, 268,
        276, 439-440, 456, 459, 470, 471, 525
    of validation 492
star plots 88
statistical investigations 1, 4, 19-28
    PDCA cycle 19-23
statistical methods 12, 15
statistical process control 8, 24, 93, 98, 103,
        189, 295, 314, 316, 318, 544
    average run length 94, 106, 322, 326, 330,
        331, 332
    chisquare control charts 329

    control charts 50, 51, 93, 106, 109, 168,
        317, 319-323, 333
    control limits 318, 321, 322, 325, 326,
        327, 331, 332, 343
    Cusum 54-58, 106, 319, 323, 327
    Cusum of Hotelling T 332
    EWMA 66, 69, 319, 325, 327
    fault diagnosis 342
    monitoring 22, 93, 334, 338, 340
    moving range 320
    moving standard deviation 320
    multivariate control charts 319, 327, 328,
        334, 341, 344
    multivariate Cusum 330
    multivariate EWMA 332
    multivariate monitoring
    optimisation 314
    out of control 330, 331, 335, 339
    range chart 320
    sampling 94
    Shewhart charts 51, 55, 106, 319-323, 327
    simulation 317, 347
    standard deviation chart 320
    univariate control charts 319, 327, 344
    Western Electric Rules 323, 325
    xbar (average) chart 319-322
statistical quality control 327, 333
statistical thinking 2-7, 30
Statistics 1, 2, 3, 4, 6, 7, 15, 26, 27, 32, 279,
        281, 291
statistics interpretation 125, 230, 431, 455-
        457, 480, 507, 509, 536
Statistics mathematical 6, 7, 289
stem and leaf diagram 161
sum of absolute differences 290
sum of squares 284, 286, 287, 423, 435, 442,
        464
    about mean 442
    about regression 442, 443, 471
    between group 57, 300
    due to the regression 292, 442
    error 287, 289, 292, 293, 295, 299
    mean residual sum of squares 299
    mean treatment sum of squares 300
    model 338
    of residuals 284, 435, 464, 469, 476
    prediction 338, 339, 344
    residual 287, 292, 295, 297, 299, 300, 304
    special causes 286, 287
    total 284, 286, 287, 292, 203, 299
    Type I 307
    Type II 307

within group 57, 299
systematic error 8, 11, 38, 39, 193, 436
transformation 407, 416, 505
    autoscaling 75, 77, 409, 410, 411
    data 166, 168, 175, 176, 183, 408
    linear 416
    linearisation 462-464
    logarithmic 408, 410
    mean scaling 334, 353, 409, 478, 497
    nonlinear 400, 401, 416
    range scaling 404, 409, 410, 411
    sigmoidal 400, 401
treatments 24, 178, 190, 296, 302, 308
trend plots 49, 50
true value 3, 7, 15, 16, 73, 126, 255, 256, 262, 288, 296, 423, 467
t-test 106, 107, 258, 310, 440
    confidence interval 100, 122, 124, 131
    paired comparison 140-142, 169, 171, 182, 310
    Student's t-distribution 122, 123, 131, 180
    two sample 6
Type I error 17, 22, 133, 169, 259, 310, 325
Type II error 17, 22, 134, 325

uncertainty 12, 37, 88, 90, 94, 99, 107, 113, 124, 130, 259, 281, 522, 527
uniform distribution 144

validation 18, 21, 22, 24, 29, 32, 90, 182, 227, 256, 257, 279, 288, 341, 478
    check samples 344
    graphical 258
    of models 6, 227, 257, 396, 405-407, 476, 480, 491
    software 32
    standard error 492
    statistical 26
variability 3, 8, 37, 38, 42, 54, 113, 117, 189, 216, 217, 279, 281, 285, 302, 308, 431
    additive 9
    causes 7-11, 281
    common cause 7, 11, 25, 279, 282, 285, 286, 287, 291, 296, 299, 300, 302, 305, 317, 319, 327, 337, 344
    in measurements 116, 126, 131, 479
    method 8, 21, 282, 289, 305
    noise 8, 38, 52, 57, 66, 74, 75, 188, 189, 193, 227, 232, 251, 253, 257, 269, 279, 286, 288, 302, 317, 353, 371, 421, 431, 454, 509, 519
    nuisance 11

predictor 295
random 3, 8, 9, 11, 39, 282, 324, 335, 436
sampling 116, 131
special cause 7, 11, 24, 25, 279, 282, 285, 287, 291, 305, 334
systematic 8, 282
total 283, 284, 286
variables 7, 21, 75, 91, 244, 281, 317, 341, 365, 375, 388, 408, 424, 487
    attribute 147
    categorical 147, 270, 272, 274, 275
    colinear 253, 273, 454, 473, 477
    continuous 21, 147, 153, 170, 191, 227, 281, 291, 347
    dependent 191, 255, 423
    discrete 21, 191, 227, 281, 346
    independent 182, 190, 255, 275, 346, 349, 422, 428
    indirect 494
    latent 319, 333, 338, 369, 370, 481, 482
    linear combinations 334, 461, 481, 482
    nominal 274
    ordinal 147, 148, 414
    predicted 424
    predictor 291, 333, 422, 431, 434, 436, 471
    process 69, 317, 320, 327, 332, 333, 335, 337, 339, 343, 353, 357, 485
    qualitative 21, 24, 147, 191, 274
    quality 317, 320, 327, 333, 334, 337, 339
    quantitative 21, 24, 147, 191, 279
    random 14
    response 182, 229, 431, 434
    selection 459, 476, 480, 483, 502
variance
    about regression 439
    components 10, 142-43, 300, 312, 421
    confidence interval 101, 125-30, 158
    explained 78, 79, 256, 260, 265, 293, 299, 334, 369, 371, 372, 411, 412, 478, 480, 483, 484, 488
    measurement 443, 445
    nonconstant 436
    population 284
    sample 284
    total 283, 284, 286

Western Electric Rules 323, 325
Wilcoxon Mann Whitney test 106, 173, 174
Wilcoxon test 158, 170, 172,
Wilcoxon rank sum test 173